IET ENERGY ENGINEERING 116

Modeling and Simulation of HVDC Transmission

Other volumes in this series:

Volume 1 **Power Circuit Breaker Theory and Design** C.H. Flurscheim (Editor)
Volume 4 **Industrial Microwave Heating** A.C. Metaxas and R.J. Meredith
Volume 7 **Insulators for High Voltages** J.S.T. Looms
Volume 8 **Variable Frequency AC Motor Drive Systems** D. Finney
Volume 10 **SF_6 Switchgear** H.M. Ryan and G.R. Jones
Volume 11 **Conduction and Induction Heating** E.J. Davies
Volume 13 **Statistical Techniques for High Voltage Engineering** W. Hauschild and W. Mosch
Volume 14 **Uninterruptible Power Supplies** J. Platts and J.D. St Aubyn (Editors)
Volume 15 **Digital Protection for Power Systems** A.T. Johns and S.K. Salman
Volume 16 **Electricity Economics and Planning** T.W. Berrie
Volume 18 **Vacuum Switchgear** A. Greenwood
Volume 19 **Electrical Safety: A guide to causes and prevention of hazards** J. Maxwell Adams
Volume 21 **Electricity Distribution Network Design, 2nd Edition** E. Lakervi and E.J. Holmes
Volume 22 **Artificial Intelligence Techniques in Power Systems** K. Warwick, A.O. Ekwue and R. Aggarwal (Editors)
Volume 24 **Power System Commissioning and Maintenance Practice** K. Harker
Volume 25 **Engineers' Handbook of Industrial Microwave Heating** R.J. Meredith
Volume 26 **Small Electric Motors** H. Moczala *et al.*
Volume 27 **AC–DC Power System Analysis** J. Arrillaga and B.C. Smith
Volume 29 **High Voltage Direct Current Transmission, 2nd Edition** J. Arrillaga
Volume 30 **Flexible AC Transmission Systems (FACTS)** Y.-H. Song (Editor)
Volume 31 **Embedded Generation** N. Jenkins *et al.*
Volume 32 **High Voltage Engineering and Testing, 2nd Edition** H.M. Ryan (Editor)
Volume 33 **Overvoltage Protection of Low-Voltage Systems, Revised Edition** P. Hasse
Volume 36 **Voltage Quality in Electrical Power Systems** J. Schlabbach *et al.*
Volume 37 **Electrical Steels for Rotating Machines** P. Beckley
Volume 38 **The Electric Car: Development and future of battery, hybrid and fuel-cell cars** M. Westbrook
Volume 39 **Power Systems Electromagnetic Transients Simulation** J. Arrillaga and N. Watson
Volume 40 **Advances in High Voltage Engineering** M. Haddad and D. Warne
Volume 41 **Electrical Operation of Electrostatic Precipitators** K. Parker
Volume 43 **Thermal Power Plant Simulation and Control** D. Flynn
Volume 44 **Economic Evaluation of Projects in the Electricity Supply Industry** H. Khatib
Volume 45 **Propulsion Systems for Hybrid Vehicles** J. Miller
Volume 46 **Distribution Switchgear** S. Stewart
Volume 47 **Protection of Electricity Distribution Networks, 2nd Edition** J. Gers and E. Holmes
Volume 48 **Wood Pole Overhead Lines** B. Wareing
Volume 49 **Electric Fuses, 3rd Edition** A. Wright and G. Newbery
Volume 50 **Wind Power Integration: Connection and system operational aspects** B. Fox *et al.*
Volume 51 **Short Circuit Currents** J. Schlabbach
Volume 52 **Nuclear Power** J. Wood
kVolume 53 **Condition Assessment of High Voltage Insulation in Power System Equipment** R.E. James and Q. Su
Volume 55 **Local Energy: Distributed generation of heat and power** J. Wood
Volume 56 **Condition Monitoring of Rotating Electrical Machines** P. Tavner, L. Ran, J. Penman and H. Sedding
Volume 57 **The Control Techniques Drives and Controls Handbook, 2nd Edition** B. Drury
Volume 58 **Lightning Protection** V. Cooray (Editor)
Volume 59 **Ultracapacitor Applications** J.M. Miller

Volume 62	**Lightning Electromagnetics** V. Cooray
Volume 63	**Energy Storage for Power Systems, 2nd Edition** A. Ter-Gazarian
Volume 65	**Protection of Electricity Distribution Networks, 3rd Edition** J. Gers
Volume 66	**High Voltage Engineering Testing, 3rd Edition** H. Ryan (Editor)
Volume 67	**Multicore Simulation of Power System Transients** F.M. Uriate
Volume 68	**Distribution System Analysis and Automation** J. Gers
Volume 69	**The Lightening Flash, 2nd Edition** V. Cooray (Editor)
Volume 70	**Economic Evaluation of Projects in the Electricity Supply Industry, 3rd Edition** H. Khatib
Volume 72	**Control Circuits in Power Electronics: Practical issues in design and implementation** M. Castilla (Editor)
Volume 73	**Wide Area Monitoring, Protection and Control Systems: The enabler for Smarter Grids** A. Vaccaro and A. Zobaa (Editors)
Volume 74	**Power Electronic Converters and Systems: Frontiers and applications** A.M. Trzynadlowski (Editor)
Volume 75	**Power Distribution Automation** B. Das (Editor)
Volume 76	**Power System Stability: Modelling, analysis and control** A.A. Sallam and Om P. Malik
Volume 78	**Numerical Analysis of Power System Transients and Dynamics** A. Ametani (Editor)
Volume 79	**Vehicle-to-Grid: Linking electric vehicles to the smart grid** J. Lu and J. Hossain (Editors)
Volume 81	**Cyber-Physical-Social Systems and Constructs in Electric Power Engineering** S. Suryanarayanan, R. Roche and T.M. Hansen (Editors)
Volume 82	**Periodic Control of Power Electronic Converters** F. Blaabjerg, K.Zhou, D. Wang and Y. Yang
Volume 86	**Advances in Power System Modelling, Control and Stability Analysis** F. Milano (Editor)
Volume 87	**Cogeneration: Technologies, Optimisation and Implementation** C.A. Frangopoulos (Editor)
Volume 88	**Smarter Energy: From smart metering to the smart grid** H. Sun, N. Hatziargyriou, H.V. Poor, L. Carpanini and M.A. Sánchez Fornié (Editors)
Volume 89	**Hydrogen Production, Separation and Purification for Energy** A. Basile, F. Dalena, J. Tong and T.N.Veziroğlu (Editors)
Volume 90	**Clean Energy Microgrids** S. Obara and J. Morel (Editors)
Volume 91	**Fuzzy Logic Control in Energy Systems with Design Applications in MATLAB®/Simulink®** İ.H. Altaş
Volume 92	**Power Quality in Future Electrical Power Systems** A.F. Zobaa and S.H.E.A. Aleem (Editors)
Volume 93	**Cogeneration and District Energy Systems: Modelling, analysis and optimization** M.A. Rosen and S. Koohi-Fayegh
Volume 94	**Introduction to the Smart Grid: Concepts, technologies and evolution** S.K. Salman
Volume 95	**Communication, Control and Security Challenges for the Smart Grid** S.M. Muyeen and S. Rahman (Editors)
Volume 96	**Industrial Power Systems with Distributed and Embedded Generation** R Belu
Volume 97	**Synchronized Phasor Measurements for Smart Grids** M.J.B. Reddy and D.K. Mohanta (Editors)
Volume 98	**Large Scale Grid Integration of Renewable Energy Sources** A. Moreno-Munoz (Editor)
Volume 100	**Modeling and Dynamic Behaviour of Hydropower Plants** N. Kishor and J. Fraile-Ardanuy (Editors)
Volume 101	**Methane and Hydrogen for Energy Storage** R. Carriveau and D.S.-K. Ting
Volume 104	**Power Transformer Condition Monitoring and Diagnosis** A. Abu-Siada (Editor)
Volume 106	**Surface Passivation of Industrial Crystalline Silicon Solar Cells** J. John (Editor)

Volume 107	**Bifacial Photovoltaics: Technology, applications and economics** J. Libal and R. Kopecek (Editors)
Volume 108	**Fault Diagnosis of Induction Motors** J. Faiz, V. Ghorbanian and G. Joksimović
Volume 110	**High Voltage Power Network Construction** K. Harker
Volume 111	**Energy Storage at Different Voltage Levels: Technology, integration, and market aspects** A.F. Zobaa, P.F. Ribeiro, S.H.A. Aleem and S.N. Afifi (Editors)
Volume 112	**Wireless Power Transfer: Theory, Technology and Application** N.Shinohara
Volume 114	**Lightning-Induced Effects in Electrical and Telecommunication Systems** Y. Baba and V.A. Rakov
Volume 115	**DC Distribution Systems and Microgrids** T. Dragičević, F.Blaabjerg and P. Wheeler
Volume 117	**Structural Control and Fault Detection of Wind Turbine Systems** H.R. Karimi
Volume 119	**Thermal Power Plant Control and Instrumentation: The control of boilers and HRSGs, 2nd Edition** D. Lindsley, J. Grist and D. Parker
Volume 120	**Fault Diagnosis for Robust Inverter Power Drives** A. Ginart (Editor)
Volume 121	**Monitoring and Control using Synchrophasors in Power Systems with Renewables** I. Kamwa and C. Lu (Editors)
Volume 123	**Power Systems Electromagnetic Transients Simulation, 2nd Edition** N. Watson and J. Arrillaga
Volume 124	**Power Market Transformation** B. Murray
Volume 125	**Wind Energy Modeling and Simulation, Volume 1: Atmosphere and plant** P. Veers (Editor)
Volume 126	**Diagnosis and Fault Tolerance of Electrical Machines, Power Electronics and Drives** A.J.M. Cardoso
Volume 128	**Characterization of Wide Bandgap Power Semiconductor Devices** F. Wang, Z. Zhang and E.A. Jones
Volume 129	**Renewable Energy from the Oceans: From wave, tidal and gradient systems to offshore wind and solar** D. Coiro and T. Sant (Editors)
Volume 130	**Wind and Solar Based Energy Systems for Communities** R. Carriveau and D.S.-K. Ting (Editors)
Volume 131	**Metaheuristic Optimization in Power Engineering** J. Radosavljević
Volume 132	**Power Line Communication Systems for Smart Grids** I.R.S Casella and A. Anpalagan
Volume 139	**Variability, Scalability and Stability of Microgrids** S.M. Muyeen, S.M. Islam and F. Blaabjerg (Editors)
Volume 145	**Condition Monitoring of Rotating Electrical Machines** P. Tavner, L. Ran, C. Crabtree
Volume 146	**Energy Storage for Power Systems, 3rd Edition** A.G. Ter-Gazarian
Volume 147	**Distribution Systems Analysis and Automation, 2nd Edition** J. Gers
Volume 152	**Power Electronic Devices: Applications, failure mechanisms and reliability** F. Iannuzzo (Editor)
Volume 153	**Signal Processing for Fault Detection and Diagnosis in Electric Machines and Systems** Mohamed Benbouzid (Editor)
Volume 155	**Energy Generation and Efficiency Technologies for Green Residential Buildings** D. Ting and R. Carriveau (Editors)
Volume 157	**Electrical Steels, 2 Volumes** A. Moses, K. Jenkins, Philip Anderson and H. Stanbury
Volume 158	**Advanced Dielectric Materials for Electrostatic Capacitors** Q Li (Editor)
Volume 159	**Transforming the Grid Towards Fully Renewable Energy** O. Probst, S. Castellanos and R. Palacios (Editors)
Volume 160	**Microgrids for Rural Areas: Research and case studies** R.K. Chauhan, K. Chauhan and S.N. Singh (Editors)
Volume 166	**Advanced Characterization of Thin Film Solar Cells** N. Haegel and M. Al-Jassim (Editors)
Volume 167	**Power Grids with Renewable Energy: Storage, integration and digitalization** A.S. Sallam and Om P. Malik
Volume 172	**Lighting Interaction with Power Systems, 2 volumes** A. Piantini (Editor)
Volume 905	**Power System Protection, 4 volumes**

Modeling and Simulation of HVDC Transmission

Edited by
Minxiao Han and Aniruddha M. Gole

The Institution of Engineering and Technology

Published by The Institution of Engineering and Technology, London, United Kingdom

The Institution of Engineering and Technology is registered as a Charity in England & Wales (no. 211014) and Scotland (no. SC038698).

First published 2020

The Institution of Engineering and Technology
Michael Faraday House
Six Hills Way, Stevenage
Herts, SG1 2AY, United Kingdom

www.theiet.org

British Library Cataloguing in Publication Data
A catalogue record for this product is available from the British Library

ISBN 978-1-78561-380-7 (hardback)
ISBN 978-1-78561-381-4 (PDF)

Typeset in India by MPS Limited

Contents

About the editors

Minxiao Han is a professor and the director of Flexible Electric Power Institute of North China Electric Power University, China. He has been the principal investigator in the key projects consigned by MOST, China and the NSF of China, is a member of CIGRE SC B4, and is a senior member of IEEE and CSEE. He has authored more than 100 referred papers, holds four national patents, and has written five books.

Aniruddha M. Gole is a distinguished professor and NSERC Industrial Research Chair at the University of Manitoba, Canada. In the past, Dr. Gole has worked in the field of DC transmission at Manitoba Hydro and Hydro Quebec in Canada. Dr. Gole is a member of the original development team for the PSCAD/EMTDC program, recipient of the 2007 IEEE PES Nari Hingorani FACTS Award, and a fellow of the IEEE and the Canadian Academy of Engineering.

Preface

Introduction of the book in brief

High-voltage direct current (HVDC) transmissions, including both the line commutated converter (LCC) HVDC and voltage source converter (VSC) HVDC, have played important role in the modern electric power system. More than 210 HVDC projects are delivering 150 GW of electric power all over the world. The reliable and efficient operation of HVDC transmission has been considered as a great contribution to the power grid. Therefore, modeling and simulation of HVDC transmission can be an essential work for the HVDC projects including the planning, the design, the commissioning, the operating and the maintenance. Modeling and simulation also serve as a platform for the engineers to study the behavior of HVDC system. With the inclusion of power electronic device, HVDC has the characteristics of nonlinearity and different timescale with traditional electromechanical system. Techniques for modeling and simulating HVDC have been developed in the past few decades and are also under investigation with the progress of HVDC technologies themselves. Hence, there is a need for a book to integrate the development and updated techniques of the modeling and simulation which is significant for the engineers, researchers and the related students majoring in the electric engineering.

The presented book focuses on the development and scenario of modeling and simulation for HVDC transmission system. Development of HVDC technologies will be introduced in brief at first and then the role of modeling and simulation in the research and development of HVDC system will be discussed. The HVDC to be studied can be LCC HVDC or VSC HVDC, the modeling covers the electromagnetic transient (EMT) model, electromechanical transient (transient stability [TS]) model, or the dynamic average model, with the electromagnetic transient model emphasized. The algorithms and their realization are explained for simulation with all of the mentioned models. The hardware-in-loop real-time simulation has become a well-recognized and developed technique for the simulation of HVDC system. The last chapter will give a sufficient explanation of the development and application of hardware-in-loop real-time simulation. Case studies and application examples will be given in the related chapters.

Chapters' organization and the contributors

Chapter 1: Development of HVDC and the needs for modeling and simulation.

The first chapter presents the general description of the development of HVDC, the classification of HVDC, the properties of HVDC, and the needs for modeling and simulation.

Contributors: Minxiao Han, North China Electric Power University; and Aniruddha M. Gole, University of Manitoba, Canada

Chapter 2: General description of modeling and simulation of HVDC

Chapter 2 gives a general description of the HVDC modeling, the classification, and properties for different modeling techniques, a general description of HVDC simulation and its development.

Contributors: Shujun Yao, Minxiao Han, North China Electric Power University; and Aniruddha M. Gole, University of Manitoba, Canada

Chapter 3: Electromagnetic modeling of LCC HVDC

This chapter discusses the modeling of the main circuit for LCC HVDC, the control system for LCC HVDC, and the newly developed HVDC projects such as UHVDC and CCC HVDC.

Contributor: Chao Hong, EPRI of China Southern Power Grid Co. Ltd., China

Chapter 4: VSC HVDC system modeling and stability analysis

Chapter 4 presents electromagnetic modeling of the main circuit and the control system of VSC HVDC and the stability transient of VSC HVDC will be analyzed and simulated.

Contributors: Hui Ding, Yi Qi, and Xianghua Shi, RTDS Technologies Inc., Canada

Chapter 5: Electromagnetic modeling of DC grid

Chapter 5 discusses the principle of multiterminal HVDC and DC grid. Some key equipment relating to DC grid will be given. The control strategies will be described. The authors will also discuss the steady and dynamic operation of DC grid.

Contributors: Minxiao Han, Guangyang Zhou, Zmarrak Wali Khan; North China Electric Power University

Chapter 6: Electromagnetic simulation of HVDC transmission

Algorithms for electromagnetic simulation and their realization are discussed in this chapter. After the introduction of principles and algorithms of electromagnetic transient simulation, the author discusses the general theories about the accuracy and numerical stability. A case study is given to illustrate the simulation strategies of the circuit with power electronic device.

Contributor: Shengtao Fan, Shengtao Fan Consulting, Canada

Chapters 7 and 8: Electromechanical modeling and simulation of HVDC, for LCC in Chapter 7 and VSC in Chapter 8

Electromechanical modeling and simulation are considered as powerful tools for the analysis of a large-scale power system with HVDC. These two chapters deal with the modeling of the main circuits, their control, and the simulating algorithms for LCC-HVDC and VSC-HVDC, respectively. The chapters also discuss the power flow calculation and stability analysis. Case studies and application examples are also illustrated in the chapters.

Contributors: Junxian Hou, Lei Wan, Jian Zhang, Electric Power Research Institute, China

Chapter 9: Dynamic phasor modeling and simulation of HVDC

The chapter starts with an introduction to dynamic phasors. Then the modeling of transmission lines using dynamic phasors is presented. It is followed by modeling of individual components of the HVDC systems. The combination of individual components is explained with the aid of block diagrams to illustrate the signal flow between components. Models are validated against electromagnetic transient simulation results.

Contributors: U.D. Annakkage, University of Manitoba, Canada; C. Karawita, Transgrid Solutions, Canada; S. Arunprasanth, H. Konara, RTDS Technologies Inc., Canada

Chapter 10: Small-signal modeling and analysis of HVDC system

This chapter is structured to interpret the small-signal modeling of both LCC and VSC-HVDC. The application and verification are given through examples. The construction of a converter input admittance as seen from the point of common coupling (PCC) is discussed. And this input admittance is expressed in matrix form and it is exemplified how the generalized Nyquist criterion can be used to analyze the stability of the converter–grid interconnection.

Contributors: Lennart Harnefors, ABB Corporate Research, Sweden; Lidong Zhang, ABB Power Grid Research, Sweden; Alan Wood, University of Canterbury, New Zealand

Chapter 11: Hybrid simulation for HVDC

Integration of electromagnetic and electromechanical hybrid modeling and simulation has been considered as an appropriate strategy to simulate large-scale power systems with HVDC. Chapter 10 is designed to introduce the idea, the methodology, and the case study of the hybrid simulation.

Contributors: Wenchuan Wu, Yizhong Hu, Tsinghua University, China

Chapter 12: Real-time modeling and simulation for HVDC

This chapter describes the modeling and simulation methodologies for HVDC systems on real-time digital simulators (RTSs). RTSs have been successfully used in factory and dynamic testing of HVDC controllers by HVDC manufactures and electrical power utilities. This chapter will focus on the special techniques used in real-time modeling of HVDC equipment and systems.

Contributor: Yi Zhang, RTDS Technologies Inc., Canada

Acknowledgments

The chief editors wish to thank the support from the Key Project of Smart Grid Technology and Equipment of National Key Research and Development Plan of China and the opportunities of discussions for the book supplied by "111" Project of China. We also express our sincere appreciation to the IET Assistant Editor Olivia Wilkins for her persistence in keeping us moving forward.

Chapter 1
HVDC and the needs for modeling and simulation

Minxiao Han[1] *and Aniruddha M. Gole*[2]

1.1 Development of HVDC

High-voltage DC (HVDC) transmission technology has been considered as the key technology for bulk electric power transmission and network interconnection [1,2]. HVDC is also referred to as the earliest and most significant application of power electronics in the electric power system. The technology is consistently active in innovation and the development for energy integration, transmission, and grid interconnection. This chapter will first provide a brief description of the configuration, classification, and development and then the discussions will be oriented to the needs for the modeling and simulation of HVDC.

1.1.1 General description of HVDC

The equipment used for HVDC can be classified as the main circuit devices and control/protection system, as shown in Figure 1.1 [3].

The main circuit devices include the converters that convert electric energy between DC and AC electricity, DC lines or wires connecting converters, converter transformers, filters, and reactive power compensator. The control and protection equipment ensure the arranged energy transmission and fault protection. The core of the HVDC transmission system is converters referred to as rectifier or inverter according to the direction of power flow. From the structural point of view, HVDC is power electronics conversion circuit with AC-DC-AC form. Most of the converters (also known as the converter valve) used in HVDC projects are composed of half-controlled device, thyristor, called line commuted converter (LCC) or LCC-HVDC which needs the support of the AC power source for the commutation. Recently, the voltage source converter (VSC)-based HVDC [2], VSC-HVDC, is developing quickly and applied widely as it can realize commutation based on the ability of its device to turn off. The converter of the VSC-HVDC transmission system adopts full-controlled self-commutated power electronic devices such as

[1]North China Electric Power University, Changping, Beijing, China
[2]University of Manitoba, Winnipeg, MB, Canada

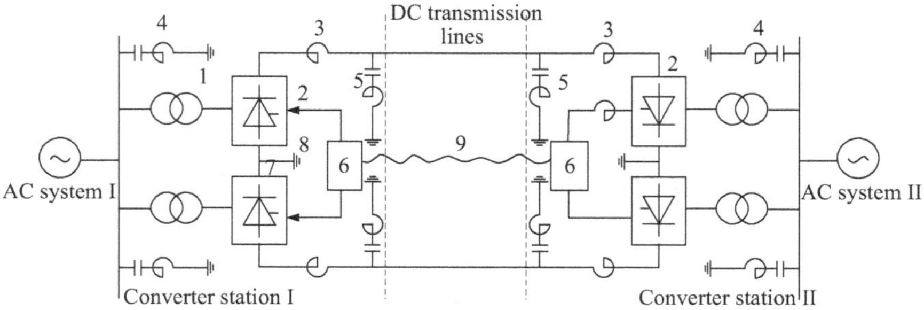

Figure 1.1 Schematic diagram of a two-terminal HVDC transmission system 1, converter transformer; 2, converter; 3, smoothing reactor; 4, AC filter; 5, DC filter; 6, control and protection system; 7, electrode lead; 8, electrode; 9, telecontrol communication system

gate-turn-off thyristor (GTO), insulate gate bipolar transistor (IGBT), and integrated gate-commutated thyristor (IGCT) with better controlling characteristics and less response time. The rapid development and commissioning of HDVC projects push a large quantity of power electronic equipment used in power system. The modern power system has the properties of multi-timescale, low inertia, and strong correlation.

1.1.2 Classification of HVDC

The HVDC transmission can be classified based on the different ways of commutation, different terminal numbers, or different connection relationships with the AC system. As illustrated above, converters can be divided into LCCs and self-commuted converters as the control characteristics of the adopted power electronic devices are different. At first, the traditional LCCs will be discussed in the following paragraphs. Categories concerned about the self-commuted HVDC are similar to that of line commuted HVDC, replacing LCCs and its accessory equipment with VSCs and the corresponding device combined with systems that will be illustrated briefly in Section 1.1.4 of this chapter.

1. Long-distance HVDC: Typical configuration shown in Figure 1.1 is the main scenario of HVDC transmitting power electricity from energy source center to load center, and the link between the mainland areas to islands through cables is achieved by HVDC.

   Long-distance HVDC can be further divided into single-direction HVDC and bidirectional DC transmission according to the power, which can be transmitted in one way or in two directions. In general, transmission from the thermal power plants and hydropower energy bases to load centers and islands with weak AC power grids is a one-way transmission. When the sending end contains a certain scale of AC system or islands are with an extensible power supply, DC power transmission usually adopts a two-way transmission mode. In this condition, the converters in the sending and receiving sides adopt the same structure used as both rectifiers and inverters.

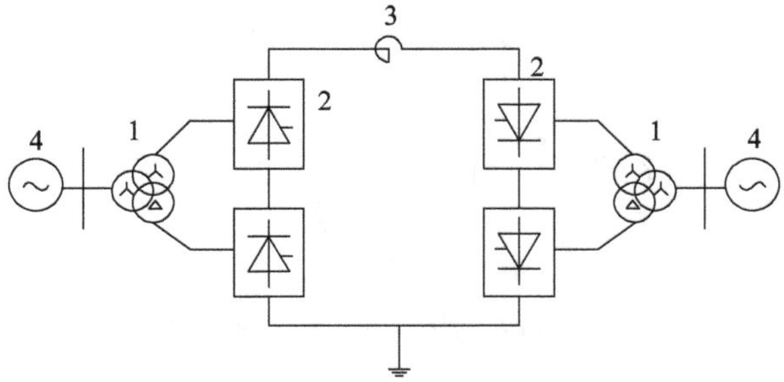

Figure 1.2 The circuit of back-to-back HVDC 1, converter transformer; 2, converter; 3, smoothing reactor; 4, AC system

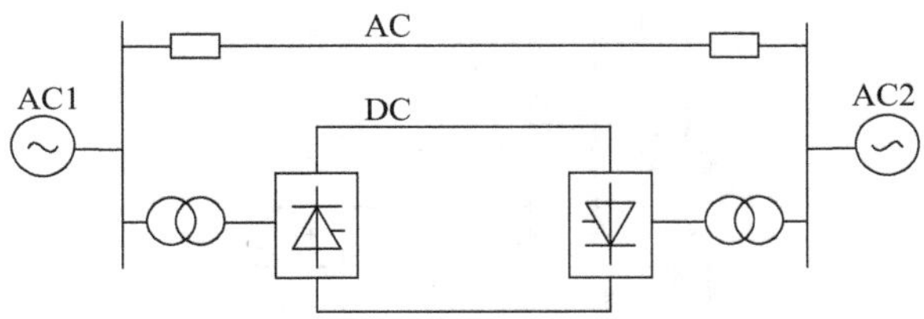

Figure 1.3 The circuit of sending power through AC and DC in parallel

2. Back-to-back HVDC: As shown in Figure 1.2, two groups of antiparallel coupling converters are connected through a smoothing reactor, so it is called as back-to-back HVDC. The converters on both sides are set in the same place with no DC lines, and this kind of HVDC transmission mode can quickly reverse the direction of power flow and can conveniently be used to control the power and frequency of the AC systems. Back-to-back HVDC can be divided into unipolar HVDC, bipolar HVDC, and HVDC with multiple sets of unipolar or bipolar in parallel. Figure 1.2 shows the main circuit of unipolar 12-pulse HVDC.
3. Sending power through AC and DC in parallel: As is shown in Figure 1.3, the two-terminal AC systems are connected by AC and DC lines. This transmission mode can be advantageous as HVDC control plays a significant role in the stability of AC systems, especially when the distance of two-terminal AC systems is relatively far [4]. The AC transmission paralleled with DC lines has the convenience of middle placement and can provide electricity for the loads from the central districts.
4. AC overlaid with DC transmission: Superimposing DC component on AC lines to make AC channel for DC transmission [5]. Both AC and DC powers are simultaneously transmitted to improve the transmission capability of lines. The mode shown in Figure 1.4 is free of the stability limit of power angle and, therefore, improves the transmission capability. According to the AC circuit number, this mode can form monopole pattern, returning current through the

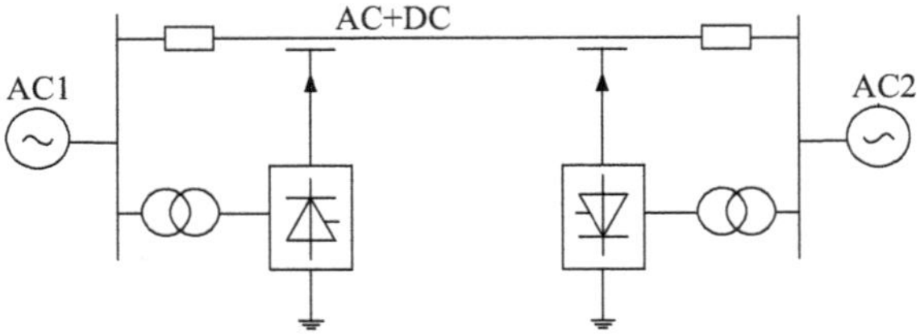

Figure 1.4 The circuit of AC overlaid with DC transmission

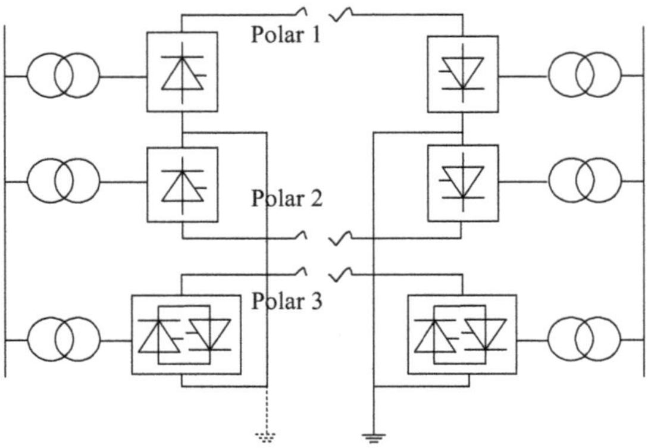

Figure 1.5 The circuit of tripolar HVDC

earth or bipolar pattern. In order to avoid the DC component flowing into the main transformer insulation, a capacitor should be added between the main transformer and the DC accessing point. In addition, the countermeasures of control and protection of pulsating voltage and current formed by the superposition of AC and DC have been important issues in this mode. Nowadays this kind of mode is still under research and has not conducted practical projects.

5. Tripolar HVDC: Using the existing AC transmission channel, the tripolar HVDC can adopt the topology of combining converters as shown in Figure 1.5 [6]. Currently, the research of this mode is still in the early stage. SIEMENS has carried out some tests and research but has not conducted practical projects.

1.1.3 Configuration of HVDC system

The HVDC construction can be divided into two-terminal HDVC and multi-terminal HVDC (MTDC), according to the number of converter station. Currently, most of the HVDC projects are of two terminals, and only a handful of MTDC projects have been put into operation. At first, the Quebec-New England project was designed as a five-terminal HVDC system but then changed into three-terminal HVDC operation because of the difficulties in controlling the coordination. With the practical utilization of VSC-HVDC, the construction of MTDC will become more flexible and relatively easier. The world has put into commercial operation more multiterminal VSC-HVDC projects.

1. Monopolar HVDC
   (a) Current return through the earth or seawater: The polar line of this mode can adopt an overhead line or cable, and the flowing back of the current can utilize the earth or seawater as the channel to reduce the transmission line investment, as shown in Figure 1.6. However, this mode requires sophisticated materials for the grounding electrode and also needs complicated commissioning. Furthermore, the return current will impose corrosion on the objects laid underground such as oil pipes, communication lines, and magnetic compass. So far, no practical project is available to demonstrate the returning current through the earth. Current returning through seawater has been applied for some projects which transmitted power crossing sea channel.
   (b) Current return through conductors: In order to avoid the problems illustrated above, a conductor is added as the recirculation channel as shown in Figure 1.7. It is apparently unreasonable economically that the monopolar converter adopts two loops of conductors; but as the HVDC project can be invested and built by stages, there are some practical projects using two conductors in monopolar HVDC as a stage in the bipolar HVDC construction. The Hokkaido–Honshu interconnection project was built in a similar manner and China's southwest water power ultra HVDC (UHVDC) project also used this mode in the commissioning stage. In case of an exceptional circumstances such as earthquakes, the bipolar scheme, as in (b), can change its operation mode from bipolar operation to monopolar operation, returning current through conductors.

2. Bipolar HVDC
   (a) Both ends of the neutral points are grounded: Figure 1.8 shows the neutral points of rectifier and inverter both accessing the earth or seawater through the grounded electrodes. This pattern is similar to two monopolar modes using the earth or seawater as the reflux channel. For the symmetry operation, the real current is relatively small as the directions of two reflux currents are opposite. When one polar quits operation as a result of faults, the other polar could still transmit 50% of the total power

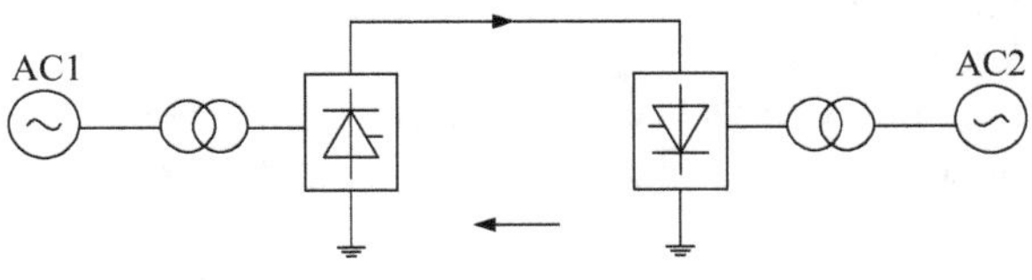

Figure 1.6 The circuit of current reflux through the earth or seawater

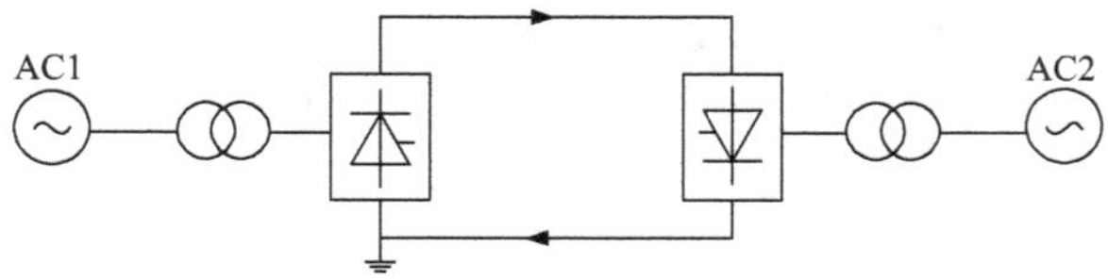

Figure 1.7 The circuit of current reflux through conductors

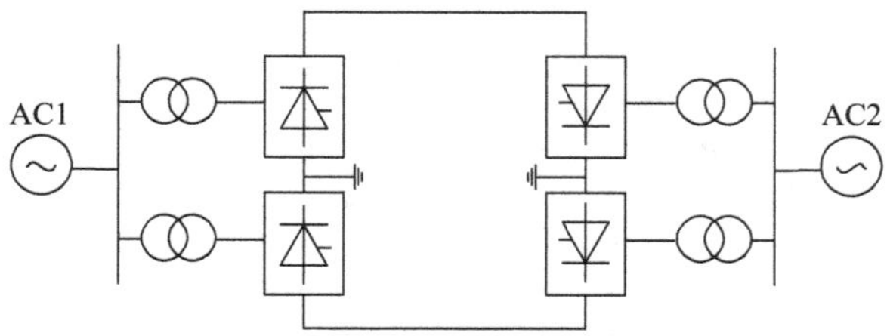

Figure 1.8 Bipolar HVDC with both ends of neutral points grounded

using the earth or seawater as the reflux channel. Therefore, this pattern greatly improves the reliability and availability of DC transmission. (Availability of DC transmission refers to the ratio of the equivalent operation hours converted under the maximum continuous capacity and the number of hours in a statistical cycle.) At present, most of the DC project building and operating adopt this mode wherein both the ends of neutral points are grounded and the majority of the bipolar HVDC projects in the world use this mode. Similarly, most of the HVDC projects in China are built in this way. During normal operation, as imperfect symmetry of the transformer parameters and the triggering angle, a certain amount of current flow will emerge in the neutral wire, and the effect caused by the current on the transformer near the grounded neutral point, the equipment laid underground, and communication should arouse enough attention.

(b) The single end of neutral points grounded: In this mode, only one end of the neutral point of the converter is grounded, as shown in Figure 1.9. This pattern can effectively avoid the grounded electrode current caused by the imbalance as noted in (a) and can greatly reduce the electromagnetic interference effect of the grounded electrode current when a monopolar fault occurs. However, the whole DC system could be out of operation when the monopolar becomes a failure, reducing the reliability and availability of DC system. The cross-channel project interconnecting Britain and France is one such instance. In the project, the system cannot operate normally due to the monopolar fault, which is corrected by building the neutral cable.

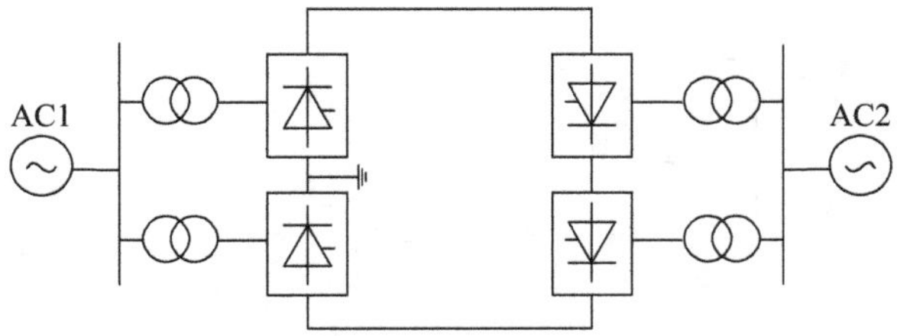

Figure 1.9 Bipolar HVDC with single end of the neutral points grounded

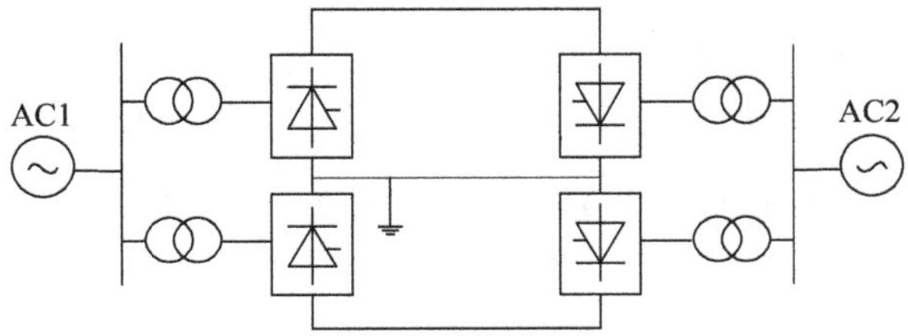

Figure 1.10 Bipolar HVDC using the neutral line mode

(c) Neutral line mode: Figure 1.10 demonstrates the bipolar HVDC using the neutral line mode. Another line can be neutral points of two-terminal converters grounded at the same time. When a monopolar fault occurs, 50% of the current will flow through the earth or seawater, so that the designed capacity of the neutral line can be reduced. In bipolar operation, smaller imbalance in current flow through the neutral line will reduce the electromagnetic interference caused by the neutral point current. The projects including Vancouver project in Canada, Hokkaido–Honshu interconnection project, and the Kii Channel project are all designed in this manner.

3. Sending power through multicircuit DC lines

   Multicircuit lines with parallel converters: As shown in Figure 1.11, both the converters and the transmission lines are in parallel; and through the connection between the two sets of DC transmission, the mutual backup can be realized, thereby improving the reliability of DC transmission and its availability.

4. MTDC and DC grid [7,8]

   The HVDC projects can be referred to as an MTDC when the number of converter stations is more than three. The MTDC can be divided into parallel mode and series mode, according to the connection pattern of converters.

   (a) Parallel MTDC: Converters share the same DC voltage as in parallel, and the two patterns called radial type and ring or mesh type are shown in (a) and (b) of Figure 1.12. Only the radial-type projects are currently in operation.

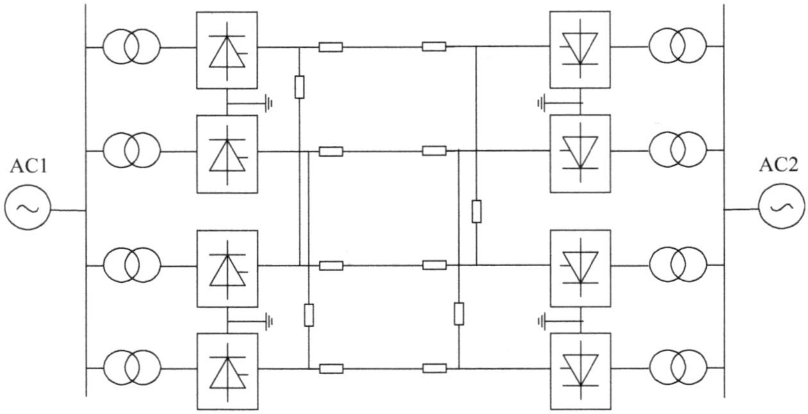

Figure 1.11 HVDC circuit with multicircuit lines having parallel converters

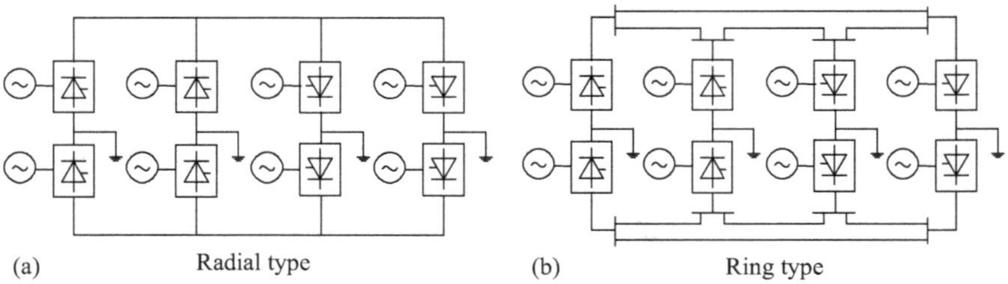

Figure 1.12 The circuit of parallel MTDC

The power can be modulated through adjusting the current, but the reverse of the power must change the connection mode between the converters and the DC lines through switching breakers.

Italy Sardinia (Sardinia) project [9] just adopts this radial type using the earth and seawater as the reflux channels, and the Corsica converter located in the middle was joined in the DC network through polarity reversal by switching breakers. Quebec – New England project [10] is also operating with a three-terminal HVDC structure (which was originally designed as a five-terminal system).

(b) Series MTDC: Converters are connected in series, so they can share the same amount of current. As shown in Figure 1.13, the power exchange between converters and AC systems can be modulated through adjusting the voltage. When one converter station stops running as a result of faults, putting it into a bypass circuit can assure the normal operation of other converters. Because of the difficulties in insulation coordination and high-power loss, the series MTDC has not yet found practical application.

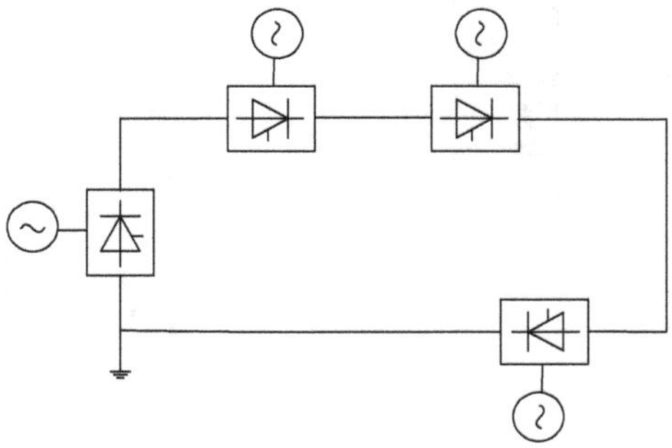

Figure 1.13 The circuit of series MTDC

The DC grid is generally considered to be the DC network with more complicated interconnections of converters, and there is no distinct line between MTDC and DC grid. If certain terminals in the MTDC system have more than one channel to reach each other, or if one or more meshed structures are included, it is referred to as a meshed DC grid. The MTDC can be considered as a radial-type DC grid. The terminology DC grid will be used in this book for both the MTDC and the meshed grid.

With the rapid development of large-scale remote area renewable generation and wide area energy transmission as in West Europe, North America, and mainland China, the scenarios call for the need of multipoint integration, multipoint consumption, and multiarea interconnection. The MTDC or DC grid has demonstrated their potential use during such scenarios.

1.1.4 Fundamental configuration and features of VSC-HVDC

Compared to the traditional LCC-HVDC transmission, the fundamental features of VSC-HVDC originated from the power electronic devices it employs. Thyristors for traditional HVDC can be triggered on but cannot be turned off unless the external circuit forces its current drop below the holding current. Therefore, traditional HVDC must rely on the external power source to achieve commutation. At present, the thyristor is the largest power electronic device applied in electric power transmission. Another group of continuously developing power electronic devices is the fully controlled ones such as metal-oxide-semiconductor field effect transistor (MOSFET), IGBT, GTO, IGCT, and injection-enhanced gate transistor (IEGT), which can be turned on or off by triggering pulse. HDVC constructed by the fully controlled device is called self-commuted HVDC. The self-commuted HVDC can be further categorized into VSC and current source converter (CSC) as shown in Figure 1.14. The structure of VSC is shown in Figure 1.14(a), and the DC voltage is maintained constant at the DC side, while the AC current is decided by the relationship between the voltage at the AC side and the output voltage of the inverter. The CSC is with a constant current source at the DC side as in Figure 1.14(b). In the aspect of output waveforms, the AC side voltage of VSC and the current of CSC can

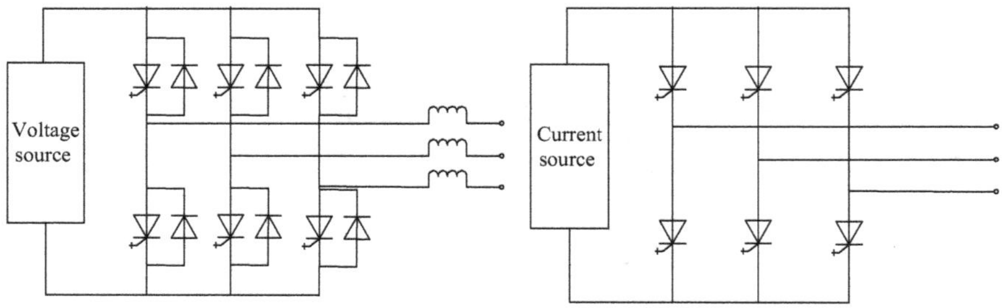

(a) Voltage source inverter (b) Current source inverter

Figure 1.14 Circuit configurations of self-commuted inverters with GTO

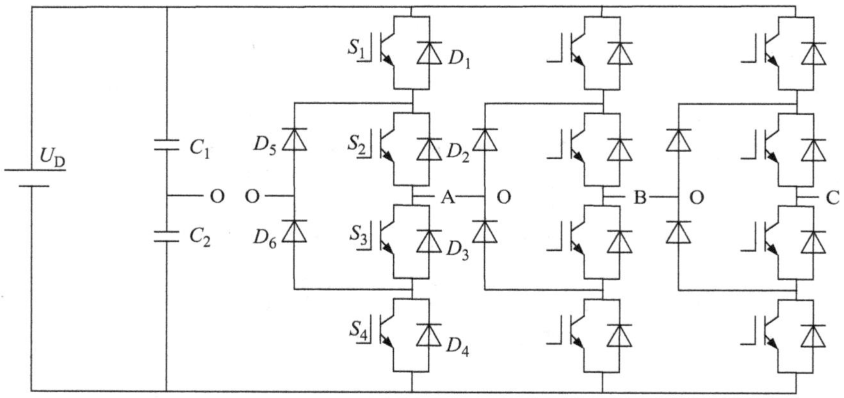

Figure 1.15 The circuit of three-level inverter with IGBT

be rectangular or pulse width modulation (PWM) waves. Both the self-commuted VSC and CSC inverters with small- and medium-sized capacities have been practical; and based on the purposes, different types of inverters can be used. VSC inverters are used in power systems with high voltage and large capacity; thus, VSC is a reasonable choice for HVDC and referred as VSC-HVDC [7].

A three-level inverter shown in Figure 1.15 has been applied in projects with small capacities and are usually used for the principle explanation. In the three-level pattern, four devices of each phase constitute three conduction modes corresponding to three DC voltage values, and the conducting time is obtained based on PWM control. The advantages of three-level mode are its simple structure, convenient control, high reliability, etc., but it lacks extension flexibility. It is mainly used for medium and small power projects. In fact, the high requirements for HVDC projects are to have a higher voltage and larger capacity for bulk energy transmission. Currently, the way for power devices in series connection to improve capacity has great limitations such as voltage sharing, harmonics, and device stress and cannot be widely and effectively used. Many novel topologies have been developed for circuits or modular connection techniques. Modular multilevel converter (MMC) technology is one of the most effective solutions to boost the capacity of VSC [11].

MMC uses submodules in series, which can avoid the direct series connection of semiconductor device and improve the harmonic characteristics of the output waveform. MMC based on the cascaded half-bridge and including public positive and negative DC bus bars was proposed as shown in Figure 1.16. Siemens utilizes this MMC topology in "Trans Bay Cable" VSC-HVDC project [12]. Since the "Shanghai-Nanhui" VSC-HVDC project in China adopted this topology, all the VSC-HVDC projects in China have employed the MMC topology [13]. MMC has no DC capacitor group between DC buses of multilevel inverters but has the structured characteristic of cascaded multilevel voltage source inverter.

MMC can generate the voltage at the multilevel AC side through controlling the on–off number of modules at each bridge arm. Modulation strategies of MMC could adopt the methods called the nearest level modulation and the carrier phase shift (CPS). The CPS method has a wide range of applications as it has features of simple control and low switching losses.

MMC based on the cascaded half-bridge has its limitations. With the development of MTDC and DC network and the adoption of DC overhead lines, VSC-HVDC has an urgent request for the ability to block the DC fault in converters. Multilevel converters based on full-cascaded H bridge modular have the ability to block the DC fault current and possess a more flexible controlling to satisfy the above requirements. Admittedly, the full H-bridge of power devices increases the cost and power loss, and the feasibility of its application in the VSC-HVDC is being paid wide attention and researched.

The VSC-HVDC control strategies will be discussed in Chapter 4 of modeling and simulation for VSC-HVDC. In general, through adopting device commutation, the DC power transmission can have better controllability and connect with weak AC systems or passive system. By decoupling the control of active and reactive

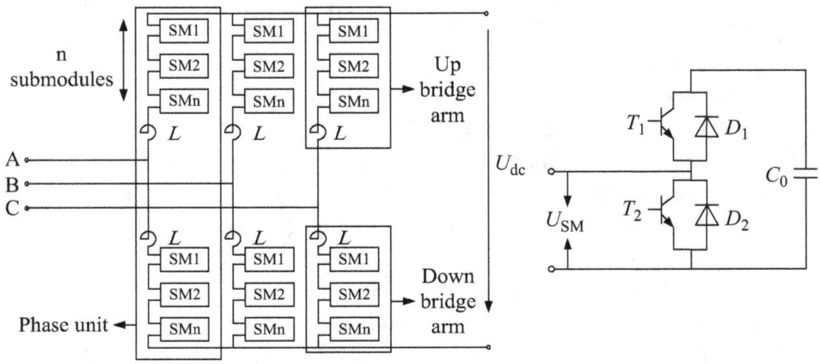

(a) Main circuit (b) Single module

Figure 1.16 The topology of MMC based on cascaded half-bridge

power, VSC-HVDC can play an active role in the stability control of AC network. With the properties of the four-quadrant operation of the VSC converters, the VSC-HVDC is prone to extend to a MTDC network. Compared with the traditional HVDC, the VSC-HVDC adopting PWM trigging is characteristic of higher working frequency and faster dynamic response, which also means some special consideration will be required for the modeling and simulation of the system including VSC-HVDC.

1.2 HVDC properties and its application

1.2.1 Economic and technical features for HVDC

(a) Economy benefits

HVDC has its rationality and applicability for long-distant bulk electric power transmission. Actually, the cost and operation expenditures of DC transmission line are lower than that of AC transmission, but the cost and operation expenditures of converter station are higher than that of AC substation. Thus, an equivalent distance is achieved when the cost of the DC transmission line and converter stations is equal to that of the AC transmission line and substations with that distant. In fact, to improve the transmission capacity of AC systems, we need to adopt various technical measures, which raise cost considerably. That is because the AC power capacity will be limited by its stability limit, with an increase in the transmission distance. At present, the equivalent distance of the overhead line is approximately 600–700 km; while for cable lines, it has been reduced to 20–40 km. On the other hand, according to the constitution of the HVDC system, HVDC project can be constructed in phases on the basis of voltage level or the number of poles, which is considered better for investment management. Following Moore's law for semiconductor power electronic devices, the economic benefits are getting more prominent.

The economic benefits of VSC-HVDC can be interpreted indirectly for more efficient and more beneficial for the renewable generation.

(b) Interconnectivity

The AC transmission capacity is limited by angle stability of synchronous generators. The longer the transmission distance, the larger the contact reactance between the synchronous generators, and the stability problem also becomes more prominent, which causes greater restrictions on AC transmission capacity. Compared with this, there is no synchronous issue in HVDC. The only limitations for HVDC power transmission capacity are the equipment capacity and receiving the AC system accepting capability.

The expansion of the AC system will lead to an increase in short circuit capacity. In many cases, the short circuit current level can be beyond the installed circuit breaker. And it is very difficult to choose an appropriate circuit breaker for all operating modes. When HVDC is used to interconnect the AC systems, it will not cause the increase in short circuit capacity and can prevent the expansion of the AC system fault as well. Thus, through dividing the huge AC systems into several relative independent subsystems, and using HVDC to interconnect, the short circuit capacity issue can be settled down and the system reliability can effectively be improved.

It is not necessary to synchronize two AC sides linked by HVDC; thus, HVDC can achieve asynchronous interconnection of the power grids. Further, HVDC can achieve the interconnection of AC systems with different frequencies. That means HVDC is operating as a frequency converter.

The development and application of VSC-HVDC provide a more effective solution for electric power system interconnection. It is unnecessary to reverse voltage to achieve power inversion; therefore, constituting a multi-terminal system or DC grid to actualize more regional interconnection seems much easier. The VSC-HVDC has no commutation failure problem and also no special requirements for the strength of AC systems. Thus, VSC-HVDC can be more flexible for the interconnection of power grids with various strengths and loads.

(c) Controllability of HVDC

HVDC, which has the characteristic of fast power flow control, can be used to control the stability and frequency of the interconnected AC systems. HVDC converter is a power control circuit composed of power electronic devices, which can rapidly and accurately control power flow. Both LCC-HVDC and VSC-HVDC can quickly reverse the power flow within the selected period. A slow power reversal can be observed in normal operation and fast power reversal for emergency control. The speed of power reversal depends on the DC power change rate, which is demanded by both sides of the AC systems and the limitation of HVDC main circuit. In normal operation, the time of power reversal usually involves a few seconds or even tens of seconds. When emergency power is needed, the time of fast power reversal is restricted to HVDC circuit parameters, especially for cable lines; its insulation property will be damaged by the fast reversal in voltage polarity, and its reverse time is usually greater than 200–500 ms. This kind of speed is significant for the stability control of the interconnected AC systems, the frequency control of load random fluctuation during the normal operation of the AC system, and the occurrence of failure. In VSC-HVDC, the power flow reversal is much easier and quicker, typically within a few milliseconds, because the power flow reversal is realized by the reverse of current but not the voltage in the case of LCC.

The VSC-HVDC converter has the characteristics of independently controlling active and reactive power, so that the VSC-HVDC plays a crucial role in the operation and control of AC/DC interconnected power systems. Active power control is more flexible and can easily achieve equivalent inertia. Reactive power can be adjusted in two directions to realize voltage support for a weak AC system.

(d) Multiple-timescale properties [14]

The application of power electronic technology is the core of HVDC. On the timescale, the operation time of power electronic devices has significant differences compared to the traditional power systems. The time constant of the boiler system ranges from tens of minutes to a few hours, the inertia time constant of turbine and generator system is a few seconds, and the time constant of power frequency quantities is tens of milliseconds. In LCC-HVDC, the trigger frequency of the devices equals to the power line frequency, but the turn-on time and turn-off time of the device are both less than 1 ms. The VSC-HVDC converter is triggered by PWM which usually ranges from hundreds to thousands of Hertz. The dynamic response time of the fully controlled devices is shorter, that is, usually tens of microseconds.

As shown above, the power system with HVDC included will be a multiple timescale system. This kind of system is particular in the modeling and simulation processes. In the modeling process, some basic problems include how to use the models to describe the dynamic process with different timescales, how to link different models and how to consider the influence of fast process in a slow process and vice versa. AC/DC hybrid power system simulation refers to the selection of time step, the improvement in the numerical method and the interfaces of subsystems with different timescales. These constitute the core content of this book as well.

(e) Disadvantages of HVDC

The disadvantages of HVDC are with more equipment, complex structure, high cost, high loss, and lower reliability of HVDC converter station. The converter will produce a lot of harmonics in the working process, especially in LCC-HVDC. If it is not handled properly, the harmonics that get into the AC system will cause a series of problems in the AC system operation. So it is necessary to set up a large number of filters to eliminate these harmonics. Second, when the HVDC adopts line commutation to transfer active power, it absorbs a large amount of reactive power that can reach 50–60% of the active power. This requires a large number of reactive-load compensation equipment and the corresponding control strategies. Moreover, the HVDC earth electrode and DC circuit breaker still have some unsolved technical problems. The disadvantages discussed above must be fully considered in the modeling and simulation of HVDC in order to correctly evaluate the feasibility of the HVDC application. With the application of VSC-HVDC and the development of power electronic technology, the issues discussed above gradually improve.

1.2.2 HVDC application

1. Pertinent cases for the application of HVDC

Due to its characteristics, HVDC has its application scope, which are discussed as follows.

(a) Submarine cable power transmission: From a global perspective, the submarine cable power transmission project accounts for one third of HVDC projects. As mentioned above, the cable line equivalent distance of HVDC reduced 20–40 km; therefore, the submarine cable power transmission has been widely used in the following two cases.
   (i) Load power supply and power delivery: Forty percent of submarine cable power transmission is used for load power supply and power delivery. Gotland project was the first operation project to solve the problem of power transmission from the mainland to the islands, after which three similar projects were constructed. After 1997, submarine cable power transmission was used to deliver the power generated by heat, water, fuel, wind, and so on from islands to mainland in other projects.
   (ii) AC systems' interconnection: Sixty percent of submarine cable power transmission is used for AC systems' interconnection. This kind of interconnection not only has an economic advantage but also has benefits on the operation and control of AC systems. Hokkaido–Honshu HVDC link and Kii Channel HVDC link demonstrate this kind of project. Kii channel HVDC link plays an important role in the stability control of Honshu and Shikoku AC systems.

(b) Long-distance overhead transmission: Research showed that with regard to 10 GW, 300 km power delivery, HVDC overhead transmission already proves its advantage. Based on this conclusion, HVDC overhead transmission capacity will account for more than 26% of the global total transmission capacity. Two applications are represented.
   (i) Power transmission

   The obvious advantages in using HVDC overhead transmission are to deliver a large amount of power from the power center to the load center with a certain distance. Even if the economic distance is not considered, it can also bring the advantage of stable control by sending electrical energy through DC or AC/DC parallel. Nelson River project, Québec-New England project in North America, Gezhouba-Nanqiao project, Sanxia-Guangzhou project, and Guizhou-Guangzhou project in China demonstrate this kind of system.
   (ii) AC systems' interconnection: About 20% of the long-distance overhead transmission is used for AC systems' interconnection to improve the stability and frequency control of the AC systems. Pacific inertia is one of the few typical applications. With an increase in the short-circuit current limitation of the AC system and an improvement in power flow control requirements, this kind of application will increase in the future. In addition, the development

of the VSC-MTDC can realize the interconnection of multiple asynchronous power grids.

(c) Back-to-back (BTB): The BTB project accounts for about 40% of HVDC projects in the world, which is principally used for power exchange and emergency power support. The application of BTB can be divided into the following two categories.
   (i) AC systems' interconnection: This kind of BTB is used to connect the AC systems with the same frequency to realize asynchronous interconnection of two AC systems. This kind of project accounts for about 80% of all BTB. Eel River project and Lingbao project demonstrate such a system.
   (ii) AC system interconnection of different frequencies: This kind of BTB is used to connect AC systems with different frequencies, which accounts for about 20% of all BTB. It is used in Japan and Arabia mainly to interconnect the continents with the different frequency systems.

(d) Countermeasure of short-circuit capacity: With a constant increase in load, generator, the scale of the AC system, becomes larger. Under this circumstance, the short-circuit current increases. HVDC, as a countermeasure to limit short-circuit current, received great attention:
   (i) Load power supply: To supply power to the concentrated urban areas, the underground cables must be used. In this case, choosing the circuit breaker will be difficult because of the large short-circuit current passing through a small equipment that is limited by small space, and the use of DC transmission at this time will be advantageous. The VSC-HVDC presented later is more advantageous.
   (ii) System partition: Dividing the existing large-scale AC system into a number of relatively small and independent operation systems, using BTB or other methods can effectively reduce the short-circuit current. The engineering application has been explored in North America. Japanese scholars studied the serial system between Kansai, Chugoku, Kyushu, and Shikoku, which shows that if DC connection is adopted between Kansai and Chugoku, Chugoku and Kyushu, Kyushu and Shikoku, and Shikoku and Kansai, the short-circuit current will greatly inhibit and can also realize the power supply from the small system to the large system.

(e) Collection and delivery of renewable energy generation: Due to the limitation of the global environment and changes in the form of energy development, renewable energy generation rapidly expanded. Offshore wind power, wind power in remote areas, and desert photovoltaic power station far from the load centers are becoming much bigger and have a more urgent need for power delivery. This kind of power supply has the characteristics of wide distribution, strong power volatility, and weak AC power grid; therefore, it is difficult to adopt AC system or LCC-HVDC to connect and deliver. As a new type of transmission, VSC-HVDC has obvious advantages. It has active and reactive power regulation and fault blocking abilities which can reduce the influence of renewable energy generation uncertainty on AC system stability. Under the same condition, compared with AC transmission, the transmission capacity of VSC-HVDC is larger and the transmission distance is longer. It is conducive to collect and deliver renewable energy in a large area. VSC-MTDC is convenient for collection and renewable energy delivery in a large area. With the application of large capacity and high ratio of DC/DC converter, VSC-HVDC can be directly used for DC boost, collection, and delivery, reducing the AC-DC transformation. Thus, VSC-HVDC has become an effective way for grid connection and delivery of renewable energy generation in the future.

2. Key events in the history of HVDC
   (a) Evolution of LCC-HVDC [3]: Grid controlled mercury arc valve was successfully developed in 1928, which made HVDC practically realize. The Gotland project (put into operation in 1956) in Sweden was an early HVDC project and was also the first commercial operation HVDC project. The initial design of the Gotland power delivery consists of 20 MW, 100 kV, monopole, and cable connection. Engineering demonstration showed that this kind of transmission was more economical than building a new thermal power plant on the island. Besides, AC cable transmission could not be adopted under the condition of this distance (96 km). Sakuma interconnection project (1965, 1993) in Japan was the first zero-distance HVDC, which was used for interconnecting two systems with different frequencies. In Japan, the project could realize the bidirectional power exchange between 50 and 60 Hz AC systems, with 300 MW capacity and $\pm$125 kV voltage. Pacific Intertie project (put into operation in 1970) in America and the Nelson River project (put into operation in 1973) in Canada are the examples of long-distance mercury-valve HVDC project.

      Sardinia-Italy (1967, 1987) project, which connected Sardinia and Italy mainland, adopted monopole, ground seawater return mode, rated voltage 200 kV, and transmission capacity 200 MW, can supply the frequency support for Sardinia. Transmission lines are of different types due to the different environments, with the overhead line on land and submarine cable under the sea, and the total cable distance is 121 km, which hinted HVDC as the economic choice. In 1987, the system capacity extended to 300 MW, and a new branch converter station was built on Corsica, making the project become a three-terminal HVDC, which was the first MTDC example in the world.

      Eel River BTB project commissioned in 1972 connected two asynchronous AC systems of Québec and New Brunswick in Canada. Since then, all the new HVDC projects adopted the thyristor valve that caused great development of HVDC. New England Region in America and Canada Hydro-Quebec agreed to purchase 30 billion kWh electricity within 11 years, so an HVDC project, bipolar 690 MW ±450 kV, was constructed. After that, the project was expanded with Sandy Pond, Nicolet, and Radisson converter stations, forming a multiterminal system. Its transmission distance consisted of 1500 km length. However, due to the complexity of the operation control, Des Cantons and Comerford converter stations did not join the multiterminal system. By the end of 2015, more than 210 HVDC projects were put into operation in the world, including 11 mercury arc valve HVDC projects which were built in the first 25 years of the HVDC technology development. Most of the remaining were LCC-HVDC projects, and more than ten were VSC-HVDC. In all the HVDC projects were about 58 BTB, which accounted for nearly one third of all the HVDC projects, and two third were long-distance HVDC projects.

Since the 1950s, China got engaged in the research of HVDC transmission technology. In the 1960s, China Electric Power Research Institute built the first thyristor valve simulation device. In 1977, an AC cable line in Shanghai was transformed into a 31-kV DC transmission test line for researching HVDC technology. Since the late 1980s, considering the power delivery from Three Gorges, the research and development of HVDC technology in China improved by leaps. At present, 26 HVDC projects are put into operation with about 70 GW total capacity, which is close to half of the world HVDC capacity. China plans to put more than 30 HVDC projects into operation before 2020. In particular, to realize the West–East electricity transmission project, China is energetically expanding the construction of UHVDC, including ±660, ±800, and ±1,100 kV. In 2010, China built the world's first ±800 kV UHVDC project, which had the highest DC voltage level at that time. Chinese power grid has become the world largest AC-DC hybrid power grid, which causes modeling and simulation analysis a challenging research topic.

(b) Development of VSC-HVDC engineering projects [15]: In March 1997, the world's first VSC-HVDC Hellsjön project was put into operation. Since then, a number of VSC-HVDC transmission projects have been put into commercial operation. The representative applications of the VSC-HVDC project are shown as follows.

Hellsjön project in Sweden was the first project to test device commutation. This transmission line whose capacity is 3 MW, voltage ±10 kV, and the length 10 km was transferred from a temporary unused AC 50 kV line, which connected Hellsjön and Grangesberg in the middle of Sweden.

Gotland project, located in Gotland of Sweden, was the world's first commercial operation VSC-HVDC system which was put into operation in December

1999. Its voltage was rated at ±80 kV, converter capacity 65 MVA, and power devices adopted IGBT (2.5 kV/700 A). The purpose of the Gotland project was to deliver surplus wind power in the southern part of Gotland to the North load center. The output power fluctuation generated by wind farms on the island (about 10%) would affect the power quality if the power was sent directly to the load side. Thus, by adopting VSC-HVDC, the power quality was controlled. In the transmission mode, considering the transmission line corridor, the cable transmission method was adopted. The cable line length was about 70 km and the buried depth was about 50–70 cm. In order to save the cost of laying lines, a 40-km line was used from the previous AC transmission line. In the reactive power compensation, stabilizing the voltage in this project could be realized through controlling the VSC, compensating reactive power of asynchronous wind turbines and load consumption.

Directlink project connected Queensland and New South Wales power grids which were asynchronous through a 59-km cable line in Australia. Directlink project used the previous AC line as the DC line to reduce the impact on the environment. In addition, three pieces of the same equipment (±80 kV and 60 MW) operated parallelly, and the total capacity reached 180 MW, and its cable was 2 × 3.

Tjæreborg demonstration project in Western Denmark was mainly used to deliver offshore wind power into the AC system. At present, Denmark is actively promoting the offshore wind power technology. The purpose of this project was to verify the transmission technology of large capacity offshore wind power generation, which was 50 km to land and 100 MW of capacity. Its wind turbines are of four kinds, and the total capacity was 65 MW. The project was an AC/DC hybrid system and had three operation modes (AC, DC, and AC/DC hybrid). To achieve the highest efficiency of wind turbine operation, the general operation was maintained between 32 and 52 Hz. Therefore, in order to isolate from the AC system and maintain the power quality of the AC system, the VSC-HVDC system is usually used for power supply.

The Eagle Pass project, located in the west US border with Mexico, adopted the BTB method with power rated at 36 MW, asynchronously interconnecting the Texas power grid of America and Mexico power grid through two transmission lines with 138 kV. However, if a fault happens during the peak time due to the increasing load, it may lead to voltage instability. At this point, the load could be switched to the 138 kV transmission line of the American side; but in order to switch, it must first cut off the load, which in turn causes the interruption to some users. To solve the voltage stability problem and ensure continuous power supply between America and Mexico, this project was equipped with a VSC-HVDC system whose power was rated at 36 MW using the BTB method in the 138 kV system. This system can operate not only as a BTB system but also as STATCOM. Through switching, the system can be used as two 36 MVA STATCOMs putting into 138 kV system on the American or Mexican side. The project can also be used for black start and power transmission to the passive system. Meanwhile, it can realize the interconnection with the main grid when it gives power to the load.

The Cross-Sound project was the first long-distance submarine cable transmission project, connecting the Connecticut power grid and Long Island power grid in America, which was put into operation in May 2002. Its power is rated at 330 MW, and voltage at ±150 kV.

Nanhui VSC-HVDC is the first VSC-HVDC project in China. It was commissioned in 2011 for collecting the wind power in Nanhui, Shanghai China. The system is with a DC voltage of ±30 kV, transmission capacity of 18 MVA, and transmission distance of 8.6 km. Since then, VSC-HVDC has been rapidly developed and applied in China [16]. The following are brief introductions to representative projects commissioned in the past few years.

(i) Three terminal VSC-HVDC project in Nan'ao Island

In December 2013, the world's first commercial application of multi-terminal VSC-HVDC, which was three terminal, was put into operation in China's southern power grid. This project delivers the power from wind farms in Nan'ao Island to the inverter station in the mainland. The designed capacity was 200 MW, DC voltage level was ±160 kV, the capacity of three stations was 50, 100, and 200 MW, respectively, and hybrid transmission line (with both cable and overhead lines) was of 40.7 km length. The construction of the project is shown in Figure 1.17.

(ii) Five-terminal VSC-HVDC project in Zhoushan Islands

In July 2014, ±200 kV five-terminal VSC-HVDC science and technology demonstration project was put into operation in Zhoushan Islands of Zhejiang, which had the largest terminals and each terminal with the maximum capacity. This marks Chinese VSC-HVDC technology in the forefront of the world. The capacity of the converter stations and spacing between them are shown in Figure 1.18. Operating of the project strengthens the power interconnection between islands and improves the power supply capacity and flexibility of operation. Recently, the successful operation of the DC breakers based on this system provides favorable support for the development of the DC power system.

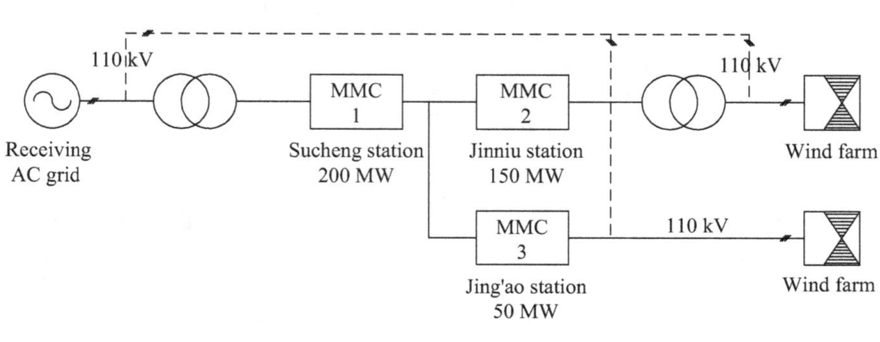

Figure 1.17 Three-terminal VSC-HVDC project in Nan'ao Island

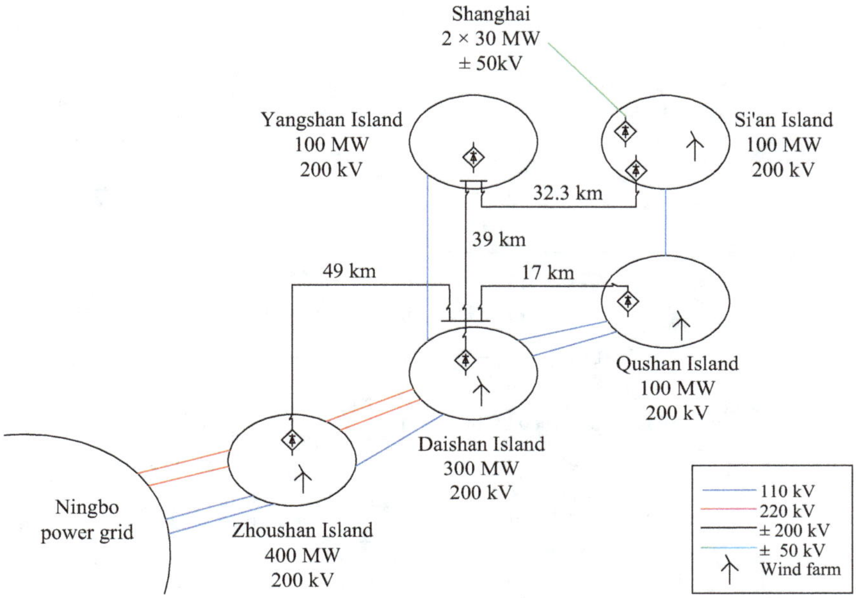

Figure 1.18 Five-terminal VSC-HVDC project in Zhoushan Islands

(iii) VSC-HVDC project in Xiamen

Xiamen flexible HVDC transmission project, with a voltage of 320 kV and the transmission capacity of 1,000 MW, had the world highest voltage level and the largest capacity of flexible DC transmission project at the commissioning time. Two new converter stations were built inside and outside the Xiamen Island. DC transmission line consisted of 10.7 km length and all land cables through the Xiang'an Xiamen subsea tunnel. Once implemented, the project provides a more flexible and effective means for voltage control and operation mode selection of the Xiamen AC grid.

(iv) VSC-HVDC DC grid project in Zhangbei

China's Zhangbei region is rich in wind and photovoltaic resources, which is a key position connecting the Inner Mongolia power grid and the North China power grid. A large-capacity pumped storage power project was constructed in this area. Therefore, the favorable characteristics of VSC-HVDC are complex energy forms, a more interconnected power grid, and high requirements of flexible operation mode, which are favorable for. A VSC-HVDC project, short-term four-terminal and long-term six-terminal with loops, is being designed, as shown in Figure 1.19. A group of state-of-the-art technologies are employed in the system design and operation, such as overhead line for interconnection, DC breaker for fault isolation, and coordinating control for meshed DC grid. The detailed modeling and simulation will be given in Chapter 5.

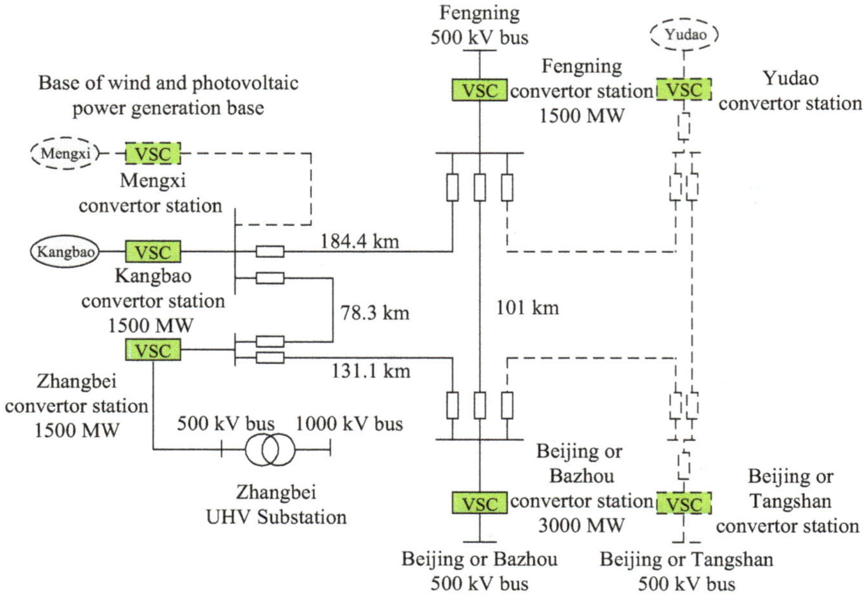

Figure 1.19 VSC-HVDC grid project in Zhangbei

1.3 Modeling and simulation of power system with HVDC

1.3.1 Properties and development of HVDC modeling and simulation

Modeling and simulation of HVDC transmission can be an essential work for the HVDC project planning, designing, commissioning, and also for the analysis and operation of the power system with HVDC links. The modeling and simulation also serve as a platform for engineers and researchers to study the behavior of the HVDC system [17].

The properties of the HVDC electric power grid are different from the traditional electromechanical power system with regard to the multi-timescale, low inertia, nonlinearity, and time varying. The basis for the difference is the inclusion of power electronics devices, converter circuits, and control systems. The successfully simulation of power system with HVDC is based on the pertinent modeling of power electronics system [18].

Both physical modeling and digital modeling have been developed for HVDC simulation. The physical models are built with real-world physical devices, according to the emulating equivalent principle. The physical models have played a key role in the early stage of HVDC development and even today they are also used in some key laboratories all over the world. With the development of computing capability and the need for more flexible simulation of a variety of HVDC operating modes, digital modeling and simulation have been developed in the past few decades. The digital modeling for a power system with HVDC means a group of differential and or algebraic, Boolean equations will be established and simulation is the solution of the equation by means of digital calculation. Modeling needs a full understanding of the steady and dynamic behavior of the HVDC system, and the models developed must be complete without any redundancy. The simulation procedure usually resorts to the pertinent numerical computation method [19,20].

The digital modeling for HVDC can be further classified into electromagnetic modeling (EMT) [21] and electromechanical modeling or transient stability [22], according to the selection of circuit variables. Hybrid modeling [23] combining EMT modeling and the electromechanical modeling has been a hot research topic for HVDC power system. With respect to simulation speed, digital simulation can fall into non-real-time simulation and real-time simulation. The principle study, design, or planning of system with HVDC may need no real-time simulation [24]. However, the test of control and protection devices for HVDC needs a real-time simulation, which is in the form of a hardware-in-loop system [25].

1.3.2 HVDC modeling and simulation in the presented book

Techniques for HVDC modeling and simulation have been developed in the past few decades and are also under investigation with the progress of HVDC technologies themselves. Hence, there is a need for a book to integrate the development and updated techniques of the modeling and simulation which is significant for the engineers, researchers, and the related students majoring in electric engineering.

The book will focus on the development and scenario of modeling and simulation for the HVDC transmission system. The development of the HVDC technologies will be introduced in brief at first, followed by a discussion on the role of modeling and simulation in the research and development of HVDC system. Both LCC-HVDC and VSC-HVDC will be studied, and the modeling covers the electromagnetic transient model, electromechanical transient model, or the dynamic average model, with the electromagnetic transient model emphasized. The algorithms and their realization are explained for simulation with all of the mentioned models. A platform based on small-scale physical equipment can also be an effective strategy for the simulation of HVDC system, which will be introduced in brief. The hardware-in-loop real-time simulation has become a well-recognized and -developed technique for the simulation of HVDC system. The last chapter in this book will give a sufficient explanation of the development and application of hardware-in-loop real-time simulation. Cases studies and application examples will be given in the related chapters.

References

[1] C. Adamson and N. Hingorani, *High voltage direct current power transmission*. London: Garraway, 1960. [Online]. Available: http://books.google.com/books?id=mQkjAAAAMAAJ (accessed Feb 2014).

[2] J. Arrillaga, Y. H. Liu, and N. R. Watson, *Flexible power transmission: the HVDC options*. Chichester: John Wiley, 2007.

[3] K. R. Padiyar, *HVDC power transmission systems*. 2nd ed. New Delhi: New Age International, 2012.
[4] J. Arrillaga and B. D. Smith, *AC-DC power system analysis*. Stevenage: Institution of Electrical Engineers, 1998.
[5] CIGRE Brochure 244, *Conductors for the uprating of overhead lines*[R]. [S.l]: CIGRE Brochure 244, WG B2.12, 2004.
[6] CIGRE Brochure 127, *Guide for upgrading transmission systems with HVDC transmission*[R]. [S.l]: CIGRE Brochure 127, WG 14.11, 1998.
[7] N. R. Chaudhuri, B. Chaudhuri, R. Majumder, and A. Yazdani, Multi-terminal direct-current grids: modeling, analysis, and control. New Jersey: Wiley-IEEE Press, 2014.
[8] D. Jovcic and K. Ahmed, *High-voltage direct-current transmission: converters, systems and DC grids*. Chichester: John Wiley & Sons, Ltd, 2015.
[9] J. P. Taisne, V. Arcidiacono, and F. Mazzoldi, 'The Corsican tapping: from design to commissioning tests of the third terminal of the Sardinia-Corsica-Italy HVDC'. *IEEE Transactions on Power Delivery*, vol. 4, no. 1, pp. 794–799, 1989.
[10] G. Morin, L. X. Bui, S. Casoria, and J. Reeve, 'Modeling of the Hydro-QUEBEC—New England HVDC system and digital controls with EMTP'. *IEEE Transactions on Power Delivery*, vol. 8, no. 2, pp. 559–566, 1993.
[11] Sanchez D and Green T,, 'Control of a modular multilevel converter-based HVDC transmission system[C]'. *Proceedings of Power Electronics and Applications Conference.* Birmingham, UK: *IEEE*, 2011: 1–10.
[12] T. Singh, *10 EU countries pledge to create North Sea renewable energy grid.* 2010 [Online]. Available: http://inhabitat.com/10-eu-countries-pledge-to-create-north-searenewable- energy-grid/ (accessed Feb 2014).
[13] B. Gemmel, J. Dorn, D. Retzmann, *et al.,* 'Prospects of Multilevel VSC technologies for power transmission[C]'. *Transmission and Distribution Conference and Exposition*. 2018: 1–6.
[14] X. Chen, M. Han, C. Liu, S. Ma, and X. Guo, 'Modeling of a large scale UHVAC/DC power networks based on PSCAD/EMTDC'. *Asia-Pacific Power and Energy Engineering Conference, APPEEC*, Mar. 2010.
[15] A. Yazdani and R. Iravani, *Voltage-sourced converters in power systems: modeling, control, and applications*. Oxford: Wiley, 2010.
[16] *Planned HVDC projects in China*. [Online]. Available: www.siemens.com/energy/hvdc.
[17] Jovcic, D., Pahalawaththa, N., and Zavahir, M. 'Analytical modeling of HVDC-HVAC systems'. *IEEE Transactions on Power Delivery*, vol. 14, no. 2, pp. 506–511, 1999a.
[18] A. M. Gole, A. Keri, C. Kwankpa, *et al.*, 'Guidelines for modeling power electronics in electric power engineering applications'. *IEEE Transactions on Power Delivery*, vol. 12, no. 1, pp. 505, 514, 1997.
[19] J. L. Hay and N. G. Hingorani, 'Dynamic Simulation of Multi-converter HVDC Systems by Digital Computer[J]'. *IEEE Transactions on Power Apparatus and Systems*, vol. PAS-89, no. 2, 1970.

[20] A. Kurita, H. Okubbo, D. B. Klapper, *et al.*, 'Multiple time scale power system dynamics simulation[J]'. *IEEE Transaction on Power System*, vol. 8, no. 1, pp. 216–223, 1993.

[21] H. W. Dommel, 'Techniques for analyzing electromagnetic transients'. *IEEE Computer Application in Power System*, vol. 10, no. 3, pp. 18–21, 1997.

[22] A. A. Fouad and V. Vittal, *Power system transient stability analysis using the transient energy function method*. Englewood Cliffs, NJ: Prentice-Hall, 1992.

[23] V. Jalili-Marandi, V. Dinavahi, K. Strunz, J. A. Martinez, and A. Ramirez, 'Interfacing techniques for transient stability and electromagnetic transient programs, IEEE Task Force on Interfacing Techniques for Simulation Tools'. *IEEE Transactions on Power Delivery*, vol. 24, no. 4, pp. 2385–2395, 2009.

[24] T. L. Maguire, 'Real time digital electromagnetic transients simulation of power systems'. *2000 Power Engineering Society Summer Meeting* (Cat. No. 00CH37134), vol. 2, 2000.

[25] Zhe Zhu, Xiaolin Li, Hong Rao, Weihua Wang, and Wei Li, 'Testing a complete control and protection system for multi-terminal MMC HVDC links using hardware-in-the-loop simulation'. *IECON 2014 – 40th Annual Conference of the IEEE Industrial Electronics Society*, pp. 4402–4408, Oct. 29, 2014–Nov. 1, 2014.

Chapter 2

General practices of HVDC modeling and simulation

Shujun Yao[1], *Minxiao Han*[1] *and Aniruddha M. Gole*[2]

2.1 General description of high-voltage DC (HVDC) modeling and simulation

Modeling and simulation of HVDC transmission have been considered essential for the HVDC project planning, commissioning, and operation. The modeling and simulation also serve as a platform for the engineers and researchers to analyze the behavior of the HVDC system [1]. The fundamental difference in the modeling and simulation of the power grid with HVDC is the inclusion of the power electronic systems [2]. First, this chapter will discuss about the modeling and simulation for power system with HVDC, which basically depend on the successful modeling and simulation of power electronic system including the power electronic devices, converter circuits, and control systems. Then the general practices for HVDC modeling and simulation will be introduced.

With the development of computing capability and the need for more flexible simulation of a variety of HVDC operating modes, the mathematical modeling and digital simulation have been developed in the past few decades [3,4]. The mathematical modeling for a power system with HVDC means a group of differential and or algebraic, Boolean equations will be established. Digital simulation is the solution of the equations by means of digital calculation. Modeling needs a full understanding of the steady and dynamic behavior of the HVDC system, and the models developed must be complete, without any redundancy. The simulation procedure usually resorts to the pertinent numerical computation method.

According to the selection of circuit variables, the modeling for HVDC can be further classified into electromagnetic modeling (EMT) [5] and electromechanical modeling or transient stability (TS) [6]. Hybrid modeling [7] combining EMT and electromechanical modeling has been a hot research topic for HVDC power system. With respect to simulation speed, digital simulation can fall into the nonreal-time simulation and real-time simulation [8]. The modeling and simulation

[1]North China Electric Power University, Changping, Beijing, China
[2]University of Manitoba, Winnipeg, MB, Canada

techniques are described in brief in this chapter; however, the detail discussion will be given in the following chapters.

2.2 Modeling and simulation of power electronics

The properties of the electric power grid with HVDC are different from the traditional electromechanical power system: with the small transient time constant, nonlinearity, and time varying. The basis for the difference is the including of power electronic devices, converter circuits, and control system. The general description of the modeling and simulation related to the three aspects are described in brief as follows.

2.2.1 Modeling and simulation of power electronic devices

The typical power electronic devices used for HVDC projects include the power diode, thyristor, GTO, IGBT, and IGCT. A detailed nonlinear i-v modeling is needed for converter design and heat dissipation and electromagnetic compatibility (EMC) analysis, while a piece-wise linearity i-v modeling is satisfied for the input–output analysis and the interaction behavior study with the power grid. The on-state voltage drop, typically from 2 to 5 V [9], will be used for device modeling which helps in the efficiency study. The switching transient speed of the devices is typically limited by the turning off transient that defines a period of time for the devices to recover their blocking capability. The switching speed is one of the key parameters for the operation of the converter, thus the choice of the simulation strategy.

The switching behavior of power electronic devices can be described with various equivalent power sources (independent or controlled) and R, L, and C elements (linear or nonlinear). The equivalent elements constitute a series of conventional circuits that vary with the switching mode [9]. In each mode, the general circuit simulation method can be employed to solve the problem, and the coherent solution of each mode constitutes the overall simulation of the circuit. In the actual simulation, the power electronic device model can be divided into three categories according to different simulation objectives and timescales:

1. The physical model of the device: This model is used to study the mechanism and switching process of the device, considering the semiconductor materials, the fabrication structure, the designed parameters, the manufacturing processes, and environmental factors. This model is suitable for the design and research of power electronic devices but is seldom used for the simulation of power electronic circuits such as for HVDC power transmission.
2. Micro model: This model uses a detailed equivalent circuit to describe the external characteristics of the device switching process. Although the model also involves some of the physical parameters within the device, the main purpose is to obtain the corresponding equivalent parameters. Such a model usually takes into account the influence of peripheral circuits, such as snubber circuits and their parameters, and is suitable for studying the voltage and current stress, switching loss, and electromagnetic compatibility in the transient process of the device switch.

3. Macro model: The accuracy of the power electronic model mainly affects the switching process of the device itself and has little effect on the performance of the whole power electronic circuit or system (such as HVDC transmission), and so for the research of power electronic circuits/systems, a macro model can meet its requirements. The core role of power electronic devices in the circuit or a system is a switch, which can be described simply by a switch model. According to certain external conditions, the switch turns on or off, and the transient process of the switch can be simulated by combining a small number of passive components. This power electronic device model is called the macro model, which can be further divided into variable topology model and constant topology model [10]:
   (a) Variable topology model: If the power electronic devices are regarded as ideal switches, which means the resistance is zero during on-state and zero conductance when switched off. Then, corresponding to these two states, the topology of the circuit will change, and the corresponding circuit equation will also be changed. When there are n power devices in the circuit, the number of maximum possible circuit topology will be 2^n. If the power electronic circuit is complex, and the number of switching devices is large, using the variable topology model makes the problem become very complex and difficult to solve. In this case, the following constant topology model is practically employed.
   (b) Constant topology model: This model treats power electronic devices as nonideal switches (which is also more appropriate). The topology of the circuit remains the same in different states and is described by different parameters. Not only will this model greatly reduce the difficulty of the simulation software to solve the equation but also improve the precision and efficiency of the simulation. The binary resistance switch model is the most widely used one. This model describes the device's on-state with a very small value of resistance (0.01–1 Ω) and with a very high resistance (0.1–5 MΩ) to describe the blocking state of the device. The choice of resistance depends on the external control conditions, so that the circuit topology remains the same. If only binary resistance is used to simulate the switching process, the transition of the circuit time constant may change because of the jump in the resistance value, which will result in the instability of the numerical calculation when the switch state changes. To overcome this problem, a binary R-L circuit or binary resistance plus C/L circuit is frequently used to simulate the switching process. The principle of parameter determination for the binary R-L circuit is that the time constant before and after the switching remains the same.

2.2.2 Modeling and simulation of power electronic circuit

Power electronic converters are the basic building block for HVDC projects. Rectifier (line commutating and device commutating), inverter (line commutating and device commutating), and possible DC/DC converter in the future are some of the typical converters for HVDC application. Three typical models have been employed for the simulation of power electronic converters for different proposes.

Depending on the range of frequencies and the timescale to be represented, three simulation models are widely used to describe the behavior of power electronic converters integrated into power grids.

1. Detailed model or EMT model: The detailed model includes a detailed representation of the power electronic device switching transient. In order to achieve an acceptable accuracy with the hundreds and thousands of Hertz of PWM switching frequencies used in the converters, the model must be discretized at a relatively small time step (typical, a few microseconds). This model is well suited for observing harmonics and control system dynamic performance over relatively short periods of time (typically hundreds of milliseconds to one second). The *dv*/*dt*, *di*/*dt*, and the overshoot of voltage and current are to be investigated for the converters. The power loss and the related EMC can also be calculated through the simulation.
2. Average model: In the average model, the voltage-sourced converters are represented by equivalent voltage sources generating the AC voltage averaged over one cycle of the switching frequency. This model does not emphasize harmonics, but the dynamics resulting from the control system and the power system interaction is preserved. This model allows using much larger time steps (typically 50 μs), thus allowing simulations to several seconds. The harmonic can be included with the modified average model. The dynamic phasor model [11] can be considered as a kind of modified average model and used to approximate different levels of harmonics according to the problem to be investigated. The dynamic phasor modeling theory will be systematically discussed in Chapter 9.
3. Phasor model or TS model: The phasor model is also referred to as the electromechanical model in the following chapters. In the phasor simulation method, the sinusoidal voltages and currents are replaced by phasor quantities (complex numbers) at the system nominal frequency (50 or 60 Hz). The information on harmonics and unbalance is not included in the phasor model. The model is better adapted to simulate the low-frequency electromechanical transient over long periods of time (tens of seconds to minutes) for a large-scale power grid. The same technique is successfully used in TS software. The pseudo-steady model for LCC-HVDC can be considered as a phasor model which describe the relationship between the steady-state AC active and reactive power, the DC voltage and current, and the AC root mean square (RMS) value, firing angle, commutation angle, and distinguish angle. The triple-pulse model for LCC-HVDC can be used to calculate the harmonics, considering the AC background distortion.

2.2.3 Modeling and simulation of control system

The control system is an important part of the power electronic circuit to complete the power conversion. The main circuit structure and parameters of the HVDC transmission system are relatively simple, and the method discussed above can be used to build the model. In contrast, the control system for HVDC transmission (together with protection called the secondary system) is much more complex, and the modeling and simulation of the control system become the core of the HVDC analysis. Although the control system in the project is also realized through specific circuits, such as digital signal processing (DSP) and field programmable gate array (FPGA), for modeling and simulation, the more concerning issue is about its logical relationship, rather than the circuit voltage and current signal behavior. Therefore, the modeling and simulation of the control system is mainly a description of the logical relationship. A differential equation, transfer function, or state equation is the basic description of the control system. Discrete processing is a necessary part of computer digital simulation. Logic blocks such as selection, delay, limiting, etc., are also involved in the logic processing description. The switching behavior of power electronic devices in the main circuit is decided by the logic of the control system. The function of the power electronic circuit is realized by combining the model of the control system with the main circuit model of the power electronics by triggering pulse logic.

2.3 Modeling and simulation of power system with HVDC

2.3.1 Classification of HVDC modeling and simulation

Modeling and simulation for HVDC can be defined and classified according to the platform and timescale [12]. Both physical and digital modeling have been developed for HVDC simulation. The physical models are built with real-world physical devices according to the emulating equivalent principle. The physical models have played a key role in the early stage of HVDC development, and even today they are also used in some key laboratories all over the world. With the development of technology and hardware platform, the digital modeling and simulation have been proposed and developed in the past few decades. A group of differential and or algebraic equations are used to describe the dynamic characters of power system with HVDC for the digital modeling. The solutions of those equations are usually found by means of discretion technology and numerical computing method.

As discussed above, the modeling for power system with HVDC can usually be classified into EMT modeling and TS modeling for the different research destination and variable choice. The EMT modeling, also referred to as a detailed model, with the instantaneous voltage and current has been developed for small-scale, short-term behavior study, whereas the TS modeling with the phasor voltage and current has been developed for large-scale and long-term dynamic study. The hybrid modeling, which combines EMT and electromechanical modeling has been developed, which in turn has become a hot research topic for the HVDC power system. The hybrid modeling can make full use of the advantages of the two types of modeling that identify the power electronic devices with EMT and the traditional generation and transmission system with electromechanical modeling. The above will be described in brief in the following paragraphs and the detailed discussion will be given in the following chapters.

The digital simulation usually involves the following simulations: a nonreal-time simulation and real-time simulation. The principle study, design, or planning of system with HVDC may not need real-time simulation; while the test of control and protection devices for HVDC needs a real-time simulation which is in the form of a hardware-in-loop system.

2.3.2 HVDC EMT modeling and simulation

EMT studies the behavior of the instantaneous electric field and magnetic field and the related voltage and current in the time domain. The transient of overvoltage and overcurrent after switching, the process of commutation failure, the dynamics of harmonics are typical scenario for EMT study. EMT goes to the detail of the circuit dynamics, so the instantaneous voltage and current will be used in the modeling studies.

To illustrate the transient details of EMT, the nonlinearity, the electric–magnetic coupling, and the distributed parameters of the transmission line must be considered in the modeling. The unbalance in the three-phase system and the frequency-dependent characteristics of the device parameters should also be taken into account for the EMT study.

Considering the nonlinearities, a general modeling for EMT study is as follows:

$$\begin{cases} \dot{x} = f(x(t), u(t), t) \\ y = g(x(t), u(t), t) \end{cases} \tag{2.1}$$

where x is the state variables, $\boldsymbol{u}$ is the input vector, and $\boldsymbol{y}$ is the output vector. In power electronics, the circuit can be changed as a function of time due to the action of switches. But the circuits are usually linear for each interval without switching. A set of linear differential equations describes the system for each circuit state as follows:

$$\begin{cases} \dot{x} = A_k x(t) + B_k u(t) \\ y = C_k x(t) + D_k(t) \end{cases} \tag{2.2}$$

where A_k, B_k, C_k, and D_k are the linear matrixes for switch-mode k. The matrixes can be modified according to the variable topology or variable parameters as described in 2.2.1.

In general, the differential equations in (2.1) can be solved by the implicit trapezoidal rule as follows:

$$x(t) = x(t - \Delta t) + \frac{\Delta t}{2}[f(x(t), u(t), t)] + f(x(t - \Delta t), u(t - \Delta t), t - \Delta t) \tag{2.3}$$

Equation (2.3) has $x(t)$ at both sides and can only be solved by iterative procedures. Newton–Raphson is the normal method employed for the iterative procedure. The solution of the linear counterpart as in (2.2) can be easily found by matrix manipulation.

The widely used algorithm for EMT programming is the Dommel's method [13]: with the selected integral algorithm, all electric elements are depicted by the equivalent company circuit consisting of resistor (or conductor) with a parallel current source (or series voltage source) at each discrete time step. The circuit is algebraic description of $v(t)$, $i(t)$, and $v(t-\Delta t)$, $i(t-\Delta t)$, where $v(t)$ and $i(t)$ are the variables for the new step, $v(t-\Delta t)$ and $i(t-\Delta t)$ are the variables that have been found in the previous step. The equation for the circuit above is as follows:

$$G_{eq}V_n(t) = I_s(t-\Delta t) \tag{2.4}$$

where G_{eq} is the equivalent conductor matrix that keeps constant if no switch or parameter change, and G_{eq} will be modified according to different treatment when switching or circuit parameter changes take place. $V_n(t)$ is the node voltage variable vector at time t. $I_s(t-\Delta t)$ is the equivalent current source vector at $t-\Delta t$, which is a known vector from the previous step.

The typical time step for EMT is 10–50 μs, and much smaller maybe needed for the study of fast transient such as power electronic switching and traveling wave performance. Limited by the modeling and simulation, EMT is difficult for large-scale power system simulation. The following enhancement can be explored to boost the capability of simulation:

(a) The conductor matrix for power system circuit is a typical sparse matrix, which can be effectively simulated with nonzero element treatment.
(b) With on–off equivalent circuits, the switch behavior can be described by two groups of different circuit parameters but with the same topology. The change of parameters can be integrated into a group of inversed conductor matrix before the start of the simulation. The time simulation can be boosted effectively with the prestored inversed matrix.
(c) Network partitions [14] and parallel computation are an effective way for large-scale network simulation. The technique for the data interface between different partitions must be carefully studied.

The modeling for thyristor can be an on–off equivalent circuit or equivalent circuit plus switch, the modeling for six-pulse LCC-HVDC converter valve can be an on–off equivalent circuit (as in PSCAD/EMTDC) or equivalent circuit with switching function (as in MATLAB®).

The modeling for VSC-HVDC can be different for different circuit topology. The on–off circuits or switching function can be used for a two-level converter or three-level converter but normally not for MMC. Dynamic averaged models with a controlled source are generally employed for MMC modeling for electromagnetic transient [15]. Some more efficient method are also utilized for MMC modeling [16].

2.3.3 HVDC electromechanical modeling and simulation

Electromechanical modeling and simulation also referred to as TS modeling and simulation is used to study power system transient behavior in the time domain after one or a series of large interruptions. The variables in the electromechanical model employ the fundamental phasor (mostly, the positive sequence for a three-phase system). A general description of the electromechanical model is as follows:

$$\begin{aligned} \dot{x} &= f(x, V) \quad \text{(a)} \\ g(x, V) &= 0 \quad \text{(b)} \end{aligned} \tag{2.5}$$

where x is the state variable, and V is the node voltage phasor vector. Equation (2.5a) is a differential equation for dynamic elements in the power system and (2.5b) is an algebraic equation for static elements.

Two strategies have been generally used to find the solution for (2.5): alternating method and simultaneous method.

With given initial $x(0)$, the alternating method finds $V(0)$ at first from (2.5b), by employing step integral, such as Euler's method, implicit trapezoidal integral scheme, etc., x of next time step will be solved from (2.5a). And then in (2.5b), the next step V will be fixed possibly by a few rounds of iteration. Again, the implicit trapezoidal integral scheme is the most widely used integral approach because of its numerical stability [9]. The estimation–correction alternating algorithms can be used for the solution. The iteration algorithm for the implicit trapezoidal integral is:

$$x_{n+1}^{k} = x_n + \tfrac{\Delta t}{2}[f(x_n, V_n) + f(x_{n+1}^{k-1}, V_{n+1}^{k-1})] \tag{2.6}$$

We can use $x_{n+1}^{(0)} = x_n$, $V_{n+1}^{(0)} = V_n$ as the estimated initials for the iteration and V_{n+1}^{k} can be found by (2.5b) with x_{n+1}^{k} found in (2.6). And the iteration process will stop when $\left|V_{n+1}^{k} - V_{n+1}^{k-1}\right| = \varepsilon$ is reached for a given tolerance. At the right side of the equation are also the variables that need to be fixed. They can only be solved by alternating calculation with the network algebra equation as in (2.5b).

The principle for simultaneous solution method is to differentiate (2.5a) and to combine with (2.5b) to find the solution simultaneously as:

$$\begin{cases} x_{n+1} = x_n + \dfrac{\Delta t}{2}[f(x_n, V_n) + f(x_{n+1}, V_{n+1})] \\ g(x_{n+1}, V_{n+1}) = 0 \end{cases} \tag{2.7}$$

The Newton–Raphson method is generally used for the solution of (2.7). Jacobi matrix has to be found for the calculation which makes the process complex and lacks flexibility for an extension.

The time step for electromechanical simulation is a few milliseconds, typically, 10 ms. The electromechanical simulation can be used to study relatively a large-scale network but cannot penetrate the detail of a device, especially for HVDC or FACTS equipment. Following is a brief description of the models for HVDC converters in the EM simulation.

In the electromechanical transient simulation model of the power system, the quasi-steady model is commonly used in DC converter stations. Assuming that the commutating bus voltage is a three-phase symmetrical sine wave, the AC positive sequence voltage is generally used. The converter main circuit model is depicted by a controlled voltage source and an equivalent impedance on the DC side. The rectifier and inverter are combined with the DC line with dump RLC. On the AC side, the equivalent active and reactive power injections are considered for the integration of the converter to the grid.

When the commutation failure occurs, the valve group on the same half-bridge arm in the inverter-side converter is simultaneously turned on, and the continuous commutation failure can lead to simultaneous turning on of valves group in a certain phase. When the continuous commutation failure occurs, the inverter side loses the adjustment capability, and the current in the DC line is completely dependent on the regulation characteristics of the rectifier-side converter, with no power exchange between the inverter and the receiving AC network.

Similar to the conventional LCC, the electromechanical transient model of VSC is also an external equivalent model. As the physical variables are demonstrated by the three-phase fundamental phasors (positive, negative, and zero sequences), the three-phase voltages and currents of AC interface nodes are not available; therefore, the topologies of the internal valve groups and the switching progress and trigger control will be ignored during the modeling.

On the AC side, the VSC is considered as an AC voltage source whose magnitude and phase angle are controllable. This source is connected to the AC system through a series of the transformer–inductor circuit. The reactor called "commutation reactor" could supply access impedance for the injecting voltage of VSC and be used as a filter at the same time. For low-level VSC, AC filters should be installed on the secondary side of the connection transformer. For multilevel VSC (e.g., MMC), filters are not needed in general. Therefore, the AC side circuit has two forms: without filters and with filters.

The AC-phase current is decomposed into the upper bridge arm current and the lower bridge arm current under the action of the valve group switching modulation. After the three-phase synthesis, it is injected into the DC side. Therefore, VSC in the DC side is equivalent to a controlled current source. According to KCL, the magnitude and direction of currents flowing into the positive and negative terminals of the converter must be equal and opposite. Therefore, the converter is represented by an equivalent current source and its two ends are directly connected to the positive and negative buses [17]. Capacitors are considered between the current and the DC transmission line. For MMC, the equivalent capacitor will be used based on the number of submodules.

2.3.4 Hybrid simulation with HVDC [7]

Hybrid simulation in this section is the integration of electromechanical (or TS) simulation with electromagnetic transient (EMT) simulation. For a large-scale power system with HVDC or other power electronic equipment included, hybrid simulation can be selected: electromechanical simulation for traditional AC sector with slower transient, electromagnetic simulation for HVDC or other power electronic equipment with the faster transient. Hybrid simulation seeks a balance between accuracy and efficiency, which has been considered as a promising strategy for the simulation of power systems with HVDC or FACTS included. The main challenges faced by the hybrid simulation are the interface point (bus) selection, equivalent circuit models in the two zones, and the data exchange between the TS and EMT zones.

1. Selection of the interface points (buses): For the simulation of large-scale power systems with HVDC or FACTS, the typical selection of interface points are the AC buses of the HVDC or FACTS equipment integrated into the power system as in Figure 2.1.

   With this first thought interface point selection, we can use relatively less points for grid partition and easy for equivalent modeling as in (2). But without considering the unbalance and distortion of interface buses, the simulation accuracy could be not well satisfied.
2. Equivalent circuit models: In the hybrid simulation, the TS and EMT operate in two different zones. Thus, each simulation zone needs the other simulation zone supplying a much smaller scale equivalent circuit offering pertinent and sufficient boundary conditions.
   (a) For EMT simulation, the TS zone can be represented by a multiport Thevenin equivalent circuit. The equivalent circuit can be line-frequency-based model or wideband frequency model [17]. The line-frequency model uses the pseudo-steady-state model for the TS zone, the equivalent circuit is relatively easy to obtain, and the simulation speed can be greatly boosted. The line frequency model can only describe the fundamental frequency, and the simulation accuracy cannot be satisfied when harmonic issues are concerned. The wideband frequency model represents a large range of frequency of the TS zone circuit. The model can be used for harmonics and other distortion study with higher accuracy, while the parameters for the wideband frequency equivalent circuit is never an easy task. A lot of efforts have been put in to decide the parameter but both theoretical and practical applications of satisfied solutions have not yet

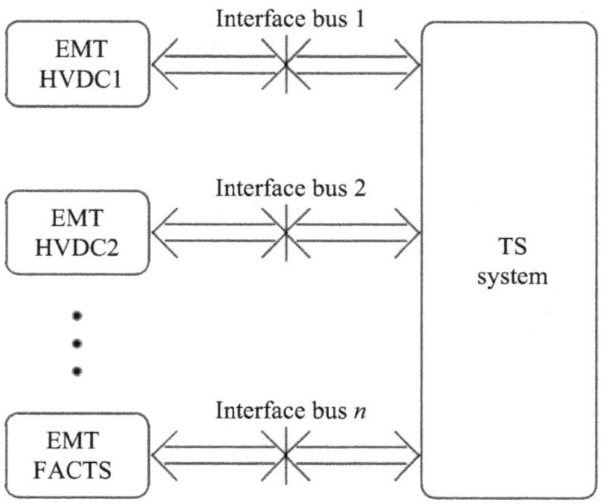

Figure 2.1 Interface points (buses) for different zones

been reached. Recently, a two-layer network equivalent has been proposed to improve the simulation efficiency with satisfying accuracy: the first layer employs the uniform transmission line model with frequency as a variable. The second layer is with a lower order admittance rational expression to represent the lower frequency properties of the TS zone.

(b) For TS simulation, the positive-sequence, fundamental component will be used to describe the EMT zone. A power source (positive or negative), current or voltage source, Norton, or Thevenin equivalent circuit can be the option to represent the EMT zone. The discrete instantaneous variables from the EMT simulation must be converted into the positive-sequence, fundamental phasor values during the modeling. The process can be finished by Fourier analysis or by fast Fourier transform (FFT) to boost the transfer speed. Some mismatching can be included when the EMT system experiences the unbalance or distortion faults. Curve fitting strategies have been proposed for extracting the fundamental phasor by filtering the nonperiodical components and high-frequency harmonics.

3. Scheme of data exchange between the TS and EMT zones:

   The EMT and TS simulations have a different time step which can be the order of microsecond versus millisecond. A data exchange rule is necessary for the information exchange and equivalent circuit update. Generally, for convenience, the step size of the TS simulation is made an integer multiple of that of EMT simulation, and exchanging of information occurs at common points in time which typically is the TS simulation time step.

The data exchange interfacing sequence can be basically categorized as serial [18] and parallel [19]. The serial sequence as illustrated in Figure 2.2, at each time instant, only one of the simulation, TS or EMT, runs, while the other one is idle. The serial sequence is obviously not satisfied with the simulation efficiency with the development of hybrid simulation.

In a parallel sequence, both simulations run simultaneously. The typical parallel scheme is illustrated in Figure 2.3.

The parallel scheme consists of much higher simulation speed. But the interface error will be unacceptable at the instances of a fault occurring or clearing. To overcome the drawbacks of the use of the only serial or parallel scheme, the combination of the two schemes has been put into use, as illustrated in Figure 2.4: the parallel sequence is used for steady simulation, and serial sequence will be employed when the system is in fault transient.

2.3.5 Parallel processing and hardware enhancement

Parallel processing techniques are one of the key strategies for efficient computation of power system simulation. And the processing capability is further enhanced by the new development of hardware configuration.

Parallel processing is defined [20] as a form of information processing in which two or more processors together with some form of an interprocessor

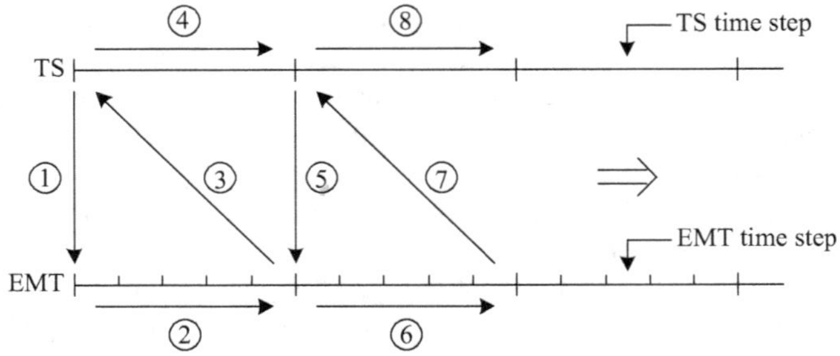

Figure 2.2 Serial sequence for data exchange

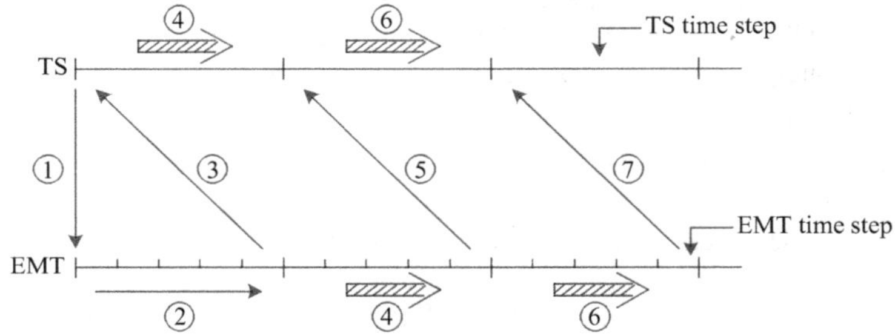

Figure 2.3 Parallel sequence for data exchange

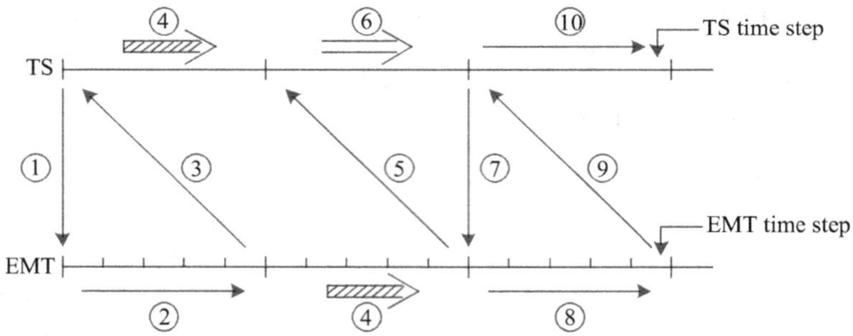

Figure 2.4 Combination of serial and parallel sequences

communication system cooperate on the solution of a problem. The method of Diakoptics introduced by Kron [21] is the earliest application of the network partitioning and parallel processing in the power system. Diakoptics divided a large power system to a group of few subsystems or subnetworks and simulated with a set of cooperative processors in parallel. Since then, parallel processing has been successfully used for the study of load flow, TS, electromagnetic transient, and security assessment.

Two types of parallel processing can be categorized according to the time step selection: the single rate and multirate. The single-rate algorithm neglects the difference in the time constants between different subsystems and uses a uniform time step for all subsystems. The obvious benefit is the easy data exchange between subsystems, and no need for interpolation. Long transmission line natural decoupling was firstly used for the single-rate parallel simulation. The power system has been considered as a stiff [22] system, especially with the inclusion of HVDC, FACTS, and many other power electronic devices. The single-rate method is very difficult to keep a balance between simulation efficiency and accuracy. The multirate strategy has been adopted for parallel processing with different time constant subsystems [23].

Data communication between different subnetworks has a great impact on parallel processing efficiency. Data exchange must be coordinated with the subnetwork calculation. A hierarchical algorithm has been developed for the processing: upper layer calculates the connection currents between the subnetworks and is used as a boundary condition for the lower layer parallel calculation. For the multiple time step processing, the W-method interpolation [24] and slack variable methods [25] are frequently used for data exchange. The W-method has the drawback of the possibility of numerical instability, whereas the slack variable has a lower efficiency because of large quantities of iteration. Latency technique [26–28] has been proposed to boost the parallel processing efficiency and accuracy by making full use of the information from the fast subnetwork during the slow subnetwork simulation.

With the development of power electronics, a large number of HVDC links and FACTS devices are included in the power system. With the HVDC link involved, the converter valves operate as a group of switches which will change the circuit topologies for many times during each cycle. The change in the topology leads to LU re-decoupling of the node admittance matrix. The computation burden will be tremendously increased with a large number of semiconductor switches included. To ease the computation burden for the total network matrix updating, the typical network partition strategy is to separate the power electronic device from the main AC system that has a fixed topology. Limited by the switches processing capability, power system with large numbers of HVDC links or FACTS devices is still a challenge for parallel simulation.

The development of computation hardware is taking the route of multikernel CPU or mass-kernel GPU [29,30]. The parallel processing capability are further enhanced by the new development of hardware configuration. Unlike the traditional parallel CPU, which is a coarse-grained processing, GPU can realize a fine-grained processing. And parallel element modeling can be practical for GPU simulation, which makes the GPU processing has a privilege for power system with a large number of power electronics. The combination of CPU/GPU system can make full use of the advantages of CPU and GPU: CPU for complex calculation and GPU for a large quantity of parallel calculation. The combination processing is illustrated as in Figure 2.5. CPU is assigned for data exchange and initialization, and GPU for subunit calculation with multithreading.

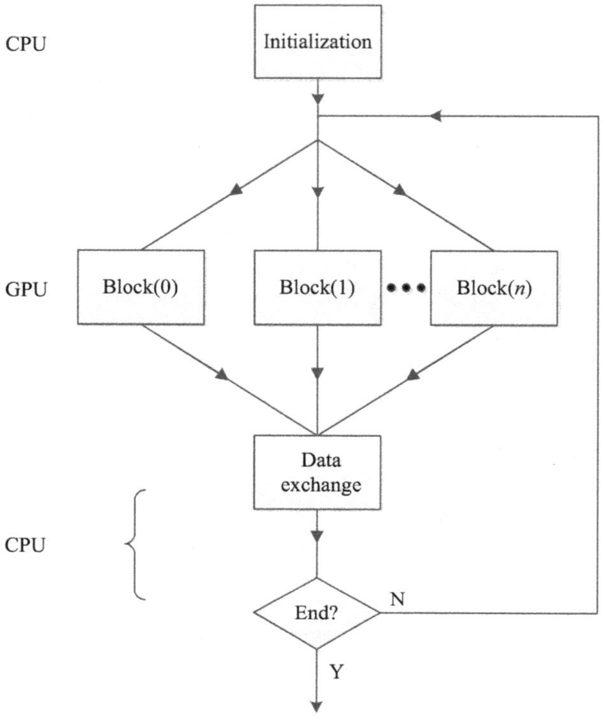

Figure 2.5 CPU/GPU combination processing

To realize the real-time simulation is another goal for the simulation efficiency study. Besides the development of the pertinent parallel algorithm, the enhanced computation hardware such as the high-clock-speed digital signal processor and reduced instruction set computer, FPGA is also a key factor for the realization of real-time simulation.

2.3.6 Selection of the modeling and simulation for the power grid with HVDC

As explained previously, the modeling and simulation of power electronics play a key role in the power system with HVDC, covering the planning, design, commissioning, and operating. During the long-term strategic planning of power system with HVDC, the power flow, TS, and benefit/investment evaluation are the most concerning issues. The behavior of a large-scale, long-term interconnection of AC/DC network needs to be studied. Because of the lack of the detailed HVDC system data, the pseudo-steady-state HVDC modeling and electromechanical simulation can be more suitable for the study. When the detailed phenomena, such as the analysis of the faults, harmonics are considered and the HVDC data are available, the electromagnetic models, or detailed model, will be the better choice. The word "hybrid" is frequently used for modeling and simulation. It will be interpreted to a different meaning for different applications. The hybrid model with electromechanical and EMT can be chosen for large-scale AC/DC network simulation with the HVDC converter studied in detail. Hybrid time-step simulation can also be a choice for large-scale system simulation with all EMT but much smaller time step for HVDC. The nonreal-time modeling and simulation are preferred to the mechanism study, design, or planning of system with HVDC. On the other hand, a hardware-in-loop system will be employed for testing the control and protection devices for HVDC. A group of new technologies has been developed in the past few years for the modeling and simulation of power system with HVDC. Their properties and application will be explained in brief in the following section.

2.4 Development and status quo of HVDC modeling and simulation

The development of modeling and simulation of HVDC is closely related to the development of HVDC technology itself. Mathematical models were built at the very beginning of the design calculation of HVDC project. A steady-state model was used for power flow calculation and no control was included within the model. And then a pseudo-steady state model [4] was employed for the calculation of the normal operating parameters such as voltage, current, and firing angles. The control function must be considered in the model, whereas the transient of the switching thyristor is neglected. The pseudo-steady-state model was successful for the main circuit parameter's decision and the analysis of steady-state operation of HVDC. While the model is not well satisfied with the analysis of transient in detail and the behavior under some interested scenarios such as unbalance faults, harmonics, etc. Before the widely accepted computer simulation era, the physical model has been developed for the HVDC study. The physical model was realized with small-scale physical devices whose parameters were chosen after the equivalent principle [31]. The physical model has been equipped in the main HVDC research institute and universities and has been successfully used for HVDC transient study. High investment and lack of flexibility were the main drawbacks of the physical model that limited its widely accepted simulation instrument. And with the rapid development of computing capability, such as multi-core CPU, graphic processing unit (GPU) and FPGA, a totally new modeling and simulation era was coming. Since then, the modeling and simulation of HVDC are focused on the digital models and numerical simulation.

In the late 1960s, scholars from North America began their dynamic simulation of HVDC systems with a multiconverter by digital computer [3]. A digital-computer program was developed, which represented many circuit configurations, different modes of control, and protective devices and provided a model that offered both accuracy and flexibility. The formation of a mathematical model that represents the converters and their associated AC systems by differential, algebraic, and Boolean equations was described in [13]. A new topological technique for the representation of converters was introduced. The technique overcomes the problems of representing the discrete switching processes that occur in converter operation. The utilization of tensor analysis and Diakoptics facilitates the construction of a flexible mathematical model of the complete systems of multi-converter stations and associated AC circuitry [21]. The techniques showed a successful simulation for the dynamics of HVDC systems. In the middle of the 1980s, the IEEE sponsored a working group for the study of dynamic performance and modeling of HVDC systems [31]. The model was widely accepted in the research activities and digital software development. With the need for a highly efficient development of the control/protection system for HVDC, the real-time simulation came to the stage [32]. From early 1990s, real-time simulation became a hot topic resorting to a variety of algorithms and computing potentials. Hardware-in-loop for real-time simulation has become the basic platform for HVDC control/ protection system development. Within the last 10 years, the small-signal model [33] for HVDC was greatly concerned and developed using the linearization technique. The dynamic phenomena such as oscillation and harmonic instability have been successfully studied with the small-signal analysis. Varieties of models of HVDC have been developed and integrated in the well-recognized digital simulation software such as EMTDC/PSCAD, BPA, MATLAB, and PSS/E. The performance of the HVDC model in commercial software has a great weight for the recognization of the software. With the applications of new devices, topologies, configuration, and control strategies for HVDC, the technology never stops innovation. Thus, the modeling and simulation of HVDC expect continuous concern and development.

References

[1] A.B.Turner, 'Modelling techniques for the study of HVDC infeeding to weak ac system'. *IEEE Transactions on Power Apparatus and Systems*. PAS-100 (7), 1981.

[2] A.M. Gole, A. Keri, C. Kwankpa, E.W. Gunther, H.W. Dommel, and I. Hassan, 'Guidelines for modeling power electronics in electric power engineering applications'. *IEEE Transactions on Power Delivery*. 12 (1), 1997, 505–14.

[3] John L. Hay, Narain G. and Hingorani, 'Dynamic simulation of multi-converter HVDC systems by digital computer'. *IEEE Transactions on Power Apparatus and Systems*. PAS-89 (2), 1970.

[4] S. Arabi, P. Kundur, and J.H. Sawada, 'Appropriate HVDC transmission simulation models for various power system stability studies'. *IEEE Transactions on Power Systems*. 13 (4), 1998, 1292–7.

[5] H.W. Dommel, 'Techniques for analysing electromagnetic transients'. *IEEE Computer Application in Power System*. 10 (3), 1997, 18–21.

[6] A.A. Fouad, and V. Vittal, *Power System Transient Stability Analysis Using the Transient Energy Function Method. Englewood Cliffs*, NJ: Prentice-Hall, 1992.

[7] V. Jalili-Marandi, V. Dinavahi, K. Strunz, J.A. Martinez, and A. Ramirez, 'Interfacing techniques for transient stability and electromagnetic transient programs, IEEE task force on interfacing techniques for simulation tools'. *IEEE Transactions on Power Delivery*. 24 (4), 2009, 2385–95.

[8] T.L. Maguire, 'Real time digital electromagnetic transients simulation of power systems'. *2000 Power Engineering Society Summer Meeting (Cat. No. 00CH37134)*. Vol. 2, 2000.

[9] N. Mohan, T.M. Undeland, and W.P. Robbins, *Power electronics – Converters, Applications and Design*, Third Edition. John Wiley & Sons, 2003.

[10] V. Rajagopalan, *Computer-aided Analysis of Power Electronic Systems*. New York: Marcel Dekker Inc., 1987.

[11] M. Daryabak, S. Filizadeh, J. Jatskevich, A. Davoudi, M. Saeedifard, and V.K. Sood. 'Modeling of LCC-HVDC systems using dynamic phasors'. *IEEE Transactions on Power Delivery*. 29 (4), 2014, 1989–98.

[12] D. Jovcic, N. Pahalawaththa, and M. Zavahir, 'Analytical modeling of HVDC-HVAC systems'. *IEEE Transactions on Power Delivery*. 14 (2), 1999, 506–511.

[13] J. Arrillaga, W. Enright, N.R. Watson, and A.R. Wood, 'Improved simulation of HVDC converter transformers in electromagnetic transient programs'. *IEE Proceedings – Generation, Transmission and Distribution*. 144 (2), 1997.

[14] A.C. Yue, X. Zhou, and R. Li, 'Node-splitting approach used for network partition and parallel processing in electromagnetic transient simulation'. *Proceedings of International Conference on Power Systems and Technology, Singapore*. November 2004, pp. 425–30.

[15] H. Saad, J. Peralta, S. Dennetière, *et al.*, 'Dynamic averaged and simplified models for MMC-based HVDC transmission systems'. *IEEE Transactions on Power Delivery*. 28 (3), 2013, 1723–30.

[16] Noman Ahmed, Lennart Ängquist, Shahid Mehmood, Antonios Antonopoulos, Lennart Harnefors, and Staffan Norrga, 'MMC modeling efficient modeling of an MMC-based multiterminal DC system employing hybrid HVDC breakers'. *2016 IEEE Power and Energy Society General Meeting (PESGM)*.

[17] H. Su, K.W. Chan, L.A. Snider, and J.C. Soumagne, 'Advancement on the integration of electromagnetic transients simulator and transient stability simulator'. *Proceedings of International Conference on Power Systems and Transients, Montreal, QC, Canada*. June 2005, IPST05-238, pp. 1–6.

[18] M.D. Heffernan, K.S. Turner, J. Arrillaga, and C.P. Arnold, 'Computation of A.C.-D.C. system disturbances: part I, II, and III'. *IEEE Transactions on Power Apparatus and Systems*. PAS-100 (11), 1981, 4341–63.

[19] H.T. Su, K.W. Chan, and L.A. Snider, 'Parallel interaction protocol for electromagnetic and electromechanical hybrid simulation'. *Proceedings of the Institution of Electrical Engineers*. 152 (3), 2005, 406–14.

[20] IEEE Task Force Report by the Computer and Analytical Method Subcommittee of the Power Systems Engineering Committee. 'Parallel processing in power systems computation'. *IEEE Transactions on Power Systems*. 7 (2), 1992, 629–38.

[21] G. Kron, 'Tensorial analysis of integrated transmission systems'. *AIEE Transactions*. 71, 1952, 814–21.

[22] C.W. Gear, *Multi-rate methods for ordinary differential equations*. Technical Report No. COO-2383-0014, UIUCDCS-F-74-880, 1974, 9.

[23] M.L. Crow, and J.G. Chen, 'The multi-rate method for simulation of power system dynamics'. *IEEE Transactions on Power Systems*. 9 (3), 1994, 1684–90.

[24] A. Bartel and M. Günther, 'A multi-rate W-method for electrical networks in state-space formulation'. *Journal of Computational and Applied Mathematics*. 147 (2), 2002, 411–425.

[25] W. Do Couto Boaventura, A. Semlyen, M.R. Iravani, *et al.* 'Robust sparse network equivalent for large systems: part I – methodology'. *IEEE Transactions on Power Systems*. 19 (1), 2004, 157–63.

[26] R.A. Saleh and A.R. Newton, 'The exploitation of latency and multi-rate behavior using nonlinear relaxation for circuit simulation'. *IEEE Transactions on Computer-Aided Design of Integrated Circuits and Systems*. 8 (12), 1989, 1286–98.

[27] F.A. Moreira and J.R. Marti, 'Latency techniques for time-domain power system transients simulation'. *IEEE Transactions on Power Systems*. 20 (1), 2005, 246–53.

[28] F.A. Moreira, J.R. Martí, L.C.J. Zanetta, *et al.*, 'Multi-rate simulations with simultaneous-solution using direct integration methods in a partitioned network environment'. *IEEE Transactions on Circuits and Systems I: Regular Papers*. 53 (12), 2007, 2765–78.

[29] J. Nickolls and W.J. Dally, 'The GPU computing era'. *IEEE Micro*. 30 (2), 2010, 56–69.

[30] S.W. Keckler, W.J. Dally, B. Khailany, *et al.*, 'GPUs and the future of parallel computing'. *IEEE Micro*. 31 (5), 2011, 7–17.

[31] IEEE Committee Report. 'Functional model of two-terminal HVDC systems for transient and steady-state stability'. *IEEE Transactions on Power Apparatus and Systems*. PAS-103, 1984, 1249–55.

[32] J.R. Marti and L.R. Linares, 'Real-time EMTP-based transients simulation'. *IEEE Transactions on Power Apparatus and Systems*. 9 (3), 1994, 1309–17.

[33] C.M. Osauskas, D.J. Hume, and A.R. Wood, 'A small signal frequency domain model of an HVDC converter'. *IEE Proceedings of Generation, Transmission and Distribution*. 148 (6), 2001, 573–8.

Chapter 3
Electromagnetic modeling of LCC-HVDC

Chao Hong[1]

3.1 Introduction

This chapter describes the electromagnetic transient (EMT) modeling of two-terminal line commutated converter-based high-voltage direct current (LCC-HVDC) transmission systems. The scope of the description is confined to modeling for the computation of EMTs associated with HVDC transmission systems using the widely spread EMT program-type method [1].

EMT programs can model AC-DC hybrid power systems in their greatest details. It is possible to investigate a very wide frequency range of transients with these tools ranging from lightning transients to electromechanical rotor oscillations [2]. It is an effective method to establish a detailed EMT simulation model for the interaction study and fault analysis of AC-DC systems. The EMT model of an HVDC transmission system generally includes the modeling of the main circuit configuration and the major components, the DC control and protection system, and the associated AC equivalent systems. Whether the model can reflect steady-state and dynamic performance of the HVDC transmission system is critical to the accuracy of simulation results [3].

From the aspects of research and engineering application, EMT models of HVDC transmission systems can be divided into typical simplified models such as generic benchmark models and detailed project-specific models. The differences in the degree of the modeling details determine the different application scenarios that two types of models support.

The generic HVDC benchmark model usually adopts a monopolar HVDC transmission configuration, and models only the major components of the main circuit, including AC filters, converter transformers, converter valve groups, smoothing reactors, DC lines, and DC filters [4,5]. From the HVDC control modeling perspective, the generic benchmark models represent only the basic control functions, thus can provide reference cases for testing simulation tools and can also be used to analyze the control strategies. For a planned new HVDC project, the

[1]EPRI of China Southern Power Grid Co. Ltd., Guangzhou, China

detailed HVDC models are usually unavailable, and the generic benchmark models may be used for studies on the project planning stage.

The detailed EMT simulation models are usually established for actual HVDC transmission projects. They should represent the relevant characteristics of the HVDC system, considering the scope of studies. The representation of HVDC transmission main circuit should be consistent with the physical configuration of the actual system. The modeling of the main circuit components includes models of AC filter/shunt capacitor banks, converter transformers, converter valve groups, smoothing reactors, DC filters, DC yard switching devices, DC lines, surge arresters, etc., models for certain applications also establish detailed representation of measuring system equipment. Since the structure and parameters of each main circuit component can be easily obtained from the actual project, the detailed EMT model of an HVDC transmission system usually adopts accurate component models. For example, the modeling of DC lines usually adopt the frequency-dependent distributed parameter model instead of the lumped parameter model adopted in the generic model. With regard to the modeling of HVDC control system, functionally accurate model of the actual control functions should be established. The detailed EMT model can then be treated like the real HVDC system, which can be used to perform various tests through simulation. However, the high level of modeling details may result in the computational burden of the HVDC system model to become significant issue [2].

The application of detailed EMT models for HVDC transmission system is very extensive, mainly including the studies of steady-state performance, AC fault response, DC fault response, system overvoltages, insulation coordination, switching surge effects, torsional damping, harmonic behavior, controller step response, optimization of control functions, design of new project-based control functions, protection coordination, support for on-site commissioning tests, and investigations for unexpected phenomena.

In summary, this chapter mainly aims at elaborating the detailed EMT modeling method of LCC-HVDC transmission systems with the experience of actual HVDC projects. In the following sections, the modeling of converter valve groups, converter transformers, transmission lines, AC and DC filters, HVDC control, and protection system are described in detail.

3.2 Electromagnetic modeling of main circuit components of an LCC-HVDC system

3.2.1 Thyristor valve groups

In an LCC-HVDC transmission system, the component tasks of the thyristor valve groups arranged in the valve towers are for converting AC to DC and vice versa. Valve-based electronics are used for receiving firing signals from the converter controls, sending trigger pulses to the thyristor valves, and reporting the status of thyristor valves to the converter control system.

The three-phase Graze bridge is the basic converter unit used in HVDC transmission projects (Figure 3.1). The six valves or valve arms of the converter are controllable if constructed from one or more thyristors in series. The valve numbers in Figure 3.1 (from 1 to 6) indicate the sequence of their conduction with reference to the positive sequence of the AC system phases.

Electric circuit configuration of the basic six-pulse valve group with its converter transformer in Wye–Delta connection is shown in Figure 3.2. When electric power flows into the DC valve group from the AC system, it is operating as a rectifier. If power flows from the DC valve group into the AC system, it is an inverter.

Two series-connected bridges constitute a 12-pulse converter group, which is shown in Figure 3.3, the most commonly used configuration in high-voltage and large-power applications.

Nearly all HVDC power converters with thyristor valves are assembled in a converter bridge of 12-pulse configuration. The most common 12-pulse configuration is the use of two three-phase converter transformers with one DC side winding as an ungrounded Wye connection and the other a Delta configuration. Consequently, the AC voltages applied to each six-pulse valve group which gives a 12-pulse valve group a phase difference of 30° which is utilized to cancel the AC side fifth and seventh harmonic currents and DC side sixth harmonic voltage, thus resulting in a significant saving in harmonic filters [6,7].

As the basic unit of the thyristor valve, a thyristor level (TL) consisted of a thyristor valve plate, a snubber circuit, and a DC voltage resistance (R_{DC}) (Figure 3.4). The snubber circuit, consisting of a series circuit of a capacitor (C_B) and a resistor (R_B), is connected in parallel with each thyristor. In interaction with the valve reactors, this arrangement serves to limit the electrical stresses on the thyristors to less than the specified levels. The DC voltage resistance, which also connected in parallel with each thyristor, can limit the commutation overvoltage stressing of the thyristor at each turnoff of the valve and linearize the voltage distribution between thyristors.

Since the voltage rating of thyristors is several kV, a high-voltage converter valve may have tens of individual thyristors connected in series groups of valve or thyristor modules [8]. In actual HVDC transmission projects, a valve section (VS)

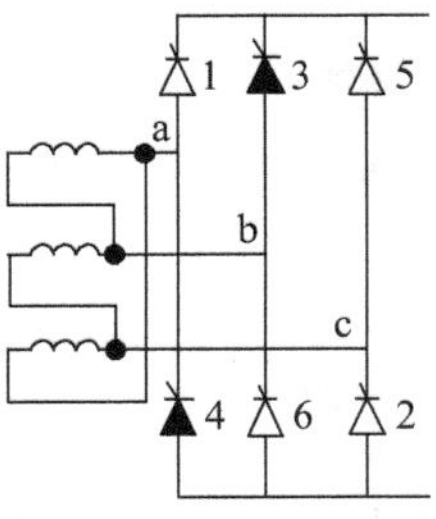

Figure 3.1 Three-phase Graze bridge

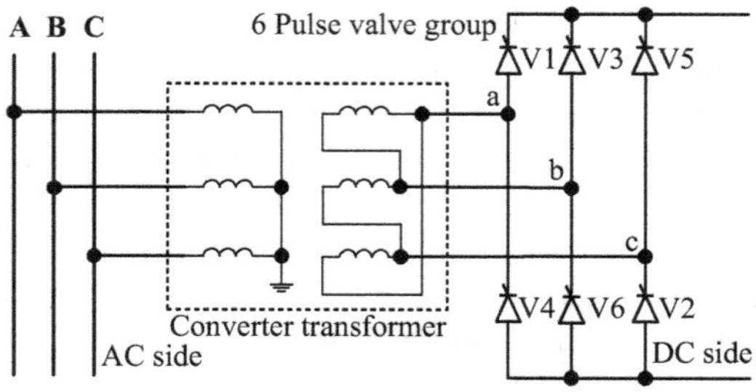

Figure 3.2 Electric circuit configuration of the basic six-pulse valve group with its converter transformer in Wye–Delta connection

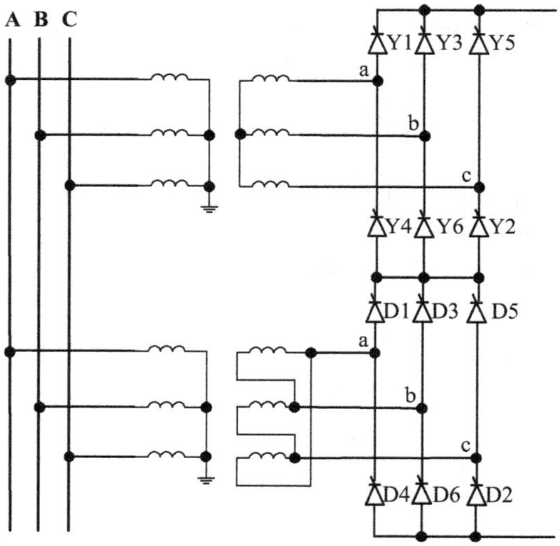

Figure 3.3 The 12-pulse converter group

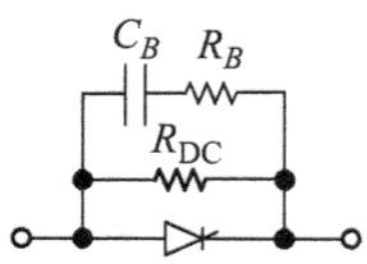

Figure 3.4 Thyristor level (TL)

illustrated in Figure 3.5 is the smallest repeatable electric unit of the thyristor valves. Each VS contains a valve reactor (L_{VD}), several or tens of series-connected TLs, and a parallel-connected voltage sharing capacitor (C_K). As the electrical characteristic of a VS and whole thyristor valves are the same, realizing the correct simulation of the VS means the correct simulation of thyristor valves.

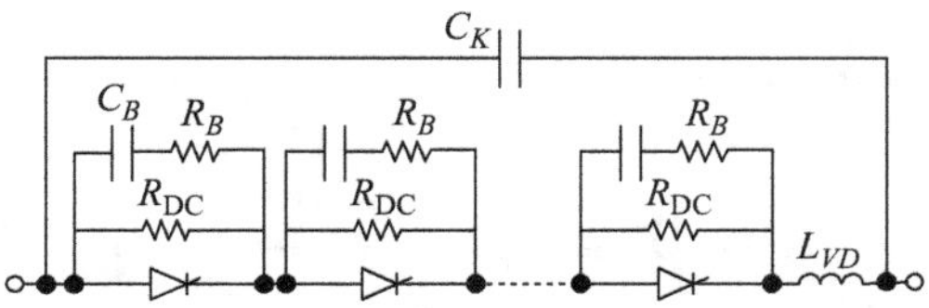

Figure 3.5 The structure of a valve section (VS)

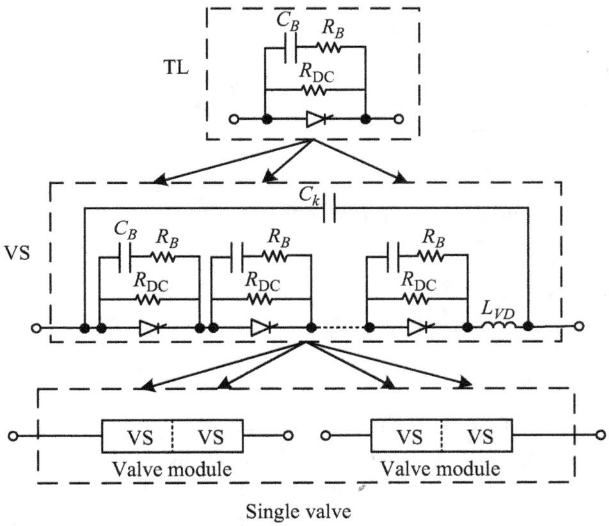

Figure 3.6 The composition structure of a single valve

Further, a thyristor or valve module is that part of a valve in a mechanical assembly of series connected VSs and their immediate auxiliaries, which include heat sinks cooled by air, water, or glycol, snubber circuits, and valve firing electronics. A thyristor module is usually interchangeable for maintenance. One valve consists of one or several serial-connected thyristor modules. Each valve is protected against overvoltage with one arrester [6]. The interrelation among TL, VS, and valve modules is shown in Figure 3.6.

Generally, in the EMT modeling of thyristor valves, L_{VD}, C_K, and R_{DC}, shown in Figure 3.6 is usually ignored to simplify the modeling process [8,9], as shown in Figure 3.7(a). The thyristor in Figure 3.7(a) can be modeled as a switching selected resistor (R_{on} or R_{off}), as shown in Figure 3.7(b). When the valve is deblocked and the thyristor conducting in the forward direction, the equivalent resistor R_{on} has a very small value (e.g., $R_{on} = 0.1\ \Omega$). The thyristor valve illustrated in Figure 3.7(b) can be modeled as a resistor connected between node k and m with snubber circuit

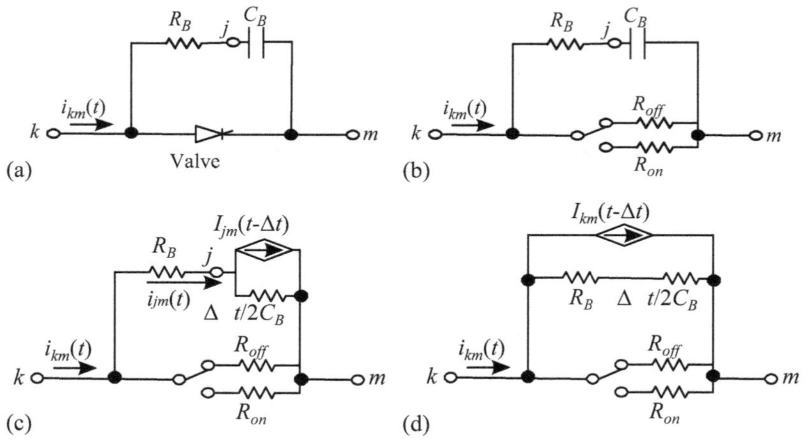

Figure 3.7 Equivalence of the thyristor valve

being ignored, that gives

$$i_{km}(t) = \frac{e_k(t) - e_m(t)}{R_{on}} \tag{3.1}$$

When the valve is blocked, the thyristor can be modeled as a resistor R_{off} with a very large resistance value (e.g., $R_{\text{off}} = 10^6\ \Omega$). The thyristor valve illustrated in Figure 3.7(b) can be modeled by the snubber circuit with the thyristor branch being ignored. Applying the trapezoidal rule, the capacitor (C_B) in Figure 3.7(b) can be modeled as an equivalent current source connected in parallel with an equivalent resistor [10], as shown in Figure 3.7(c). The following equation can be derived:

$$i_{jm}(t) = \frac{e_j(t) - e_m(t)}{\Delta t/2\mathrm{C_B}} + I_{jm}(t - \Delta t) \tag{3.2}$$

and

$$I_{jm}(t - \Delta t) = -i_{jm}(t - \Delta t) - \frac{2\mathrm{C_B}}{\Delta t}(e_j(t - \Delta t) - e_m(t - \Delta t)) \tag{3.3}$$

where Δt is the simulation time step length. From Figure 3.7(c), applying circuit transformation shown in Figure 3.7(d) gives:

$$I_{km}(t - \Delta t) = \frac{\Delta t/2\mathrm{C_B}}{R_B + \Delta t/2\mathrm{C_B}} I_{jm}(t - \Delta t) \tag{3.4}$$

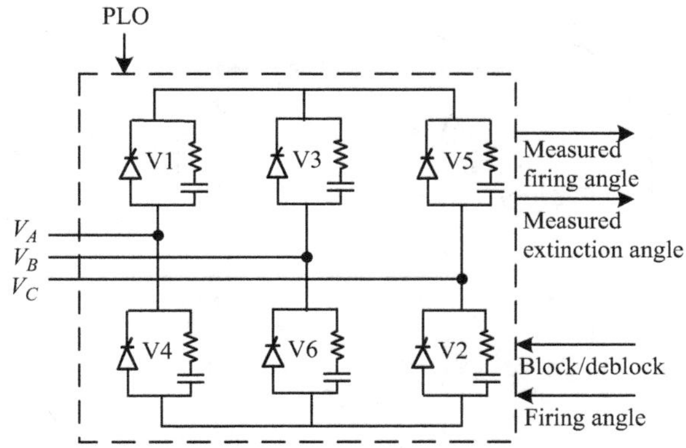

Figure 3.8 Simplified modeling of six-pulse group converter

The final simplified expression can be written as

$$i_{km}(t) = \frac{e_k(t) - e_m(t)}{R_B + \Delta t/2C_B} + I_{km}(t - \Delta t) \tag{3.5}$$

$$I_{km}(t - \Delta t) = -\frac{\Delta t/2C_B}{R_B + \Delta t/2C_B}\left(i_{km}(t - \Delta t) + \frac{2C_B}{\Delta t}(e_j(t - \Delta t) - e_m(t - \Delta t))\right) \tag{3.6}$$

where

$$e_j(t - \Delta t) = e_k(t - \Delta t) - R_B i_{km}(t - \Delta t) \tag{3.7}$$

Figure 3.8 shows the modeling component of a six-pulse valve group provided in PSCAD/EMTDC [11]. The control signals mainly include the input firing angle, block/deblock signal, the measured firing angle, and extinction angle output.

The input parameters of the converter model include thyristor resistance (R_{off} and R_{on}), forward voltage drop, forward breakover voltage, reverse withstand voltage, minimum extinction time, snubber capacitance (C_B), and resistance (R_B). The phase-locked oscillator (PLO) is used to lock the phase of the AC voltage based on the phase vector technique. The block diagram of PLO is shown as Figure 3.9 [11]. Through Clark transformation, the three-phase AC voltage is converted to V_α and V_β. In Figure 3.9, ω_0 is an estimated angular frequency of the AC voltage. This technique exploits trigonometric multiplication identities to form an error signal, which speeds up or slows down the PLO in order to match the phase.

The phase-locked loop (PLL) is also widely used to provide the phase information of the AC system to the converter. There are several analysis methods of the PLL such as the zero-crossing phase-locked method, method based on the synchronous reference frame, and method based on the virtual-balanced three-phase system [12].

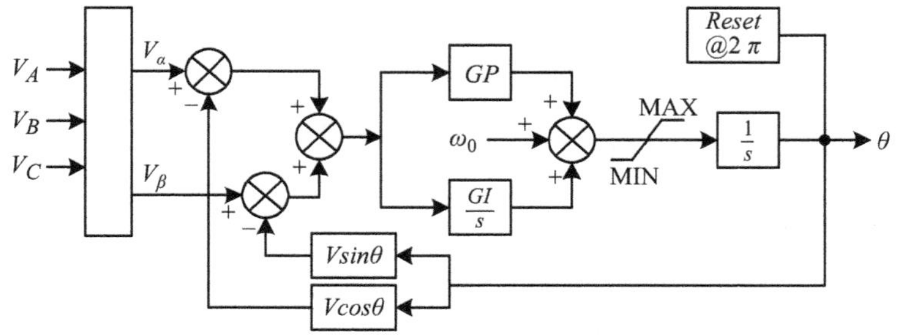

Figure 3.9 Block diagram of the PLO

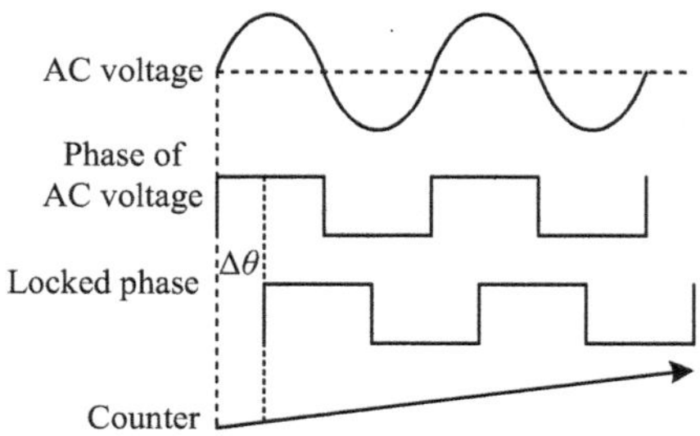

Figure 3.10 Principle diagram of the zero-crossing phase-locked method

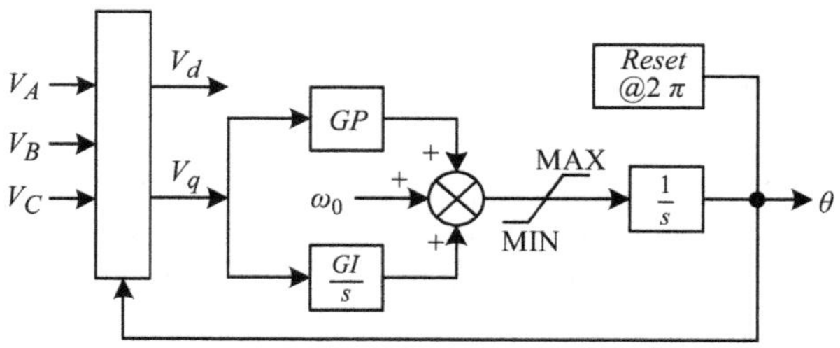

Figure 3.11 Block diagram of the PLL based on the synchronous reference frame

1. The zero-crossing phase-locked method: The principle diagram of the zero-crossing phase-locked method is shown in Figure 3.10. The AC voltage is converted to the square-wave voltage, and the rising edge moment of the voltage is noted down to compare with the phase of the PLL voltage. The AC voltage can be tracked by adjusting the measuring error. The zero-crossing phase-locked method has high precision but the dynamic performance is slow [13].
2. PLL based on the synchronous reference frame: The block diagram of the PLL based on the synchronous reference frame is shown as Figure 3.11. Through

Park transformation, the three-phase AC voltage is converted to V_d and V_q. V_q should be adjusted to zero because of the floating control characteristic of the proportional integral (PI) controller which means the AC voltage phase is locked [12]. This PLL method has high precision and fast dynamic response when the AC system is balanced.

3. PLL based on the virtual balanced three-phase system: Another PLL method is based on the virtual balanced three-phase system shown in the Figure 3.12. The single phase of the AC voltage can be decomposed into a positive and negative sequence voltage whose phase can be locked by the method based on the synchronous reference frame. Then we can get the three-phase information compounded by the single-phase information. Therefore, this method will not be affected in case of the unbalanced AC system [13].

PLO or PLL is the important base of the valve control. There are two methods of valve firing named the equidistant firing angle control and equidistant firing pulses phase control. The former may lead the harmonic instability of DC systems which has not been used in the actual HVDC projects. The equidistant firing pulses phase control aims at sending the equidistant firing pulses phase sequence to each valve of converters. For the 6- and 12-pulse converters, the firing pulses phase interval are 60° and 30°, respectively, which can effectively prevent the harmonic instability phenomena [6].

In addition, the constant simulation step such as 50 μs may cause precision problems of modeling. If the triggering time occurs in between the time points, it can only be represented at the next time step point. In order to solve the problem, different kinds of interpolation algorithms are the effective technologies used widely in the electromagnetic modeling so far [10].

3.2.2 Modeling of converter transformers

Due to equipment manufacturing and transportation restrictions, converter transformers of large-capacity LCC-HVDC transmission projects usually use single-phase transformers. This section mainly describes the EMT modeling of single-phase two-winding converter transformers, which can be represented as magnetically independent

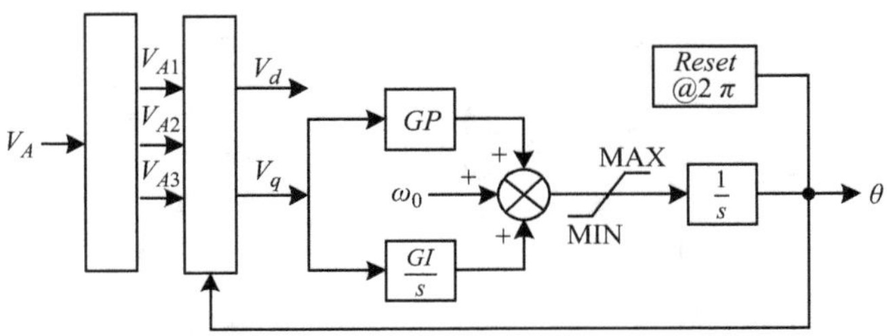

Figure 3.12 Block diagram of the PLL based on the virtual-balanced three-phase system

single-phase units with no coupling between phases. Three single-phase units are used to form Wye–Wye or Wye–Delta three-phase transformer sets according to the correct connection method. However, the three-phase multilimb converter transformers are often used in small- and medium-capacity HVDC schemes, and the detailed modeling is given in [10] and [14–17].

3.2.2.1 Single-phase, two-winding transformer model

The single-phase, two-winding transformer can be described using the two mutually coupled windings as shown in Figure 3.13, which is introduced in [17] as the classical approach. The voltages across the two windings can be expressed as

$$\begin{bmatrix} v_1 \\ v_2 \end{bmatrix} = \begin{bmatrix} L_{11} & L_{12} \\ L_{12} & L_{22} \end{bmatrix} \cdot \frac{d}{dt} \begin{bmatrix} i_1 \\ i_2 \end{bmatrix} \tag{3.8}$$

where L_{11} and L_{22} are the self-inductances of windings 1 and 2, respectively, and L_{12} is the mutual inductance between the windings [17].

Using the turns ratio "a," which is defined as the ratio of the number of turns in the two windings, (3.8) can be rewritten as

$$\begin{bmatrix} v_1 \\ a \cdot v_2 \end{bmatrix} = \begin{bmatrix} L_{11} & a \cdot L_{12} \\ a \cdot L_{12} & a^2 \cdot L_{22} \end{bmatrix} \cdot \frac{d}{dt} \begin{bmatrix} i_1 \\ i_2/a \end{bmatrix} \tag{3.9}$$

Equation (3.9) can be represented by the equivalent circuit as shown in Figure 3.14 [17], where

$$L_1 = L_{11} - a \cdot L_{12} \tag{3.10}$$

$$L_2 = a^2 \cdot L_{22} - a \cdot L_{12} \tag{3.11}$$

The parameters L_1, L_2, and aL_{12} of the equivalent circuit can be determined from open- and short-circuit tests (Figure 3.14) [17]. The nominal turns ratio is also determined from the open circuit tests.

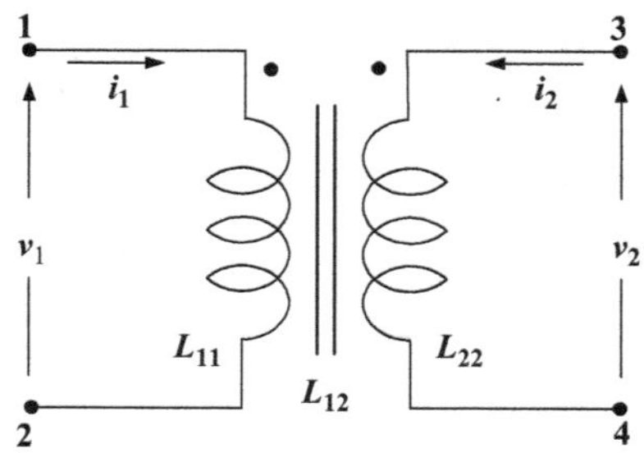

Figure 3.13 Two mutually coupled windings

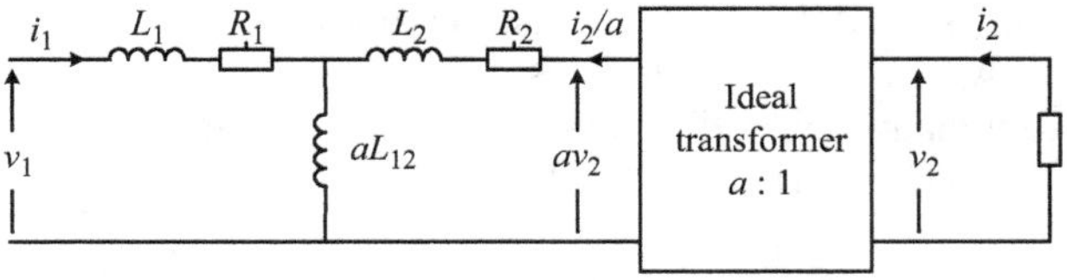

Figure 3.14 Equivalent circuit of two mutually coupled windings

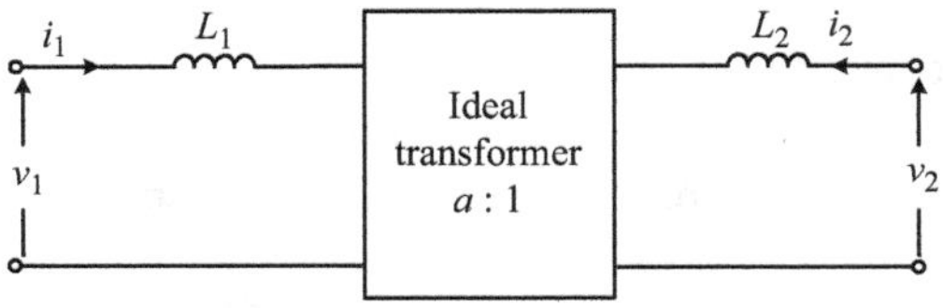

Figure 3.15 Equivalent circuit of the two-winding transformer, without the magnetizing branch

3.2.2.2 Inverting the mutual induction matrix

In order to solve for the winding currents, the inductance matrix of (3.8) needs to be inverted:

$$\frac{d}{dt}\begin{bmatrix} i_1 \\ i_2 \end{bmatrix} = \frac{1}{L_{11} \cdot L_{22} - L_{12}^2} \cdot \begin{bmatrix} L_{22} & -L_{12} \\ -L_{12} & L_{11} \end{bmatrix} \cdot \begin{bmatrix} v_1 \\ v_2 \end{bmatrix} \tag{3.12}$$

The coupling coefficient between the two coils is [17]

$$K_{12} = \frac{L_{12}}{\sqrt{L_{11} \cdot L_{22}}} \tag{3.13}$$

It can be observed that as the coupling coefficient K_{12} approaches unity, the elements of the inverse inductance matrix approach infinity. An excessively small magnetizing current also leads to such ill conditioning. In such cases, it is often advisable to model the transformer with only leakage reactances and no magnetizing branch (Figure 3.15) [17]. The relationship between the derivatives of current and the voltages can be directly expressed as in (3.14):

$$\frac{d}{dt}\begin{bmatrix} i_1 \\ i_2 \end{bmatrix} = \frac{1}{L} \cdot \begin{bmatrix} 1 & -a \\ -a & a^2 \end{bmatrix} \cdot \begin{bmatrix} v_1 \\ v_2 \end{bmatrix} \tag{3.14}$$

where $L = L_1 + a^2 \cdot L_2$ is the leakage inductance between windings 1 and 2 as measured from winding 1 terminal [17].

3.2.2.3 Modeling of transformer core saturation

Saturation can be represented with a compensating current source across the winding wound closest to the core [10,17,18]. For a two-winding, single-phase transformer, the saturation current injection placed at the secondary winding terminal is shown in Figure 3.16. This method is an approximate way to add saturation to mutually coupled windings but can provide acceptable simulation results for many power system studies [18].

3.2.3 Modeling of DC transmission line

In EMT simulations, the three basic transmission line models are the PI section model, the Bergeron model, and the frequency-dependent line model [17].

3.2.3.1 PI section model

Approximate nominal PI section models are often used for short transmission lines (less than 15 km), where the travel time is less than the solution time step (usually 50 μs). However, PI section models are unsuitable for long transmission distances. Instead, traveling waves theory is used in the development of more realistic models such as the Bergeron model and the frequency-dependent model [10,17].

3.2.3.2 The Bergeron model

Figure 3.17 shows a lossless distributed parameter line, where x is the transmission distance from node k to node m and d is the length of the transmission line. The wave propagation equations for this line are

$$-\frac{\partial v(x,t)}{\partial x} = L\frac{\partial i(x,t)}{\partial t} \tag{3.15}$$

$$-\frac{\partial i(x,t)}{\partial x} = C\frac{\partial v(x,t)}{\partial t} \tag{3.16}$$

where L is the inductance, C is the capacitance per unit length, and $i(x,t)$ and $v(x,t)$ are the current and voltage phasors at a distance x along the transmission line. The

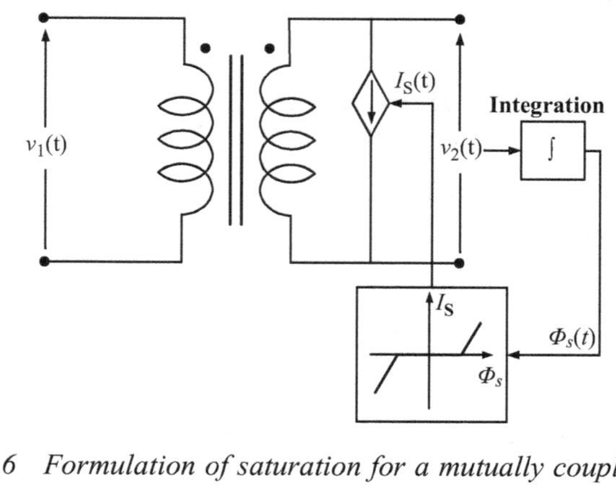

Figure 3.16 Formulation of saturation for a mutually coupled winding

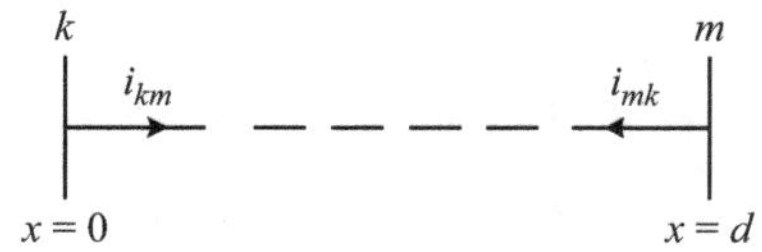

Figure 3.17 Propagation of a wave on a transmission line

general solution is

$$i(x,t) = f_1(x - wt) + f_2(x + wt) \tag{3.17}$$

$$v(x,t) = Z_C \cdot f_1(x - wt) - Z_C \cdot f_2(x + wt) \tag{3.18}$$

where $f_1(x - wt)$ represents a wave traveling at velocity w in the forward direction, $f_2(x + wt)$ represents a wave traveling in the backward direction, $Z_C = \sqrt{\frac{L}{C}}$ is the characteristic impedance, and $w = \frac{1}{\sqrt{LC}}$ is the phase velocity [19].

Let the observer leave node m at time $t - \tau$ and arrive at node k at time t, where $\tau = d/w$. Thus

$$v_m(t - \tau) + Z_C \cdot i_{mk}(t - \tau) = v_k(t) + Z_C \cdot (-i_{km}(t)) \tag{3.19}$$

Rearranging (3.19) gives

$$i_{km}(t) = \frac{1}{Z_C} v_k(t) + I_k(t - \tau) \tag{3.20}$$

where

$$I_k(t - \tau) = -\frac{1}{Z_C} v_m(t - \tau) - i_{mk}(t - \tau) \tag{3.21}$$

Similarly,

$$i_{mk}(t) = \frac{1}{Z_C} v_m(t) + I_m(t - \tau) \tag{3.22}$$

where

$$I_m(t - \tau) = -\frac{1}{Z_C} v_k(t - \tau) - i_{km}(t - \tau) \tag{3.23}$$

According to (3.20)–(3.23), the Bergeron lossless line model of the transmission line is shown in Figure 3.18. In this model, the two sections are decoupled.

When the loss is considered, there is no difference in using two sections at the terminals instead of using distributed resistances throughout the line. When $R/4 \ll ZC$, where R is the total line resistance, the lumped resistance R can be inserted in the line as shown in Figure 3.19, i.e., $R/2$ in the middle of the line and $R/4$ at each terminal [10].

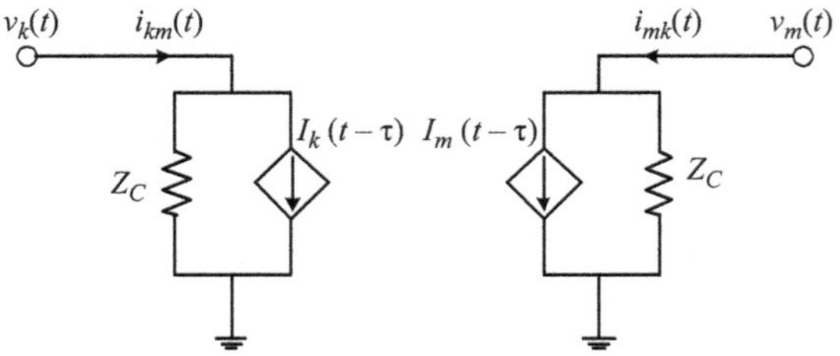

Figure 3.18 Norton equivalent for Bergeron lossless line model

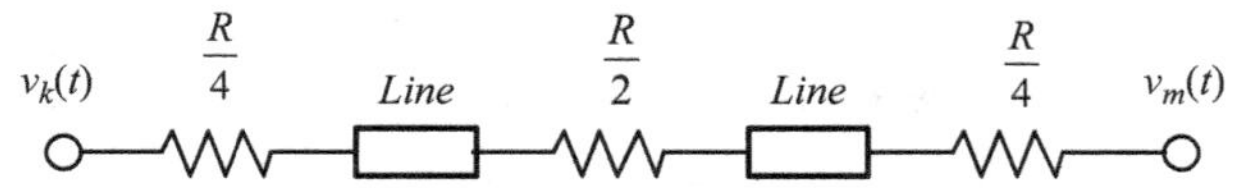

Figure 3.19 Transmission line with lumped losses

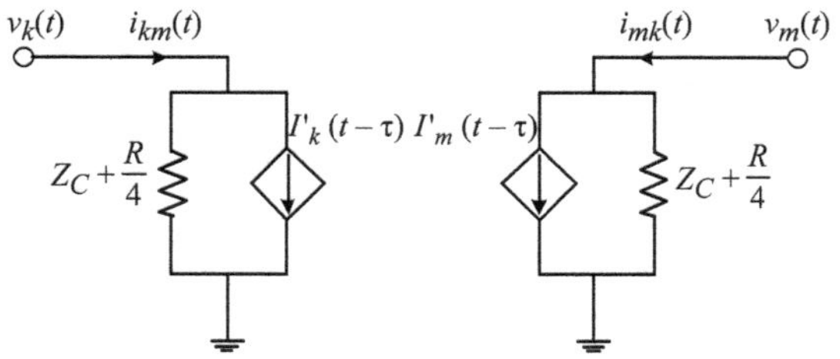

Figure 3.20 Norton equivalent for Bergeron transmission line model

Thus,

$$I_k(t-\tau) = -\frac{1}{Z_C}v_m(t-\tau) - \frac{Z_C - \frac{R}{4}}{Z_C + \frac{R}{4}} \cdot i_{mk}(t-\tau) \tag{3.24}$$

$$i_{km}(t) = \frac{1}{Z_C}v_k(t) + I_k(t-\tau) \tag{3.25}$$

The Bergeron transmission line model is shown in Figure 3.20.
And the expression for the current source at end k is [10]

$$\begin{aligned} I_k^{'}(t-\tau) \quad = \quad & -\frac{Z_C}{(Z_C+R/4)^2}(v_m(t-\tau) + (Z_C - R/4)\cdot i_{mk}(t-\tau)) \\ & -\frac{R/4}{(Z_C+R/4)^2}(v_k(t-\tau) + (Z_C - R/4)\cdot i_{km}(t-\tau)) \end{aligned} \tag{3.26}$$

The Bergeron model is a constant frequency method based on traveling wave theory and can accurately represent the fundamental frequency. Therefore, this model is suitable for fundamental frequency load flow studies but is not adequate for high-frequency studies [10,17].

3.2.3.3 Frequency-dependent line model

The frequency domain model of transmission lines is first calculated from the physical line geometry and then the convolution can be performed to obtain the time domain response [10]. As the line parameters are functions of frequency, the relevant equations should first be viewed in the frequency domain. The characteristic impedance is given by [20]

$$Z_C(w) = \sqrt{\frac{R(w) + jwL(w)}{G(w) + jwC(w)}} = \sqrt{\frac{Z(w)}{Y(w)}} \tag{3.27}$$

while the propagation constant is

$$\gamma(w) = \sqrt{(R(w) + jwL(w))(G(w) + jwC(w))} \tag{3.28}$$

Finally, $V_k(w)$ and $I_k(w)$ can be given as follows [10]:

$$\begin{cases} V_k(w) = Z_C(w)I_k(w) + A(w)(V_m(w) + Z_m(w)I_m(w)) \\ I_k(w) = Y_C(w)V_k(w) - A(w)(I_m(w) + Y_C(w)V_m(w)) \end{cases} \tag{3.29}$$

The Norton frequency-dependent transmission line model is shown in Figure 3.21.

In Figure 3.21, the $I_{kHistory}$ and $I_{mHistory}$ are given by

$$\begin{cases} I_{kHistory} = A(w)(I_m(w) + Y_C(w)V_m(w)) \\ I_{mHistory} = A(w)(I_k(w) + Y_C(w)V_k(w)) \end{cases} \tag{3.30}$$

The frequency-dependent line model should be used for studies requiring a very detailed representation of the line over a wide frequency range [17]. Thus, this model is usually used for modeling the DC lines of the HVDC transmission links.

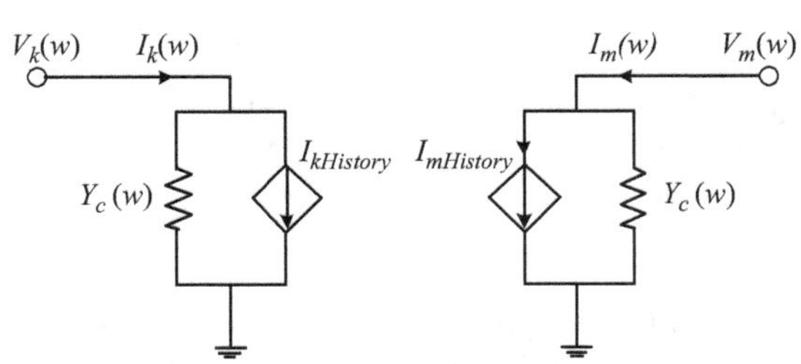

Figure 3.21 Norton equivalent for frequency-dependent transmission line

3.2.4 Modeling of AC filters and capacitor banks

In relation to an AC network, the converters of an LCC-HVDC system act as sources of harmonic currents. AC filter circuits are required to absorb the harmonic currents. In general, the tasks of AC filters and capacitor banks contain absorption of the AC side harmonic currents caused by the converters as well as compensation of the reactive power at fundamental frequency.

The characteristic AC side current harmonics generated by the 12-pulse converters are $12k \pm 1$, where k is a positive integer. The AC filters are typically tuned to 11th, 13th, 23rd, and 25th harmonics for 12-pulse converters [6,7]. The AC side harmonic filters may be switched with circuit breakers to accommodate reactive power requirement strategies. A parallel resonance is naturally created between the capacitance of the AC filters and the inductive impedance of the AC system. For the special case where such a resonance is lightly damped and tuned to a frequency between the second and fourth harmonics, then a low-order harmonic filter at the second or third harmonics may be required, even for the 12-pulse converter operation [6].

A number of suitable filter circuit configurations are available for an HVDC converter station. The typical topologies of the AC filters are shown in Figure 3.22 [3,6]. In EMT simulation, AC filters and capacitor banks are usually modeled by switched branches connected to the high-voltage bus bars on the AC side of an HVDC converter station and are generally arranged in a star (Wye) circuit which is solidly grounded, as shown in Figure 3.23.

3.2.5 Modeling of DC filters and smoothing reactors

3.2.5.1 DC Filters

The characteristic DC side voltage harmonics generated by a 12-pulse converter are of the order $12k$, where k is a positive integer [6,7]. The task of DC filters is to reduce harmonic current flow on DC transmission lines. DC filters are not required in DC transmission schemes without overhead DC lines, such as in HVDC cable connections and back-to-back links.

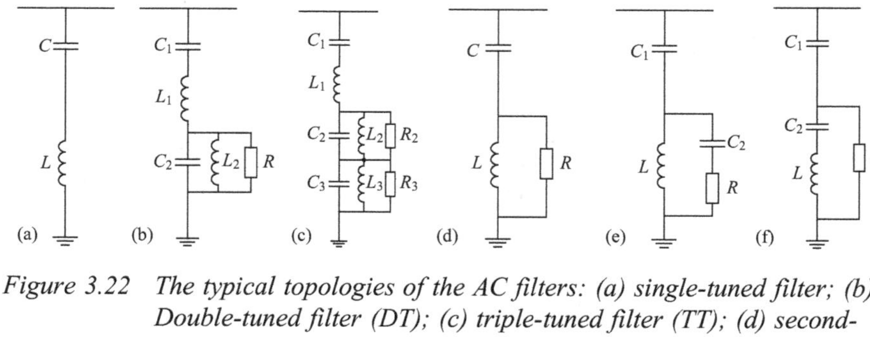

Figure 3.22 The typical topologies of the AC filters: (a) single-tuned filter; (b) Double-tuned filter (DT); (c) triple-tuned filter (TT); (d) second-order high-pass damping filter; (e) third-order high-pass damping filter; and (f) C-type damping filter

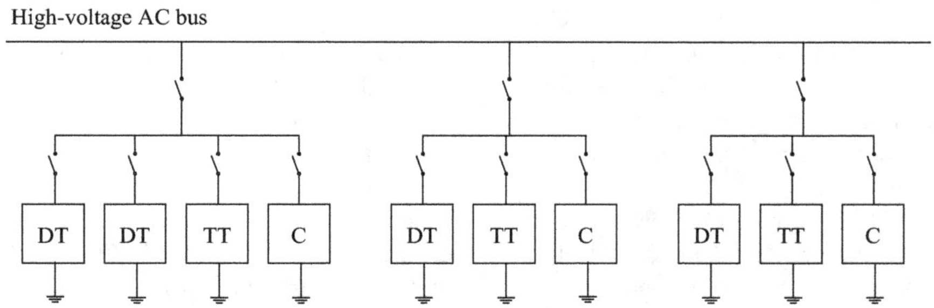

Figure 3.23 The switched branches model for AC filters and capacitor banks

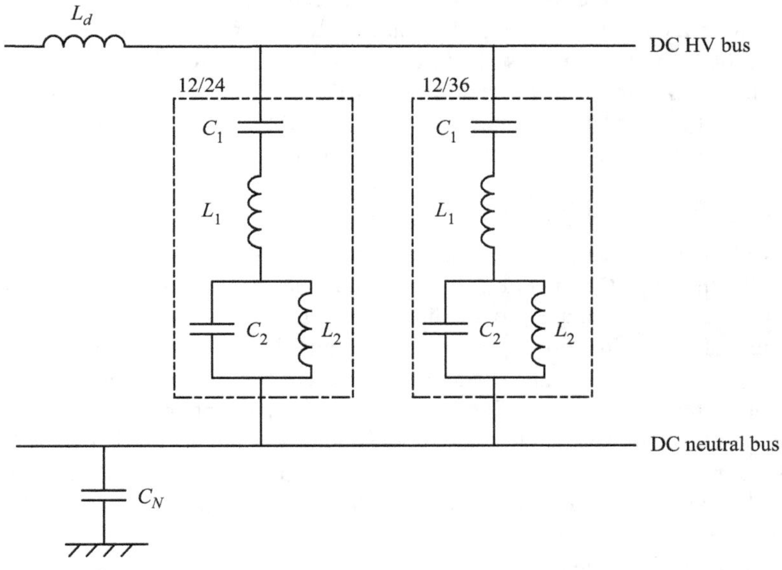

Figure 3.24 DC filter arrangement

DC filters are usually connected between the DC high-voltage bus and the DC neutral bus of the same pole. The commonly used DC filters are double-tuned filters and triple-tuned filters. For a two-terminal HVDC transmission system (12-pulse converters), a typical DC filter design for each station uses a combination of two double-tuned DC filters per pole, one filter for the 12th and 24th harmonics, and the other for the 12th and 36th harmonics, as shown in Figure 3.24.

Due to various technical and economical advantages, the triple-tuned DC filters are used in some new HVDC transmission schemes, which are typically tuned for the compensation of the harmonic currents with the order numbers 12, 24, and 36. The converter stations may use only one triple-tuned DC filter or a combination of two identical DC filters per pole.

The configuration of the DC filters very strongly resembles the filters on the AC side of the HVDC converter stations. A DC filter consists of capacitors, reactors, and current transformers (for measuring and protection). For most EMT studies, the current transformers are not modeled, and the *LC* component reduction can be performed to reduce the number of nodes and hence the computation at each time step [10].

Besides the DC filters, there are other equipment in the DC circuit, including smoothing reactor *Ld* and neutral bus surge capacitor CN, as shown in Figure 3.24, which has significant contribution for limitation of harmonic currents on the DC lines and thus shall be considered in the modeling of DC circuit components.

3.2.5.2 Smoothing reactor

Smoothing reactors are usually arranged between valve groups and DC filters of each pole in a converter station. The main functions of the smoothing reactors are to limit the rate of current rise in case of a fault in the DC circuit, to smooth the ripple in the DC, to limit the high-order harmonics in the DC lines, and to avoid resonance at low-order harmonics [6]. For most EMT studies, a smoothing reactor can be represented in the model as a nominal inductance.

3.2.6 Modeling of the connected AC system

It is well-known that the EMT model solution of the large-scale AC system connected to the HVDC converters is very time-consuming. Therefore, AC system reduction is usually employed in the modeling of AC-DC power systems. The selection of the reduction approach is determined by the research interests. For studies focusing on the transient and dynamic characteristics of an HVDC transmission system, the connected AC systems usually adopt the equivalent voltage source models. However, for studies focusing on the interaction between AC and DC systems, such as dynamic performance after the AC system faults, commutation failure and fault recovery, and the stability control functions, more detailed AC system modeling is needed.

When the connected AC system is modeled as an equivalent voltage source, Thevenin equivalent method is often used. Thevenin impedance value *Z* at fundamental frequency is obtained by short-circuit calculation at the converter bus bar, and it varies with the different operating modes of the connected AC system. The damping angle θ is usually about 75°–85° [2]. An R-R/L circuit, as shown in Figure 3.25(a), could be employed to represent the Thevenin impedance, whose impedance characteristic is displayed in Figure 3.25(b). Obviously, the R-R/L circuit possesses an additional degree of freedom compared with the Thevenin impedance, which allows for the network damping to be specified at another frequency [2]. Moreover, Thevenin source must be adjusted to guarantee that the voltage of the converter bus bar satisfies the requirement.

When the equivalent modeling of the connected AC system needs to be more detailed to meet the requirement of some studies, such as researches on post-disturbance analysis, dynamic network reduction is usually used [21–23]. For the

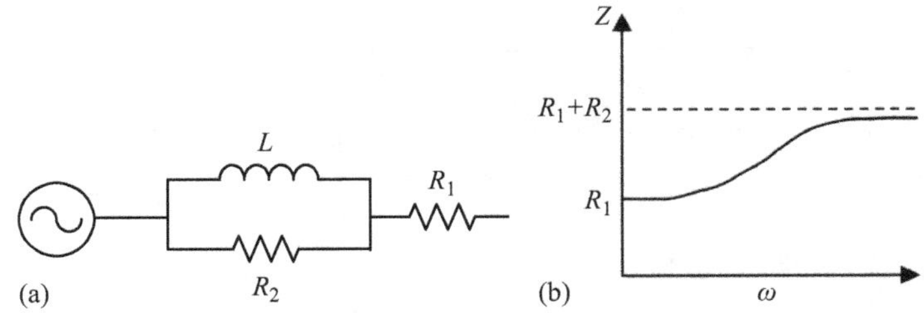

Figure 3.25 Thevenin equivalent model of the connected AC system

AC system reduction, it is necessary to define the remained parts of the AC system, which are usually of great interests and should be kept with the original network structures and components for the studies, while the other parts of the AC system will be reduced. For a two-terminal LCC-HVDC link, the following parts of the connected AC systems are likely to be remained:

1. The AC network near to converter stations of the HVDC system, including the fault and disturbance locations for simulation studies;
2. AC links in parallel with the HVDC link; and
3. Large-capacity generation units and loads near the converter stations.

Furthermore, it is necessary to evaluate the equivalent results of the dynamic network reduction. In terms of the steady-state characteristics, the power flow results and short-circuit capacities at the converter station buses of the equivalent AC systems should be comparable to those obtained from the original AC system. In addition, the dynamic response subsequent to typical DC and AC faults should be compared with the relative results obtained from the original system to check the dynamic performance of the equivalent system.

3.3 Modeling of HVDC controls

3.3.1 Introduction

HVDC transmission systems are highly controllable, and their effective operation depends on the correct implementation of this controllability. HVDC transmission systems usually transport bulk electric power that can only be accomplished under DCs and voltages being precisely controlled to realize the desired power transfer. The control system is also important for dynamic performance of the HVDC transmission system [24,25]. As well as the accurate modeling of the main circuit component, effective EMT simulation requires proper representation of HVDC control and protection functions. This section discusses the modeling of HVDC control and protection systems in EMT simulations.

The HVDC pole/converter-related control functions include determining the correct steady-state power order, control functions for the operation of the converters and the valve-firing control, high-speed control functions of DC voltage

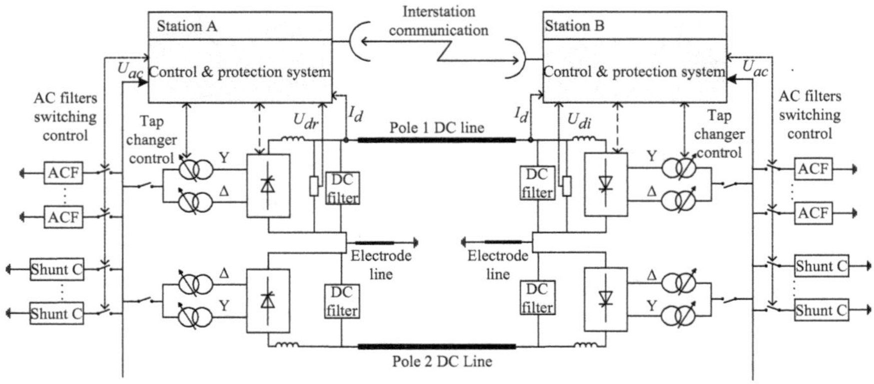

Figure 3.26 Overview of the controls of a two-terminal HVDC transmission system

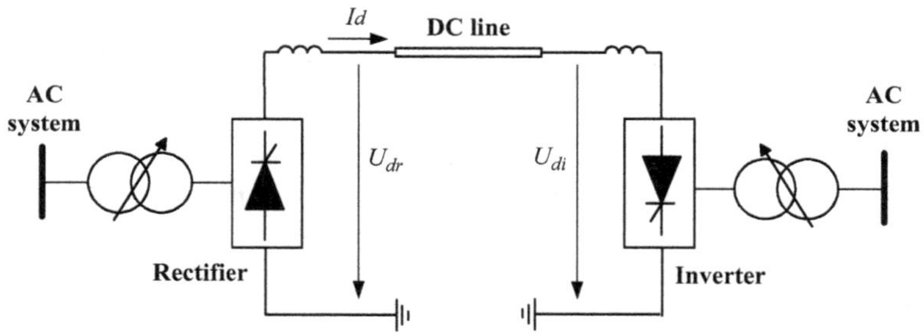

Figure 3.27 The simplified diagram of an HVDC transmission system

control, DC control, firing angle and extinction angle control, stability, modulation, and frequency control, etc. To implement the control functions, a variety of DC main circuit signals need to be continuously and precisely measured as input quantities to the control systems, such as DC side voltages and currents, three-phase AC voltages and currents, the firing angle α and for an inverter, its extinction angle γ, etc., as shown in Figure 3.26. The output signals from DC controls are used to provide firing pulses to the thyristor valves, to switch the on-load tap changers, and to open or close the controlled switches of reactive power compensation units in the converter stations. The firing control is rapid (1–10 ms), whereas the on-load tap-changer control and the reactive power compensation unit switching control are slow (5–10 s per step or per unit).

3.3.2 Principle of HVDC controls

The simplified diagram of a monopolar HVDC transmission system or one pole of a bipolar scheme is shown in Figure 3.27. The power transfer is always from the rectifier to the inverter. The control of the HVDC transmission system is essentially achieved by changing the DC voltages of the two converters, and the magnitude of DC I_d is controlled depending on the difference in terminal voltages U_{dr} and U_{di}.

$$I_d = \frac{U_{dr} - U_{di}}{R_{dc}} \tag{3.31}$$

where R_{dc} is the resistance of the DC line. The DC power of the pole at rectifier terminal P_{dc} can be calculated as

$$P_{dc} = U_{dr} I_d \tag{3.32}$$

For a two-terminal DC transmission system, one converter must be in current control and the other defines DC voltage. Normally, the inverter is assigned the task of controlling the DC voltage and the rectifier in the same pole controls the DC.

A typical static characteristic of an HVDC transmission scheme is shown in Figure 3.28. This steady-state $U_d - I_d$ characteristic shows the rectifier characteristic (ABOCDE) and the inverter characteristic (FGHJOKL). The operating point of the DC system is defined by the intersection of the two characteristics, i.e., the point O. The rectifier current order (I_{ord}) determines the constant current part of the characteristic (BOC). The DC voltage order (U_{d_ref}) maintains the constant DC voltage part of the characteristic (JOK).

The static characteristic for an HVDC transmission system is determined by the main circuit as well as the DC controls. With the determined DC main circuit structure and data, different static characteristics can be developed to determine the steady-state operation of the converter controls and the specific operation of the controls in response to system perturbations by modification of the DC control strategies and settings.

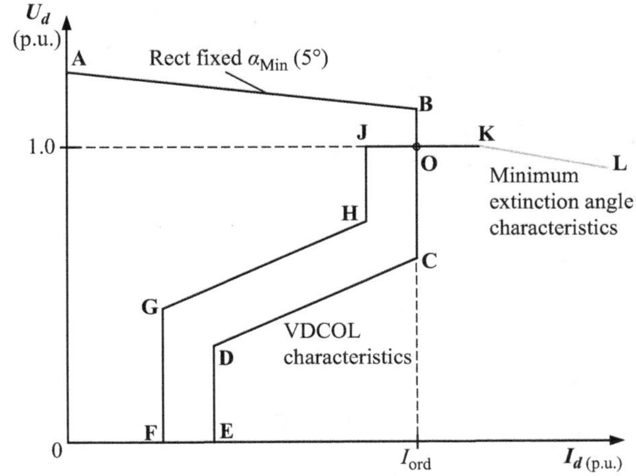

Figure 3.28 Steady-state $U_d - I_d$ characteristics

3.3.3 Basic DC control functions

3.3.3.1 DC power control

Accurate DC power control is the general operating requirement of HVDC systems. Under normal HVDC operation, a DC power has to be maintained according to the control order. The basic principle used to control the DC power in power control mode (P_{mod}) is to divide the DC power order by the measured rectifier terminal DC voltage, as shown in Figure 3.29 where the steady-state power order (P_{ord}) is normally determined by the operator. The resulting current order (I_{d_ref}) is then the control reference for the rectifier current controller.

3.3.3.2 DC control at rectifier station

The DC control is designed to keep the DC constant as long as possible. The DC is controlled according to the current order, which is calculated from DC power control functions in DC power control mode or given directly by the operator in DC control mode, as shown in Figure 3.30.

Figure 3.30 illustrates the DC control at rectifier station using a proportional-integral regulator to generate a firing angle. Actually, the DC voltage at rectifier terminal is controlled via the firing angle to maintain the desired DC line current. When the actual DC becomes lower than the ordered current (e.g., due to AC bus voltage reduction), the current control decreases the firing angle resulting in an increase in the rectifier DC voltage (U_{dr}) in order to maintain the DC. In case the actual current is too high, the controller reacts in the opposite direction. With a proportional plus integral regulator as shown in Figure 3.30, the steady-state constant current characteristic of the rectifier is the vertical section B-O-C in Figure 3.28.

3.3.3.3 Rectifier characteristics with fixed firing angles

The rectifier firing angle α cannot be reduced beyond a certain limit (typically $\alpha_{min} = 5°$). Therefore, the rectifier cannot control the DC I_d constant, if the firing angle α is at its minimum limit, and then the rectifier is operated with a fixed firing angle of $\alpha_{min} = 5°$. The uncontrolled characteristic is determined by [6]:

$$U_{dr} = \frac{3\sqrt{2}}{\pi} U_{cr} \cos(\alpha_{min}) - \frac{3}{\pi} X_{cr} I_d \tag{3.33}$$

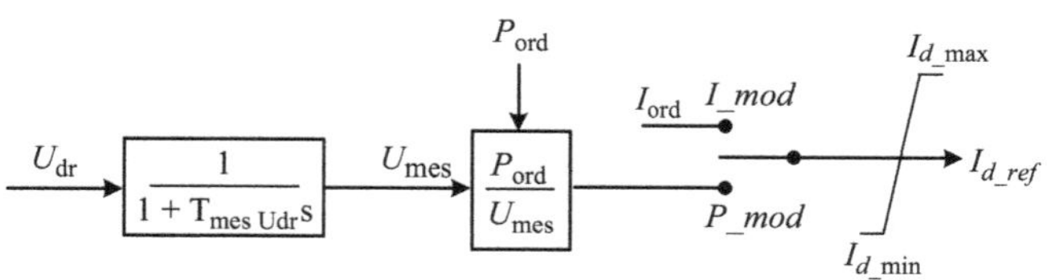

Figure 3.29 DC power control functions

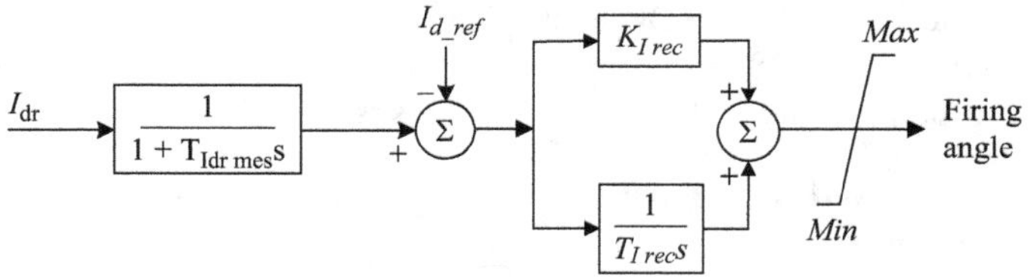

Figure 3.30 DC control at rectifier station

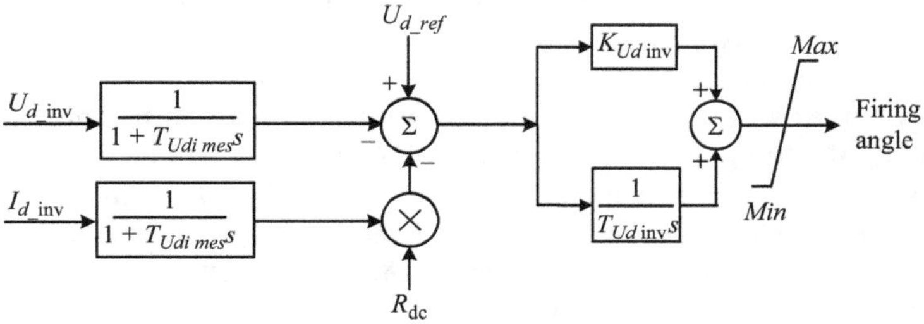

Figure 3.31 DC voltage control at inverter station

where U_{cr} is the phase-to-phase r.m.s. commutating voltages referred to the valve side of the converter transformer, and X_{cr} is the commutating reactance of the rectifier.

The steady-state fixed firing angle (α_{min}) characteristic of the rectifier is shown as the section A-B in Figure 3.28. The slope of this characteristic depends on the commutation impedance of the converter, including the impedances of the connected AC system and the converter transformers. The weaker the AC system, the steeper the slope.

3.3.3.4 DC voltage control at inverter station

Under steady-state conditions, the inverter is assigned the task of controlling the rectifier terminal DC voltage at a target value. A DC voltage controller is normally provided at the inverter station to keep the DC voltage constant as long as possible, which defines the constant DC voltage characteristic J-O-K in Figure 3.28. The DC voltage control function can be proportional plus integral controllers as shown in Figure 3.31. The measured inverter terminal DC voltage is compared with the target value, and the voltage drop on the DC line should be considered as illustrated in Figure 3.31.

When DC voltage is too high, the DC voltage control increases the PI controller resulting in a firing angle change toward 90°. This change in firing angle results in a decrease in the inverter terminal DC voltage (U_{di}). If the inverter is operating in constant U_d characteristic, the rectifier must control the DC I_d. With the decrease in the inverter terminal DC voltage, the rectifier terminal DC voltage

(U_{dr}) will be decreased by the DC controller in order to maintain the DC constant. The point O in Figure 3.28 where the rectifier and inverter characteristic intersect is the operating point of the HVDC transmission system.

3.3.3.5 The closed loop extinction angle control at inverter station

The main task of the closed loop extinction angle control at the inverter is to prevent the HVDC system from suffering commutation failures. The extinction angle reference value γ_{ref} is given as the minimum acceptable value typically in the range of 15°–18°. The actual value of the extinction angle γ is measured for each of the 12 valves. The extinction angle control function can also be proportional plus the integral controllers as shown in Figure 3.32. The extinction angle control becomes active when the measured extinction angle falls below the minimum limit. It decreases the firing angle order α toward 90° to control γ to the reference value.

Maintaining a constant extinction angle γ causes the DC voltage U_d to droop with increasing DC I_d, as shown in the minimum constant extinction angle γ characteristic, i.e., section K-L in Figure 3.28. The weaker the AC system at the inverter, the steeper the droop [24].

3.3.3.6 DC margin control at inverter

DC control at inverter is provided in order to prevent the loss of power transmission during transients, for example, AC voltage reduction at the rectifier side. The DC order initiated at the rectifier controls is also sent to inverter controls, and a margin of 0.1 p.u. is subtracted from the rectifier DC order at the inverter. The difference between the rectifier current order and the inverter current order is called the current margin denoted by I_{marg} in Figure 3.33. The current margin prevents the current controller at inverter to become active during normal operation.

The DC controller at the inverter attempts to control the DC to the value ($I_{ord} - I_{marg}$) and is forced out of action in normal steady-state operation. If a drop in the rectifier-side AC voltage occurs for some reason or other, it results in a lowering of the rectifier minimum firing angle characteristic, as shown in Figure 3.33. The operating point will shift from point O to O' on the vertical characteristic J-H, where it is intersected by the lowered rectifier minimum firing angle characteristic A'-B'. The inverter reverts to current control, controlling the DC to the value ($I_{ord} - I_{marg}$) and the rectifier is effectively controlling DC voltage

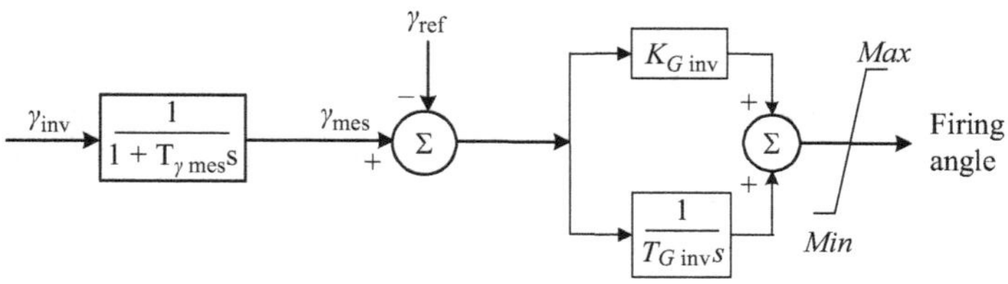

Figure 3.32 Extinction angle control

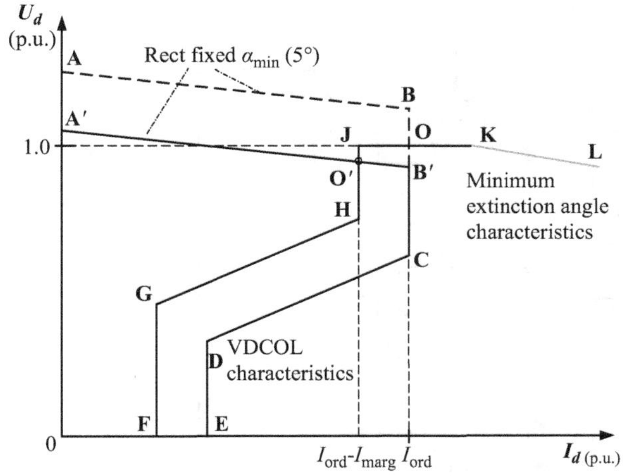

Figure 3.33 Transition to inverter current control

so long as it is operating at its minimum firing angle characteristic A'-B'. The inverter DC control ensures a stable operating point at lower DC when the rectifier minimum firing angle characteristic drops.

3.3.3.7 The selection of inverter control functions

Normally only one of the DC control, DC voltage control, or extinction angle control is active at any instant. The control angle selector is usually a "select minimum" or "select maximum," depending on whether the signal is for α or β, as shown in Figure 3.34 [8,26,27].

The change form one control mode to another can be designed such that the transition from the rectifier controlling current to the inverter controlling current is automatic and smooth.

3.3.4 Voltage-dependent current order limit

During disturbances where the AC voltage at the rectifier or inverter is depressed, it will not be helpful to a weak AC system if the HVDC transmission system attempts to maintain full DC. Especially when inverter AC voltage reduces due to a fault, the inverter may experience continuous commutation failures with some valves continuously conducting. DCs should be reduced so as not to overstress the valves.

A sag in AC voltage at either the rectifier or inverter end will result in a lowered DC voltage too. The DC control characteristics indicate the DC order is reduced if the DC voltage is lowered (Figure 3.33). This can be observed in the rectifier characteristic C-D-E and in the inverter characteristic H-G-F. The controller that reduces the maximum current order is known as a voltage-dependent current-order limit (VDCOL or VDCL) [8,27].

The VDCOL control, if invoked by an AC system disturbance will keep the DC to the lowered limit during recovery which prevents continuous commutation

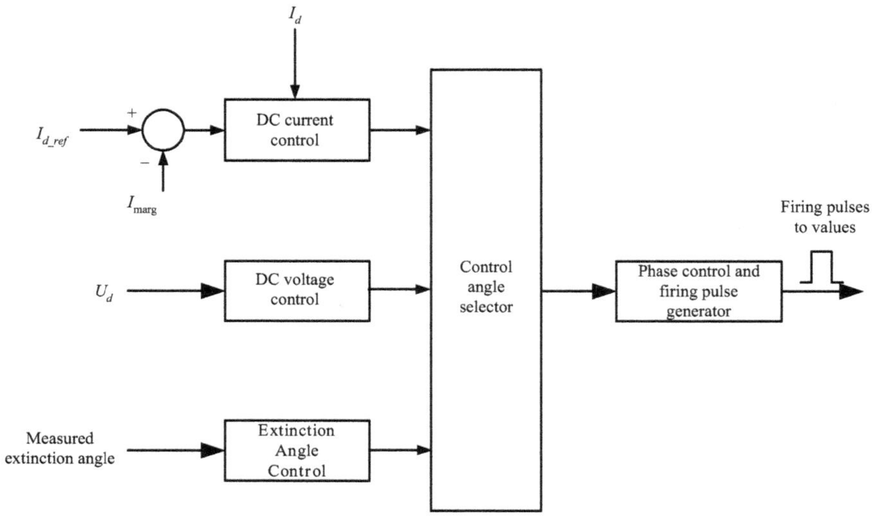

Figure 3.34 The inverter control angle selector

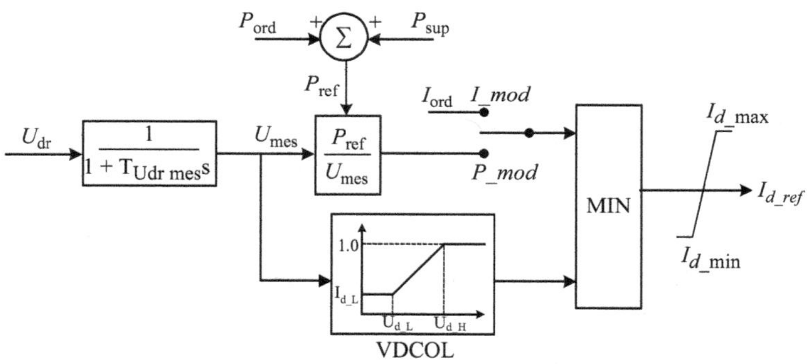

Figure 3.35 The block diagram of VDCOL control and supplementary control

failures of inverters and aids the corresponding recovery of the HVDC system. Only when DC voltage has recovered sufficiently will the DC return to its original current-order level. A schematic diagram of the VDCOL control is shown in Figure 3.35.

3.3.5 Higher level controls

In addition to the basic DC control functions discussed above, a number of supplementary controllers can also be added to HVDC controls to take advantage of the fast response of HVDC transmission system and help the performance of the AC system. These supplementary controls include AC system damping controls,

AC system frequency controls, DC power run-up/run-back change, AC voltage compensation controls, etc. The supplementary control power signals P_{sup} can be added to the current-order generation block as shown in Figure 3.35.

3.3.6 Tap changer controls of the converter transformers

The main task of tap changer control is to keep the converter transformer valve side voltage within a range suitable for the converter valves. The tap changer control is a slow acting function which only compensates slow variations in the converter station AC bus voltage.

For tap changer control, different control modes can be used depending on the operating configuration and the desired operating points. In the angle control mode, which is the standard tap changer control mode, the tap changer is controlled according to a limited steady-state firing angle range of α at rectifier and extinction angle range of γ at inverter side. If the rectifier firing angle or the inverter extinction angle goes outside the range, the corresponding converter transformer tap changer control will vary the tap position one step, to increase or decrease the converter transformer valve side voltage, and hence to bring the rectifier firing angle or the inverter extinction angle to within the steady-state range [8,25].

An alternate mode for the tap changer controls at both the rectifier and the inverter side is to keep the converter transformer valve side AC voltages (U_{di0}) within a suitable range of the nominal values. If the valve side voltage, calculated from measured primary voltage, falls below the lower limit, the tap position will be changed one step to raise the valve-side voltage. If the valve side voltage rises above the upper limit, the tap position will be changed one step to lower the valve-side voltage.

3.3.7 Summary of DC control modeling

For a simplified model based on the basic HVDC control functions, the main control mode at the rectifier is the DC control, and the main control mode at the inverter is the DC voltage control, with the backup control modes including DC margin control and extinction angel control.

For a detailed model based on the actual HVDC controls, all the functions important to the dynamic performance of the control system need to be represented. The detailed model also includes representation of the digital control sample times that are used in the real control system. This is important for accurate representation of control response and the associated stability margins and real-processing time delays in the execution of control functions. The detailed control system model for a bipolar HVDC scheme comprises the following parts.

3.3.7.1 Detailed representation of bipole and pole control functions

The DC power-order functions including ramping and reference setting, P/U_d function. The DC-order functions including current limits, current margin control, current balance control, and pole-to-pole power transfer.

3.3.7.2 Detailed representation of all converter control functions

The closed-loop controller functions include DC controller, DC voltage controller, and extinction angle controller. The functions of firing angle limits and pulse firing system should be represented in detail.

3.3.7.3 Modeling of other functions

This part includes the tap changer controls to establish correct operating points for various DC power and the reactive power controls to establish correct reactive compensation for various DC power level.

3.4 Modeling for the new developed HVDC projects

3.4.1 Modeling for the UHVDC projects

The need of cost-effective, low loss power transmission systems for huge power over very long distances promotes the application of ultra HVDC (UHVDC) transmission technology. From 2010, several ±800 kV UHVDC transmission projects have been put into commercial operation in China, India, and Brazil. Until now, all the operational UHVDC transmission systems are line commutated converter-based bipolar systems, with rated DC voltages of 800 kV and rated transmission capacity up to 10 GW. UHVDC projects with DC voltages of 1,100 kV and rated transmission capacity up to 12 GW are in the construction stage in China. Meanwhile, ±800 kV MMC-UHVDC converters with rated capacity up to 5 GW are also in the construction stage in China.

The converter configuration of 800 kV UHVDC system may be either single 12-pulse valve group per pole or two 12-pulse valve groups in a series per pole [28]. The configuration of a bipolar UHVDC system with single 12-pulse converter per pole is very similar to that of a typical HVDC bipolar system, and the modeling of the DC main circuit including main components is also similar. Therefore, this follows the focus more on the converter configuration with series valve groups in each pole.

The main circuit configuration of two 12-pulse valve groups connected in series per pole is shown in Figure 3.36, where one valve symbol represents a 12-pulse valve group [29,30]. In all the actual UHVDC transmission projects with this configuration, the DC voltage ratings of the two series valve groups are identical, which means the voltage rating is 400 kV per converter, and each converter handles half the pole power [30,31].

As shown in Figure 3.36, each of the 12-pulse converters has a high-speed bypass switch connected in parallel [30]. For a pole operating at rated DC voltage of 800 kV with the two series converters, the outage of a converter, which is the most common fault in actual operation, will result in the converter bypassed by a controlled closing of the parallel bypass switch, and the pole can continue to operate at 400 kV. Thus, the loss of one converter only represents the loss one fourth of the total transmission capacity, and balanced DC between the two poles is

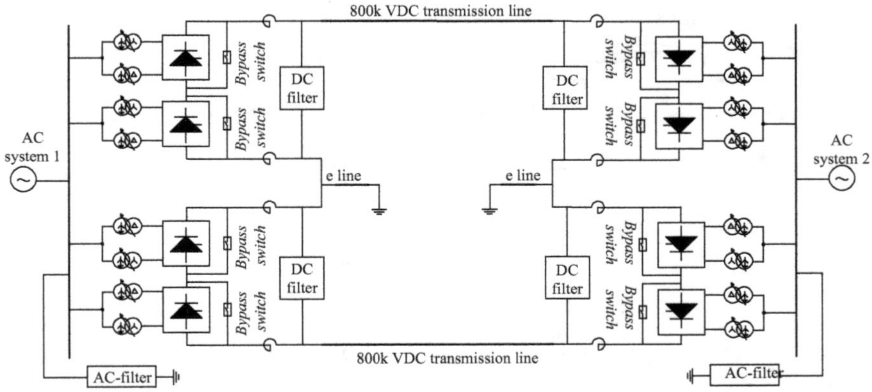

Figure 3.36 The configuration of two 12-pulse valve groups in series per pole

always kept as long as there is one valve group operating in each pole [30,31]. The modeling of a UHVDC system should consist of modeling high-speed bypass switches.

The smoothing reactor arrangement of the UHVDC system differs to that of the conventional HVDC system. Two sets of smoothing reactors are at each pole, which are, respectively, installed at the DC high-voltage bus and the neutral bus, as shown in Figure 3.36. Usually one half of the total inductance of the smoothing reactors is installed at the neutral bus [30]. The modeling of the main circuit should consider the smoothing reactor arrangement.

The UHVDC converter transformers are usually single-phase, two winding transformers [29,31]. The modeling of the converter transformer is the same as the modeling of the large-capacity HVDC converter transformer described in Section 3.2.2.

The modeling of the DC transmission line is the same as the aforementioned HVDC DC line model, and a distributed parameter model is generally used.

The principle AC filter arrangement for UHVDC schemes does not differ basically from the conventional bipolar HVDC schemes and will be modeled in a similar way. Same applies to the DC filter modeling for UHVDC schemes.

The control functionality for an 800-kV scheme is basically the same as any long-distance HVDC transmission scheme. In case of two series connected bridges per pole, a complete functional strategy will need to be developed for the control, protection, and coordination of a single-valve groups in an operating pole [28,30].

3.4.2 Modeling for the capacitor commutated converter (CCC)-HVDC systems

The CCC applies a series capacitor between the converter transformers and the valve groups, as shown in Figure 3.37 [8,32,33].

With the CCC configuration, minimum AC filtering can be applied. The filter MVAR ratings can be selected to small values (10–15% of rating), which result in a

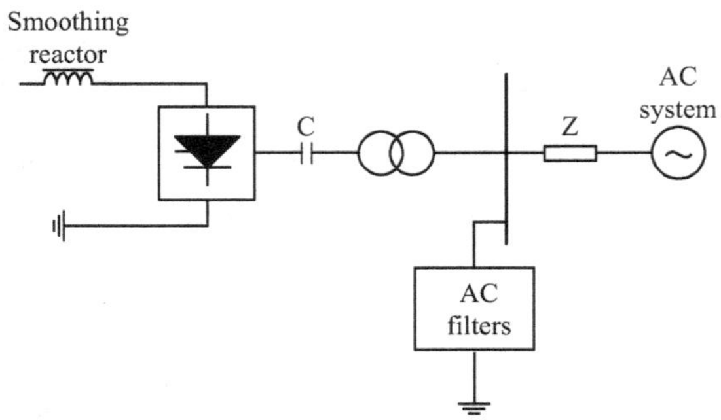

Figure 3.37 CCC DC transmission inverter

very narrow passband. Harmonics are a little higher with CCC configuration compared to the conventional configuration [32].

The inverters with series capacitor compensation are less prone to commutation failure caused from any fault in the AC power system and can also effectively operate into much lower short-circuit ratio. The incentive is to apply the series capacitor at the inverter where low short-circuit ratio and cable discharge effects are a challenge [32,33].

The control system for the CCC configuration is basically the same as for the conventional configuration except for the extinction angle controller. Extinction angle γ is modified and an effective value γ is defined. This is because the commutation voltage of a CCC is the sum of the AC line voltage and the voltage charge on the series capacitor. It is suggested that the PLOs derive their AC signals from the AC commutating bus bar. For the CCC configuration, the series reactance of the capacitor can be 0.3–0.4 per unit based on the converter transformer rating. The extinction angle order can be reduced to 2°–5° instead of the normal 15°–18°. The actual extinction angle setting to use depends upon the value for the series reactance used as well as the degree of utilization of ratings of the transformer, series capacitor, and valve group. Therefore, the maximum firing angle is larger than that of a conventional DC inverter. A CCC inverter can operate at a higher power factor than a conventional DC converter [32].

When CCC configurations are used, tests of transient overvoltages on the DC side volts, valves, the series capacitor, the converter transformer, and AC bus bar should be undertaken for various disturbances and protection sequences.

References

[1] Dommel H. 'Digital computer solution of electromagnetic transients in single- and multi-phase networks'. *IEEE Transactions on Power Apparatus and Systems*. 1969, vol. 88 (2), pp. 734–71.

[2] CIGRE 563_WG B4.38. *Modelling and Simulation Studies to be performed during the lifecycle of HVDC systems*; 2013.

[3] Han M., Ding H., Chen X., and Lin C. 'Electromagnetic transient modeling of HVDC system'. *Proceedings of the CSU-EPSA*. 2008, vol. 20 (4), pp. 7–11.
[4] Szechtman M., Wess T., and Thio C. V. 'First benchmark model for HVdc control studies'. *Electra*. 1991, vol. 135, pp. 55–75.
[5] Faruque M., Zhang Y., and Dinavahi V. 'Detailed modeling of CIGRÉ HVDC benchmark system using PSCAD/EMTDC and PSB/SIMULINK'. *IEEE Transactions on Power Delivery*. 2006, vol. 21 (1), pp. 378–87.
[6] Zhao W. *HVDC Transmission Engineering Technology*. Beijing: China Electric Power Press; 2011.
[7] Jos A. *High Voltage Direct Current Transmission (2nd Edition)*. London: The Institution of Electrical Engineers; 1998.
[8] Manitoba HVDC Research Centre Inc. *Applications of PSCAD/EMTDC*. Canada; 2007.
[9] Wang Z., Wu J., Palmer P. R., Bryant A. T., and Hudgins J. L. 'Implementation and comparison of models of conventional thyristor for system simulation'. *IEEE Power Electronics Specialists Conference*; 2006.
[10] Neville W., and Jos A. *Power Systems Electromagnetic Transients Simulation*. London: The Institution of Engineering and Technology; 2007.
[11] Le J., Hu R., Wu M., and Hu R. X. *PSCAD X4 Circuit Design and Simulation*. Beijing: China Machinery Press; 2015.
[12] Shi M., Qian L., Yue Y., and Chen X. 'Phase-locked loop designing of three phase inverter power'. *Transactions of China Electrotechnical Society*. 2015, vol. 30 (sup 1), pp. 93–97.
[13] Liu Z. *Research on Key Techniques in Cascade Multilevel STATCOM for Wind Power Generation*. Wuhan: Huazhong University of Science and Technology; 2010.
[14] Jos A., and Bruce S. *AC-DC Power System Analysis*. United Kingdom: The Institution of Electrical Engineers; 1998.
[15] Arrillaga J., Enright W., Watson N. R., *et al.* 'Improved simulation of HVDC converter transformers in electromagnetic transient programs'. *IEE Proceedings-Generation, Transmission and Distribution*. 1997, vol. 144 (2), pp. 100–6.
[16] Juan A. M., and Bruce A. M. 'Transformer modeling for low- and mid-frequency transients—a review'. *IEEE Transactions on Power Delivery*. 2005, vol. 20 (2), pp. 1625–32.
[17] Manitoba HVDC Research Centre Inc. *EMTDC User's Guide*. Winnipeg; 2005.
[18] Salimi M., Gole A. M., and Jayasinghe R. P. 'Improvement of transformer saturation modeling for electromagnetic transient programs'. *IPST 2013 Proceedings*; Vancouver, Canada, July 2013.
[19] Snelson J. K. 'Propagation of traveling waves on transmission lines - frequency dependence parameters'. *IEEE Transactions on Power Apparatus and Systems*. 1972, vol. PAS-91 (1), pp. 85–91.
[20] Marti J. R. 'Accurate modelling of frequency dependent transmission lines in electromagnetic transient simulations'. *IEEE Transactions on Power Apparatus and Systems*. 1982, vol. PAS-101 (1), pp. 147–55.

[21] Newell R. J., Risan M. D., Allen L., Rao K. S., and Stuehm D. L. 'Utility experience with coherency-based dynamic equivalents of very large systems'. *IEEE Transactions on Power Apparatus and Systems*. 1985, vol. PAS-104 (11), pp. 3056–63.

[22] Liang Y., Lin X., Gole A. M., and Yu M. 'Improved coherency-based wideband equivalents for real-time digital simulators'. *IEEE Transactions on Power Systems*. 2011, vol. 26 (3), pp. 1410–7.

[23] Zhou Y., Duan Y., Zhang B., *et al.* 'Study of oherency-based dynamic equivalent method for power system with HVDC'. *2010 International Conference on Electrical and Control Engineering*; Wuhan, China, 2010. pp. 2013–6.

[24] 'IEEE guide for planning DC links terminating at AC locations having low short-circuit capacities'. *IEEE standard 1204-1997*. New York: IEEE Power & Energy Society; 1997.

[25] CIGRE 001_WG 14.02. 'A summary of the report on survey of controls and control performance in HVDC schemes'. *CIGRE*. 2000, vol. 15 (2), pp. 704–9.

[26] Sato M., Honjo N., Yamaji K., and Yoshino T. 'HVDC converter control for fast power recovery after AC system fault'. *IEEE Transactions on Power Delivery*. 1997, vol. 12 (3), pp. 1319–26.

[27] Hong C. 'Improving fault recovery performance of an HVDC link with a weak receiving AC system by optimization of DC controls'. *IEEE 2018 International Conference on Power Systems Technology (Powercon 2018)*; Guangzhou, China, November 2018, pp. 4474–9.

[28] CIGRE 417_WG B4.45. *Technological Assessment of 800kV HVDC Applications*; 2010.

[29] Marcus H., and Dietmar R. 'Review on UHVDC equipment design aspects'. *Southern Power System Technology*. 2008, vol. 2 (4), pp. 1–12.

[30] Hong C. 'Characteristics analysis of deblocking simultaneously both 12-pulse valve groups in a pole of Yunan-Guangdong ±800 kV DC project'. *Southern Power System Technology*. 2010, vol. 4 (4), pp. 1–7.

[31] Nayak R. N., Sasmal R. P., Victor L., *et al.* 'Current status of design, engineering, manufacturing and testing of 800KV HVDC equipment'. *IEC/CIGRE Second International Symposium on Standards for UHV*; New Delhi, India, 2009.

[32] CIGRE 352_WG B4.34. *Capacitor commutated converters (CCC) HVDC interconnections digital modeling and benchmark circuit*; 2008.

[33] Sadek K., Pereira M., Brandt D. P., Gole A. M., and Daneshpooy A. 'Capacitor commutated circuit configurations for DC transmission'. *IEEE Transactions of Power Delivery*. 1998, vol. 13 (4), pp. 1257–64.

Chapter 4
VSC system modeling and stability analysis

Hui Ding[1], *Yi Qi*[1] *and Xianghua Shi*[1]

4.1 Introduction

The application of the voltage-sourced converter (VSC)-based high-voltage DC (HVDC) system has been growing fast worldwide. It has a significant advantage in controllability and flexibility, e.g. the independent control of active and reactive powers, the fast dynamic performance, and so on. Most significantly, it can avoid the commutation failure that is present in line-commutated converter (LCC)-based HVDC systems. These advanced features greatly promote the applications of VSC-based HVDC transmission systems in renewable energy integration, offshore energy transmission, multi-terminal DC grid interconnections, etc. As the two-level converter is an essential part of VSC systems and was first adopted in a real project, this chapter will only focus on the two-level pulse-width modulation (PWM)-based topology. The design and modeling of multilevel modular converter (MMC)-based HVDC is very different from the two-level or three-level VSC. The detail modeling of MMC can be found in [1] and the related topics are also covered in Chapters 8, 9, 10, and 12 of this book.

For the modeling of VSC-based HVDC systems with electromagnetic transient (EMT)-simulation tools, the parameters must first be properly designed before any further investigation. This chapter will first discuss all the constraints of the VSC electrical systems, and the corresponding criteria are given to calculate the system parameters. Then the AC dynamics of the VSC system are developed with space-phasor representation in the phasor domain. Using the dq0 transformation, the VSC space-phasor model is transformed from the phasor domain to the dq0 domain, in which the current-mode control is designed to decouple the VSC system in d-axis and q-axis. With this typical dual closed-loop decoupling control, the VSC system can be controlled in d-axis and q-axis independently.

However, given the commissioning and operating of recent VSC systems, increasing system resonance occurred in a wide-band range, from several hertz to several kilohertz. In order to give an insightful explanation of these oscillations, the Bode plot of the eigenvalues of the system open-loop transfer matrix is used. With the phase margin and gain margin analyses, the constraints of the proportional-integral (PI) controller parameters are discussed. These conclusions can guide all VSC system modeling and design.

[1]RTDS Technologies Inc., Winnipeg, MB, Canada

4.2 Design of electrical system parameters for VSC

Figure 4.1 shows the schematic of a typical two-level converter system, where an LC filter is used to filter out the high-frequency characteristic harmonics generated by the converter. To operate the converter reasonably, the LC filter and the DC-bus capacitor should be properly designed.

4.2.1 LC filter design

Several constraints are considered in the design of the LC filter of the converter:

- AC current ripple
- DC voltage utilization
- Reactive power limit
- Resonant frequency

4.2.1.1 AC current ripple constraint

In general, from the AC side, the converter can be considered as a voltage source. The AC voltage is generated from the sinusoidal pulse-width modulation (SPWM), which includes a fundamental component and high-frequency harmonics. The AC current is excited by this voltage and consequently includes all the components with the same frequencies. In order to optimize the DC voltage utilization, the third harmonic is superimposed upon the modulation waveforms. Figure 4.2 shows the switching SPWM waveform with third harmonic injection. The third harmonic injection amount is generally set to approximately 15% of the fundamental component.

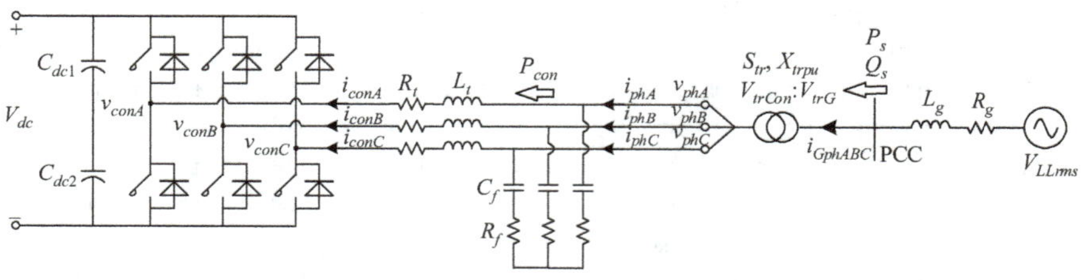

Figure 4.1 Two-level VSC systems configuration

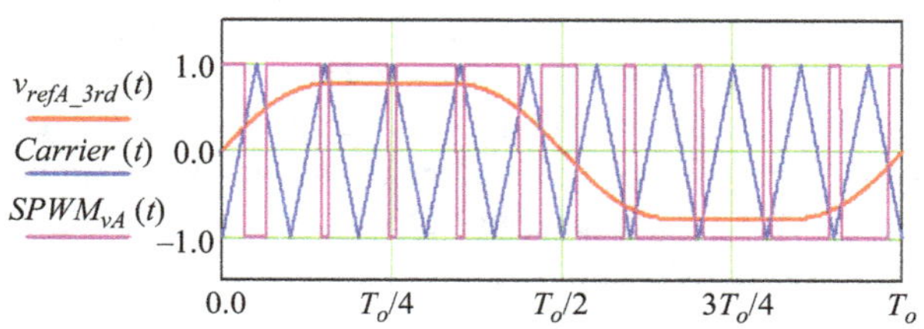

Figure 4.2 Switching waveforms under SPWM and third harmonic injection

Applying fast Fourier transform (FFT) analysis for the SPWM waveform, the harmonic components of the SPWM waveform are shown in Figure 4.3. The modulation index is set as unity.

It is observed that the harmonics are centered around the integer times of the carrier frequency

$$N_h = k \cdot N \pm j, j = \begin{cases} 2, 4, 6, ..., \text{if} \quad k = 1, 3, 5, ... \\ 1, 3, 5, ..., \text{if} \quad k = 2, 4, 6, ... \end{cases} \tag{4.1}$$

where $N = f_{sw}/f_o$. The dominant harmonic of the phase voltage generated by the two-level converter is at carrier frequency. Therefore, the dominant current ripple flowing through the filter inductor is the harmonic centered around the carrier frequency.

Figure 4.4 shows the magnitude of the voltage versus modulation index at the carrier frequency. It shows that the third harmonic injection has a slight influence on the harmonic component at the carrier frequency. This figure remains valid for other high carrier frequencies, as the proportion of the characteristic harmonic maintains a very constant value once the carrier frequency is much larger than the fundamental frequency, e.g. $>10f_o$.

The component of the per-unit SPWM harmonic voltage at the carrier frequency can be obtained approximately from Figure 4.4 (denoted as $V_{con\_Npk\_pu}$), from which the harmonic current ripple can be calculated.

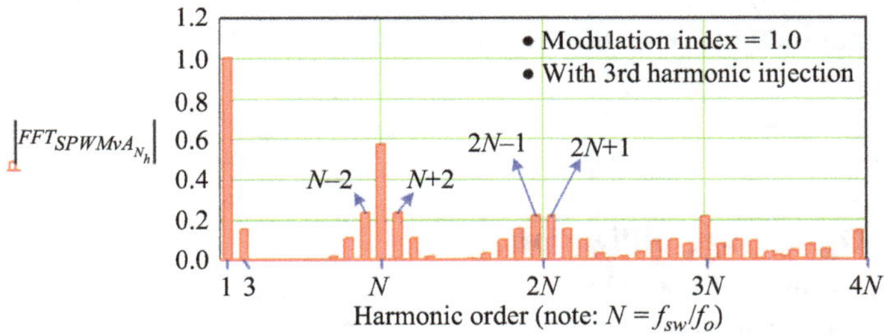

Figure 4.3 Harmonics magnitude of the SPWM waveform

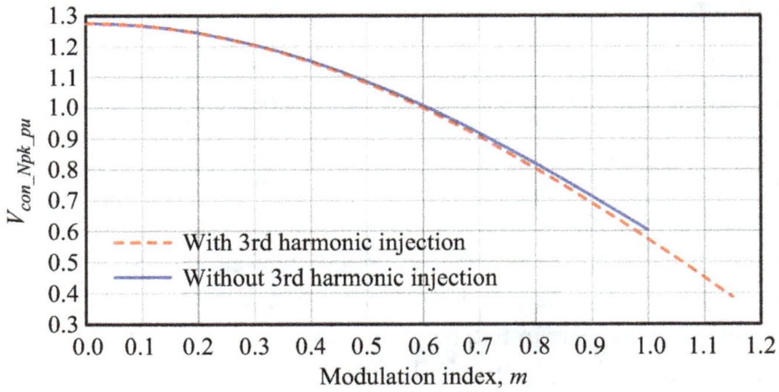

Figure 4.4 SPWM harmonic component at the carrier frequency

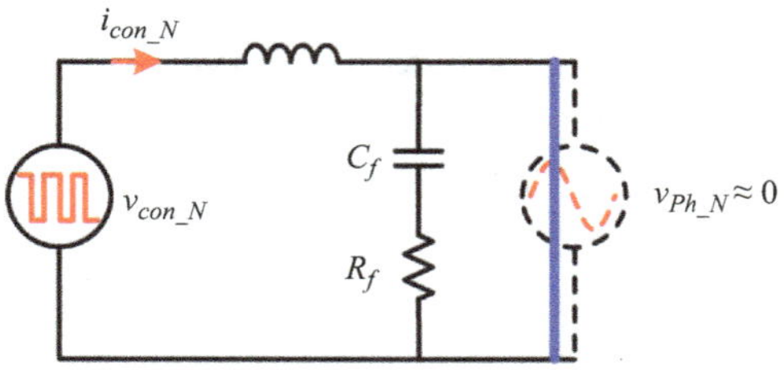

Figure 4.5 Equivalent single-phase circuit for N-th harmonic

Figure 4.5 shows the equivalent single-phase circuit with the voltage source of the *N-th* harmonic. The equivalent impedance of the AC system at the carrier frequency is very low and can be treated as a short circuit.

From Figure 4.5, the current ripple at the carrier frequency is calculated as

$$\begin{aligned} I_{con\_ripple} &\approx 2I_{con\_Npk} = 2\frac{V_{con\_Npk\_pu} \cdot V_{dc}/2}{2\pi f_{sw} L_t} \\ &\leq \Delta I_{con}\% \cdot I_{con\_rated} \end{aligned} \tag{4.2}$$

where $\Delta I_{con}\%$ is the current ripple percentage. In order to limit the current magnitude and reduce the switching losses, the maximum AC current ripple of the converter is generally limited to 20–30% of its rated current. With this constraint, the minimum inductance of the LC filter should meet the requirement

$$L_t \geq \frac{V_{con\_Npk\_pu} \cdot V_{dc}}{2\pi f_{sw} \Delta I_{con}\% \cdot I_{con\_rated}} \tag{4.3}$$

4.2.1.2 DC voltage utilization

To obtain the highest possible DC-link voltage utilization, the voltage drop across the filter inductor is typically limited to less than 10% of the rated phase voltage.

The fundamental component of the voltage across the filter inductor is calculated as follows:

$$V_{Trms} = \omega_o L_t \cdot I_{con\_rated} = \Delta V_{Tdrop}\% \cdot V_{LLcon}/\sqrt{3} \tag{4.4}$$

where $\Delta V_{Tdrop}\%$ is the voltage drop across the filter inductor. With this constraint, the maximum inductance of the LC filter should meet the requirement below

$$L_t \leq \frac{\Delta V_{Tdrop}\% \cdot V_{LLcon}/\sqrt{3}}{\omega_o \cdot I_{con\_rated}} \tag{4.5}$$

According to the limits of the filter inductance calculated in (4.3) and (4.5), a proper inductance can be selected. If the inductance cannot meet the two constraints simultaneously, compromise must be made to choose the proper inductance, i.e., slightly increase either the current ripple limit or the voltage drop limit, or increase both of the limits.

4.2.1.3 Reactive power limit

The reactive power consumption by the filter capacitor at fundamental frequency is usually limited to be less than 5% of the rated real power. The reactive power absorbed by the filter capacitor is calculated below

$$Q_{Cf} = 3\omega_o C_f \left(V_{LLcon}/\sqrt{3} \right)^2 \leq \Delta Q_{Cf}\% \cdot P_{con} \tag{4.6}$$

where $\Delta Q_{Cf}\%$ is the maximum reactive power consumption by the filter capacitor. With this constraint, the maximum capacitance of the LC filter should meet the requirement below

$$C_f \leq \frac{\Delta Q_{Cf}\% \cdot P_{con}}{\omega_o V_{LLcon}^2} \tag{4.7}$$

4.2.1.4 Resonant frequency

In order to avoid the system resonance problems introduced by the LC filter, the resonance frequency of the filter itself normally should be greater than ten times the grid frequency and less than half of the carrier frequency. With this design, the LC filter can highly suppress the harmonics around the carrier frequency, and, at the same time, maintain the fundamental frequency impedance characteristics of the AC system.

$$10f_o \leq f_{res} = \frac{1}{2\pi\sqrt{L_t C_f}} \leq \frac{f_{sw}}{2} \tag{4.8}$$

With the selected filter inductance, the filter capacitance can be calculated as follows:

$$\frac{1}{(2\pi \times f_{sw}/2)^2 L_t} \leq C_f \leq \frac{1}{(2\pi \times 10f_o)^2 L_t} \tag{4.9}$$

If the capacitance cannot meet the two constraints simultaneously, compromise must be made to choose the proper capacitance, i.e., slightly increase the reactive power limit or slightly change the resonant frequency limit range.

4.2.2 DC voltage selection

In order to prevent over modulation occurring in the designed power operation range, the DC voltage should be properly selected. Since the fundamental component of the current flowing into the filter capacitor is typically much smaller than the current of the filter inductor, the former term can be neglected when calculating the voltage drop across the filter inductor. According to the current direction defined in Figure 4.1, the output voltage of the converter is calculated as follows:

$$V_{conPH\_rms} \approx \sqrt{(V_{PHrms} + \omega_o L_t I_{con\_rated} \sin(\phi))^2 + (\omega_o L_t I_{con\_rated} \cos(\phi))^2} \tag{4.10}$$

where ϕ is the power factor of the converter system. The actual modulation index is expressed as below

$$m = \frac{\sqrt{2}V_{conPH\_rms}}{V_{dc}/2} \tag{4.11}$$

If the modulation index m is larger than 1.0 without the third harmonic injection, or is larger than 1.15 with the third harmonic injection, it will be over modulated in nominal system operation. Under this circumstance, the DC-link voltage should be increased to avoid over modulation. Considering a 5% modulation margin, the minimum DC-link voltage should be set as

$$V_{dc} \geq \frac{2\sqrt{2}V_{conPH\_rms}}{95\% m_{max}} \tag{4.12}$$

where

$$\begin{cases} m_{max} = 1.00 & \text{(without 3rd harmonic injection)} \\ m_{max} = 1.15 & \text{(with 3rd harmonic injection)} \end{cases}$$

4.3 Controller design and simulation

A VSC converter system typically controls the reactive power (or AC voltage) and the DC voltage (or active power). In this section, only the current-mode control in the rotating dq-frame is introduced. Through controlling the VSC terminal voltages at the AC side, the VSC output currents can be regulated by a dedicated current controller.

There are two control loops needed in the current-mode controls. The outer control loop generates the current reference for the inner control loop. This current-mode control can limit the VSC current output to protect against overcurrent. In addition, it is more robust against system parameter variations.

4.3.1 Space-phasor representation for balanced systems

For a balanced system, the three-phase signals can be written as sinusoidal function below [2]

$$\begin{cases} f_a(t) = \hat{f}\cos[\theta(t)] \\ f_b(t) = \hat{f}\cos\left[\theta(t) - \frac{2}{3}\pi\right] \\ f_c(t) = \hat{f}\cos\left[\theta(t) + \frac{2}{3}\pi\right] \end{cases} \tag{4.13}$$

where

$$\theta(t) = \theta_0 + \int_0^t \omega(\tau)d\tau \tag{4.14}$$

In the above three-phase functions, $\hat{f}$, θ_0, and ω are the amplitude, the initial phase angle, and the angular frequency of the sinusoidal function, respectively.

The exponential format of the above three-phase function can be rewritten as below

$$\begin{cases} f_a(t) = \dfrac{\hat{f}}{2}\left\{e^{j\theta(t)} + e^{-j\theta(t)}\right\} \\ f_b(t) = \dfrac{\hat{f}}{2}\left\{e^{j\left[\theta(t)-\frac{2}{3}\pi\right]} + e^{-j\left[\theta(t)-\frac{2}{3}\pi\right]}\right\} \\ f_c(t) = \dfrac{\hat{f}}{2}\left\{e^{j\left[\theta(t)+\frac{2}{3}\pi\right]} + e^{-j\left[\theta(t)+\frac{2}{3}\pi\right]}\right\} \end{cases} \tag{4.15}$$

By introducing the transformation matrix from the three-phase frame to the space-phasor frame

$$T_p = \frac{2}{3}\begin{bmatrix} e^{j0} & e^{j\frac{2}{3}\pi} & e^{-j\frac{2}{3}\pi} \end{bmatrix} \tag{4.16}$$

The balanced three-phase function can be represented by the space phasor below

$$\vec{f} = T_p \begin{bmatrix} f_a(t) \\ f_b(t) \\ f_c(t) \end{bmatrix} = \hat{f} e^{j\theta(t)} \tag{4.17}$$

If the angular frequency is an independent state variable and not a function of time, the system phase angle can be represented as below

$$\theta(t) = \theta_0 + \int_0^t \omega(\tau)d\tau = \omega t + \theta_0 \tag{4.18}$$

Therefore, the space phasor of the balanced three-phase function can be rewritten as

$$\vec{f} = \hat{f} e^{j[\omega t+\theta_0]} = \hat{f} e^{j\theta_0} e^{j\omega t} = \underline{f}\, e^{j\omega t} \tag{4.19}$$

where $\underline{f} = \hat{f} e^{j\theta_0}$. In the complex plane, the complex quantity $\underline{f}$ can be represented by a vector.

4.3.2 VSC systems' representation with space phasor

For VSC systems with current-mode controls, the converter itself can be represented by a controlled voltage source V_{c_abc}. The converter interfaces with the voltage V_{t_abc} of the point of common coupling (PCC) through three decoupled inductors L_t and resistors R_t (or through three independent single transformers).

When only the fundamental component of the VSC output is considered in the phase domain, the VSC AC system can be represented by the equation below

$$L_t \frac{di_{v\_abc}}{dt} + R_t i_{v\_abc} = v_{t\_abc} - v_{c\_abc} \tag{4.20}$$

The space phasor of the above VSC systems is

$$L_t \frac{d\vec{i_v}}{dt} + R_t \vec{i_v} = \vec{v_t} - \vec{v_c} \tag{4.21}$$

The space phasor of the VSC system can be represented by the vectors in the complex rotating dq-frame with angular frequency ω.

Assume the VSC system space phasors are represented by the quantities in the rotating complex plane as below

$$\begin{cases} \vec{i_v} = e^{j\theta} \underline{i_{vdq}} \\ \vec{v_t} = e^{j\theta} \underline{v_{tdq}} \\ \vec{v_c} = e^{j\theta} \underline{v_{cdq}} \end{cases} \tag{4.22}$$

The VSC systems can be represented in the dq domain as

$$L_t \frac{d\underline{i_{vdq}}}{dt} + R_t \underline{i_{vdq}} = \underline{v_{tdq}} - \underline{v_{cdq}} - j\omega L_t \underline{i_{vdq}} \tag{4.23}$$

By introducing the vector $\underline{u_{dq}}$ as the new control input

$$\underline{u_{dq}} = \underline{v_{tdq}} - \underline{v_{cdq}} - j\omega L_t \underline{i_{vdq}} \tag{4.24}$$

The above equation can be rewritten as

$$L_t \frac{d\underline{i_{vdq}}}{dt} + R_t \underline{i_{vdq}} = \underline{u_{dq}} \tag{4.25}$$

This is the VSC AC system model in the rotating dq-frame. The VSC AC system dynamics can be represented by the first-order linear transfer function in the dq-frame.

$$\begin{cases} L_t \frac{di_{vd}}{dt} + R_t i_{vd} = u_d \\ L_t \frac{di_{vq}}{dt} + R_t i_{vq} = u_q \end{cases} \tag{4.26}$$

where

$$\begin{cases} u_d = v_{td} - v_{cd} + \omega L_t i_{vq} \\ u_q = v_{tq} - v_{cq} - \omega L_t i_{vd} \end{cases} \tag{4.27}$$

Equations (4.26) and (4.27) can be represented by a control block diagram shown in Figure 4.6.

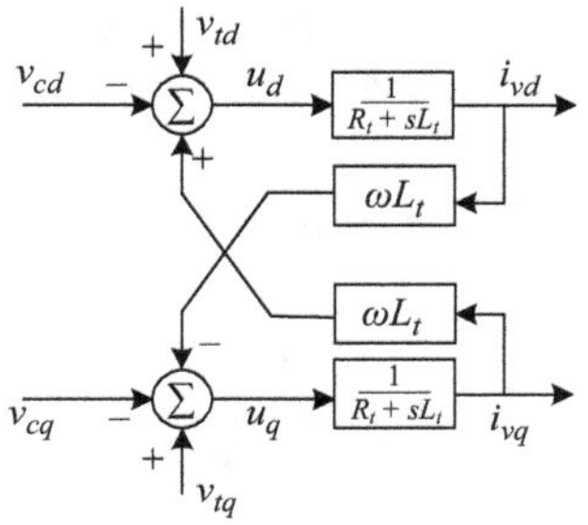

Figure 4.6 Control block diagram of VSC AC-side system

Table 4.1 Base values for VSC systems

Variable	Symbol and expression	Description
Power	P_b	VSC-rated capacity
AC voltage	$V_b = \hat{V}_n$	Peak value of the line-to-neutral voltage
AC current	$I_b = \frac{2}{3}\frac{P_b}{V_b}$	Peak value of the rated line current
DC voltage	$V_{dcb} = V_{dc0}$	Rated DC capacitor voltage
Frequency	$f_b = f_0$	System nominal frequency
Angular speed	$\omega_b = 2\pi f_b$	
Impedance	$Z_b = V_b/I_b$	
Inductance	$L_b = Z_b/\omega_b$	
Capacitance	$C_b = 1/(Z_b\omega_b)$	

4.3.3 Current-mode control design for VSC systems

The VSC system current-mode control is designed in the dq-frame. The d-axis component of the outer-loop regulates the VSC DC capacitor voltage or the DC power, and the q-axis component of the outer-loop control regulates the VSC system AC-side reactive power or the AC bus voltage. The outer control loop generates the current control reference $i^*_{v\_dq}$ for the inner current control loop. There are couplings between the d-axis and q-axis of the VSC converter dynamic model. Therefore, a corresponding compensation term should be included in the controller to eliminate these couplings. This is the main idea of the current-mode control in VSC system control design.

4.3.3.1 Base values selection for per-unit VSC system

The controller of VSC systems is often designed in a per-unit term. The base values for per-unit VSC systems follow the rules below.

For the VSC system, the base value of voltage is chosen as the peak value of the line-to-neutral voltage. The rated three-phase power is selected as the base power. With these two restrictions, the other base value can be derived as shown in Table 4.1.

For the VSC controller, per-unit values are used. With the above base value selection, the VSC system can be per-unitized as below

$$\begin{cases} L_{t\_pu}\dfrac{di_{vd\_pu}}{\omega_b dt} + R_{t\_pu}i_{vd\_pu} = u_{d\_pu} \\ L_{t\_pu}\dfrac{di_{vq\_pu}}{\omega_b dt} + R_{t\_pu}i_{vq\_pu} = u_{q\_pu} \end{cases} \tag{4.28}$$

where

$$\begin{cases} u_{d\_pu} = v_{td\_pu} - v_{cd\_pu} + \omega_{\_pu}L_{t\_pu}i_{vq\_pu} \\ u_{q\_pu} = v_{tq\_pu} - v_{cq\_pu} - \omega_{\_pu}L_{t\_pu}i_{vd\_pu} \end{cases} \tag{4.29}$$

To increase clarity, all the subscripts with *_pu* will be omitted in the following sections.

4.3.3.2 Ideal VSC system inner-loop design

An ideal inner-loop current control in dq-frame should eliminate the coupling between the d-axis and the q-axis for the VSC dynamics.

From the VSC ac dynamic function, the VSC output can be calculated by (4.30).

$$\begin{cases} v_{cd} = v_{td} - u_d + \omega L_t i_{vq} \\ v_{cq} = v_{tq} - u_q - \omega L_t i_{vd} \end{cases} \tag{4.30}$$

With the assumptions below:

(a) The control reference of VSC is exactly equal to the VSC output

$$\begin{cases} v_{cd}^* = v_{cd} \\ v_{cq}^* = v_{cq} \end{cases} \tag{4.31}$$

(b) Using the proportion-integral (PI) in the inner-loop control to represent the AC dynamic of VSC

$$\begin{cases} u_d' = \left(k_{pd} + \dfrac{k_{id}}{s}\right) \cdot \Delta i_{vd} \\ u_q' = \left(k_{pq} + \dfrac{k_{iq}}{s}\right) \cdot \Delta i_{vq} \end{cases} \tag{4.32}$$

where Δi_{vd} and Δi_{vq} are the current differences between the reference and actual measured current on the d-axis and q-axis, respectively.

The inner-loop control reference of VSC can be calculated as below

$$\begin{cases} v_{cd}^* = v_{td} - u_d' + \omega L_t i_{vq} \\ v_{cq}^* = v_{tq} - u_q' - \omega L_t i_{vd} \end{cases} \tag{4.33}$$

The above representation of inner-loop control block is shown in Figure 4.7.

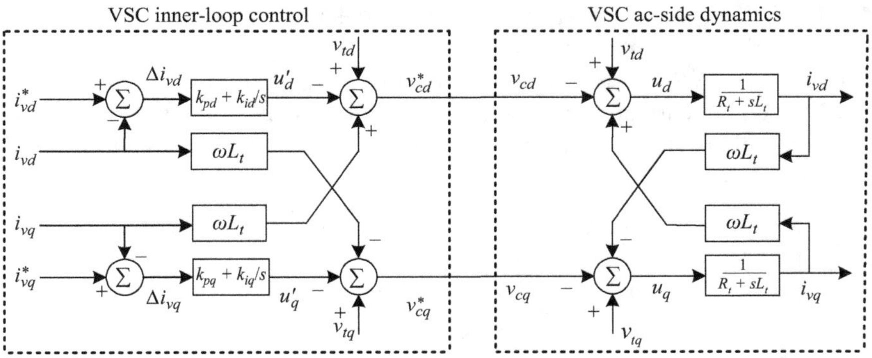

Figure 4.7 Ideal inner-loop control block diagram of VSC systems

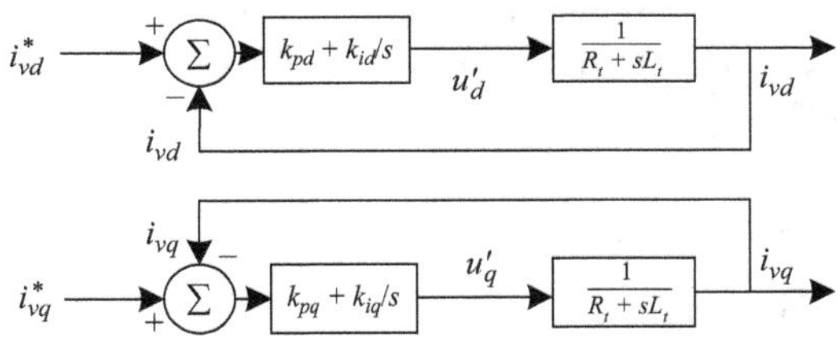

Figure 4.8 Equivalent control block diagram of VSC systems

With the above design of the inner-loop controls, the VSC system can be completely decoupled. The dynamics of VSC systems can be simplified as shown in Figure 4.8.

For the inner-loop control and the VSC AC system, the closed-loop transfer function is

$$H(s) = \frac{i_v}{i_v^*}(s) = \frac{1}{1 + \frac{R_t + sL_t}{k_i + sk_p}s} \tag{4.34}$$

If the closed-loop system is purposely designed as a first-order system with time constant τ_i

$$H(s) = \frac{i_v}{i_v^*}(s) = \frac{1}{1 + \frac{R_t + sL_t}{k_i + sk_p}s} = \frac{1}{1 + \tau_i s} \tag{4.35}$$

Then, the PI parameters can be selected by (4.36)

$$\begin{cases} k_p = \dfrac{L_t}{\tau_i} \\ k_i = \dfrac{R_t}{\tau_i} \end{cases} \tag{4.36}$$

4.3.3.3 VSC control system inner-loop design with considering the measurements

For a practical VSC system inner-loop control design, the PCC terminal voltages and VSC output currents are measured through the potential transformer (PT) and the current transformer (CT). These measurements are modeled by the first-order low-pass filters.

$$\begin{cases} H_{vm}(s) = \dfrac{1}{1 + T_{vm}s} = \dfrac{1}{1 + s/\omega_{cv}} \\ H_{im}(s) = \dfrac{1}{1 + T_{im}s} = \dfrac{1}{1 + s/\omega_{ci}} \end{cases} \tag{4.37}$$

In order to somewhat attenuate the carrier frequency harmonic and fast track the current references, the cutoff frequency should be properly selected, which is usually at several hundred hertz.

For typical current measurements, the time constant 1.0 ms is chosen. With a carrier frequency of 2.0 kHz, the magnitude attenuation and the cutoff frequency of current measurements are

$$\begin{cases} H_{im}(s)\Big|_{2.0kHz} = \left| \dfrac{1}{1 + 0.001 \cdot 2.0 \cdot \pi \cdot 2000j} \right| = 0.079 \\ f_{c\_im} = 1/(0.001 \cdot 2.0 \cdot \pi) = 159.155(Hz) \end{cases} \tag{4.38}$$

For the voltage measurements, the time constant of 10.0 ms is chosen. With a carrier frequency 2.0 kHz, the magnitude attenuation and the cutoff frequency of voltage measurements are

$$\begin{cases} H_{vm}(s)\Big|_{2.0kHz} = \left| \dfrac{1}{1 + 0.02 \cdot 2.0 \cdot \pi \cdot 2000j} \right| = 0.004 \\ f_{c\_vm} = 1/(0.01 \cdot 2.0 \cdot \pi) = 15.92(Hz) \end{cases} \tag{4.39}$$

For the grid-side converters, the system frequency variation is limited to a very narrow range (i.e. ±1.0 Hz). If the variation of system frequency is not considered in the inner-loop control design, the coupled components in the d-axis and q-axis can be represented by the rated system angular frequency ω_0.

Considering the above factors, the current-mode control cannot completely eliminate the coupling components between the d-axis and the q-axis. The representation of inner-loop control can be achieved with the equation below.

$$\begin{cases} v_{cd}^* = v_{td}H_{vm}(s) - u_d' + \omega_0 L_t i_{vq} H_{im}(s) \\ v_{cq}^* = v_{tq}H_{vm}(s) - u_q' - \omega_0 L_t i_{vd} H_{im}(s) \end{cases} \tag{4.40}$$

With (4.40), the VSC system inner-loop control can be represented by the blocks shown in Figure 4.9.

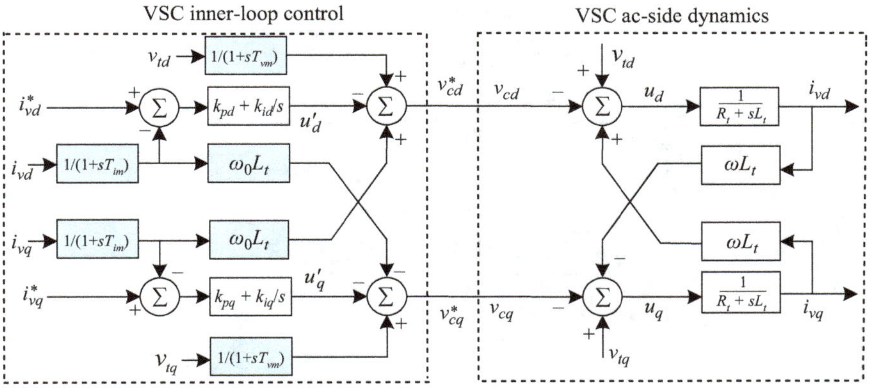

Figure 4.9 Inner-loop control block diagram of VSC systems with the measurements considered

4.3.3.4 VSC control system outer-loop design

The VSC system outer-loop control generates the current references for the inner-loop control. The grid-side converter regulates the DC voltage (or AC active power) and AC voltage (or AC reactive power). For the d-axis, the DC voltage is measured with the first-order filter (time constant 10 ms). For the q-axis, the AC voltage is also measured with the first-order filter (time constant 10 ms).

For the DC voltage and AC voltage control, the errors between the references and measurements are passed to the PI controller

$$\begin{cases} i^*_{vd} = \left(k^1_{pd} + k^1_{id}/s\right) \cdot \Delta u_{dc} \\ i^*_{vq} = \left(k^1_{pq} + k^1_{iq}/s\right) \cdot \Delta u_{ac} \end{cases} \tag{4.41}$$

In the rotating dq-frame, based on (4.22), the instantaneous active power and reactive power can be calculated with the dq quantities

$$\begin{cases} P_{ac} = v_{td} i_{vd} + v_{tq} i_{vq} \\ Q_{ac} = -v_{td} i_{vq} + v_{tq} i_{vd} \end{cases} \tag{4.42}$$

If $v_{tq} = 0$, the real power and reactive power are proportional to i_{vd} and i_{vq}, respectively. Note that, the d-axis and q-axis components are per-unit value in (4.42). If the reactive power is regulated by the q-axis component i_{vq}, the output of the PI controller should multiply –1.0 to obtain the i^*_{vq}.

$$i^*_{vq} = -\left(k^1_{pq} + k^1_{iq}/s\right) \cdot \Delta Q_{ac} \tag{4.43}$$

where $\Delta Q_{ac} = Q^*_{ac} - Q_{ac}$. The minus sign can also be put into the reactive power error, which will be $\Delta Q_{ac} = Q_{ac} - Q^*_{ac}$.

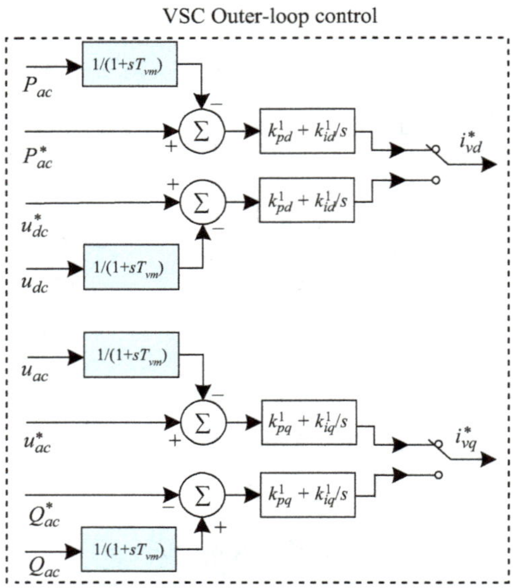

Figure 4.10 Outer-loop control block diagram of VSC systems

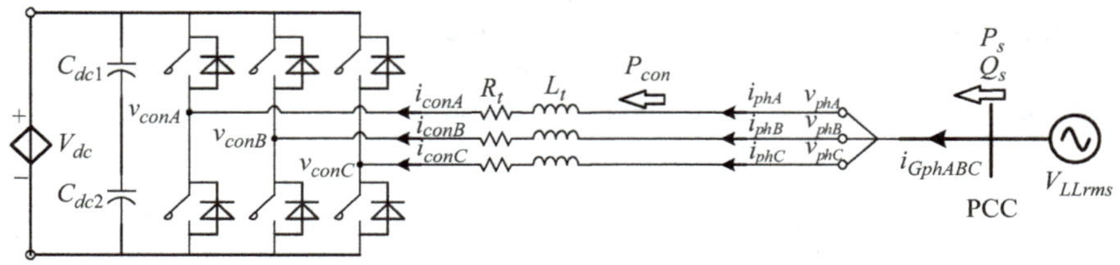

Figure 4.11 Basic structure of the converter

Based on (4.41) and (4.43), the control block of the outer loop is shown in Figure 4.10.

4.4 VSC systems stability analysis

This section studies the characteristics of VSC stability by analyzing a simplified converter system. From the simplified system, the origin of the oscillation can be investigated using bode plots. In addition, case examples of oscillation phenomena are created by adjusting the controller parameters.

4.4.1 Simplified VSC systems

The basic structure of a VSC system is shown in Figure 4.11. It includes the PCC bus, the AC inductor, and the control system. In this section, the AC system and the DC link connected to the converter are both assumed infinite buses, i.e. an ideal AC voltage source connected to the PCC, and an ideal DC voltage source connected to the DC terminal of the converter. For the control system, the current references are set as constant values. Only the inner loop of the decoupled controller is taken into account.

4.4.2 Closed-loop modeling of the VSC systems

The differential equations of the VSC electrical system are shown below

$$L_t \frac{d}{dt}\begin{bmatrix} i_{vd} \\ i_{vq} \end{bmatrix} + \begin{bmatrix} R_t & -\omega L_t \\ \omega L_t & R_t \end{bmatrix}\begin{bmatrix} i_{vd} \\ i_{vq} \end{bmatrix} = \begin{bmatrix} v_{td} \\ v_{tq} \end{bmatrix} - \begin{bmatrix} v_{cd} \\ v_{cq} \end{bmatrix} = \begin{bmatrix} u_d \\ u_q \end{bmatrix} \tag{4.44}$$

The transfer matrix of the VSC system dynamic is

$$\mathbf{H_1(s)} = \frac{\mathbf{i_{vdq}}(s)}{\mathbf{u_{dq}}(s)} = \frac{1}{(sL_t + R_t)^2 + (\omega L_t)^2}\begin{bmatrix} sL_t + R_t & \omega L_t \\ -\omega L_t & sL_t + R_t \end{bmatrix} \tag{4.45}$$

The measurements of VSC AC currents and PCC bus voltages are represented by first-order filters.

For the current measurement

$$\begin{bmatrix} i_{vd\_f} \\ i_{vq\_f} \end{bmatrix} = \mathbf{H_2(s)}\begin{bmatrix} i_{vd} \\ i_{vq} \end{bmatrix} \tag{4.46}$$

where

$$\mathbf{H_2(s)} = 1/(1 + sT_{im})\begin{bmatrix} 1 & 0 \\ 0 & 1 \end{bmatrix} \tag{4.47}$$

For the voltage measurement

$$\begin{bmatrix} v_{td\_f} \\ v_{tq\_f} \end{bmatrix} = 1/(1 + sT_{vm})\begin{bmatrix} 1 & 0 \\ 0 & 1 \end{bmatrix}\begin{bmatrix} v_{td} \\ v_{tq} \end{bmatrix} \tag{4.48}$$

The AC reference voltage provided from the controller is made up of three parts: filtered PCC voltage, the decoupled terms, and the output of the PI controller. It can be calculated as below

$$\begin{aligned} \begin{bmatrix} v_{cd}^* \\ v_{cq}^* \end{bmatrix} &= \begin{bmatrix} v_{td\_f} \\ v_{tq\_f} \end{bmatrix} - \begin{bmatrix} k_p + k_i/s & 0 \\ 0 & k_p + k_i/s \end{bmatrix}\begin{bmatrix} i_{vd}^* - i_{vd\_f} \\ i_{vq}^* - i_{vq\_f} \end{bmatrix} \\ &+ \begin{bmatrix} 0 & \omega_0 L_t \\ -\omega_0 L_t & 0 \end{bmatrix}\begin{bmatrix} i_{vd\_f} \\ i_{vq\_f} \end{bmatrix} \end{aligned} \tag{4.49}$$

Note that the parameters of PI-controllers on d-axis and q-axis are assumed identical.

By introducing the transfer matrix $\mathbf{H_3}$ and $\mathbf{H_4}$, (4.49) can be rewritten as

$$\begin{bmatrix} v_{cd}^* \\ v_{cq}^* \end{bmatrix} = \begin{bmatrix} v_{td\_f} \\ v_{tq\_f} \end{bmatrix} + \mathbf{H_3(s)}\begin{bmatrix} i_{vd\_f} \\ i_{vq\_f} \end{bmatrix} + \mathbf{H_4(s)}\begin{bmatrix} i_{vd}^* \\ i_{vq}^* \end{bmatrix} \tag{4.50}$$

where

$$\mathbf{H_3(s)} = \begin{bmatrix} k_p + k_i/s & \omega_0 L_t \\ -\omega_0 L_t & k_p + k_i/s \end{bmatrix} \tag{4.51}$$

$$\mathbf{H_4(s)} = -\begin{bmatrix} k_p + k_i/s & 0 \\ 0 & k_p + k_i/s \end{bmatrix} \tag{4.52}$$

With the ideal DC voltage, the output of the inner-loop controller is equal to the output of the VSC converter.

$$\begin{bmatrix} v_{cd} \\ v_{cq} \end{bmatrix} = \begin{bmatrix} v_{cd}^* \\ v_{cq}^* \end{bmatrix} \tag{4.53}$$

By replacing the input of the VSC system dynamic with the output of the inner-loop controller, the VSC system can be represented as

$$\begin{bmatrix} u_d \\ u_q \end{bmatrix} = \frac{sT_{vm}}{(1+sT_{vm})}\begin{bmatrix} v_{td} \\ v_{tq} \end{bmatrix} - \mathbf{H_3(s)H_2(s)}\begin{bmatrix} i_{vd} \\ i_{vq} \end{bmatrix} - \mathbf{H_4(s)}\begin{bmatrix} i_{vd}^* \\ i_{vq}^* \end{bmatrix} \tag{4.54}$$

The above equation and the VSC dynamic can form a closed loop as shown in Figure 4.12.

The above control blocks that represent the VSC AC dynamic and inner-loop controller clearly show that the voltage measurement filter does not influence the stability of the simplified system. The VSC dynamic performance cannot be improved by adjusting the time constant of the voltage measurement. Note that this is true if the outer-loop controller is not included.

The transfer matrices $\mathbf{H_1}$, $\mathbf{H_2}$, and $\mathbf{H_3}$ determine the closed-loop stability. From the above definitions, $\mathbf{H_1}$ is associated with the AC inductance and resistance. $\mathbf{H_2}$ is associated with the VSC current measurement filter, which is represented by a first-order filter. $\mathbf{H_3}$ is associated with the PI parameters of the inner-loop controller, and includes the coupling term between the d-axis and q-axis.

The open-loop transfer matrix is the matrix product of the transfer matrices of the three blocks, i.e. $\mathbf{H_1}$, $\mathbf{H_2}$, and $\mathbf{H_3}$.

$$\mathbf{H_{0L}(s)} = \mathbf{H_3(s)H_2(s)H_1(s)} \tag{4.55}$$

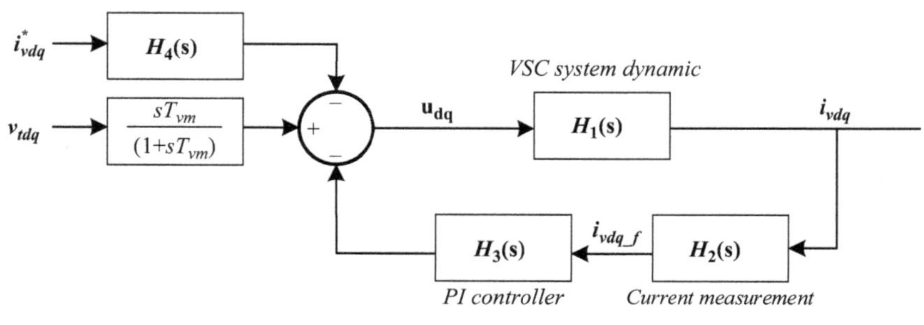

Figure 4.12 Control block representation of the simplified VSC system

Their eigenvalues are denoted as $\boldsymbol{\lambda}_{H1}$, $\boldsymbol{\lambda}_{H2}$, and $\boldsymbol{\lambda}_{H3}$, and are all conjugate pairs. For the transfer matrix $\mathbf{H_1}$, the eigenvalues can be calculated as

$$\begin{cases} \lambda_{\mathbf{H}1,1}(s) = \dfrac{sL_t + R_t + j\omega L_t}{(sL_t + R_t)^2 + (\omega L_t)^2} \\ \lambda_{\mathbf{H}1,2}(s) = \dfrac{sL_t + R_t - j\omega L_t}{(sL_t + R_t)^2 + (\omega L_t)^2} \end{cases} \tag{4.56}$$

For the current measurement transfer matrix $\mathbf{H_2}$, the eigenvalues can be calculated as

$$\lambda_{\mathbf{H}2,1}(s) = \lambda_{\mathbf{H}2,2}(s) = \frac{1}{1 + T_{im}s} \tag{4.57}$$

For the PI controller transfer matrix $\mathbf{H_3}$, the eigenvalues can be calculated as

$$\begin{cases} \lambda_{\mathbf{H}3,1}(s) = k_p + \dfrac{k_i}{s} + j\omega_0 L_t \\ \lambda_{\mathbf{H}3,2}(s) = k_p + \dfrac{k_i}{s} - j\omega_0 L_t \end{cases} \tag{4.58}$$

For the three transfer matrices $\mathbf{H_1}$, $\mathbf{H_2}$, and $\mathbf{H_3}$, the diagonal elements are equal and the non-diagonal elements have opposite signs. The eigenvalue $\boldsymbol{\lambda}_{oLi}$ can be calculated as the product of the eigenvalues $\boldsymbol{\lambda}_{H1i}$, $\boldsymbol{\lambda}_{H2i}$, and $\boldsymbol{\lambda}_{H3i}$ ($i = 1$, and 2)

$$\lambda_{\mathbf{oL}i}(s) = \lambda_{\mathbf{H}1i}(s)\lambda_{\mathbf{H}2i}(s)\lambda_{\mathbf{H}3i}(s) \tag{4.59}$$

With the above calculations, the bode plots of the eigenvalues of the system transfer matrices can be obtained by

$$\begin{cases} |\lambda_{\mathbf{oL}i}(s)| = |\lambda_{\mathbf{H}1i}(s)||\lambda_{\mathbf{H}2i}(s)||\lambda_{\mathbf{H}3i}(s)| \\ \angle\lambda_{\mathbf{oL}i}(s) = \angle\lambda_{\mathbf{H}1i}(s) + \angle\lambda_{\mathbf{H}2i}(s) + \angle\lambda_{\mathbf{H}3i}(s) \end{cases} \tag{4.60}$$

From the above equations, the open-loop system eigenvalues can be handled separately. By analyzing the impact of each part's parameters on the eigenvalues, the nature of the stability of the VSC system and its inner-loop controller can be revealed.

4.4.3 Parametric studies of VSC AC system and inner-loop controller

For the bode plot of the VSC system transfer matrix eigenvalues, if the phase versus frequency curve crosses over −180°, the system is marginally stable once the amplitude at this crossing frequency is equal to one.

(a) For the phase responses of the transfer matrices at frequency zero:

The phase response of $\boldsymbol{\lambda}_{H1i}(\boldsymbol{s})$ can be calculated as

$$\begin{cases} \angle\lambda_{\mathbf{H}1,1}(s)\Big|_{f=0} = \tan^{-1}\left(\dfrac{\omega L_t}{R_t}\right) \\ \angle\lambda_{\mathbf{H}1,2}(s)\Big|_{f=0} = -\tan^{-1}\left(\dfrac{\omega L_t}{R_t}\right) \end{cases} \tag{4.61}$$

The phase response of $\boldsymbol{\lambda}_{H2i}(\boldsymbol{s})$ can be calculated as

$$\begin{cases} \angle\lambda_{\mathbf{H}2,1}(f)\Big|_{f=0} = 0^\circ \\ \angle\lambda_{\mathbf{H}2,2}(f)\Big|_{f=0} = 0^\circ \end{cases} \tag{4.62}$$

The phase response of $\boldsymbol{\lambda}_{H3i}(\boldsymbol{s})$ can be calculated as

$$\begin{cases} \angle\lambda_{\mathbf{H}3,1}(f)\Big|_{f=0} = -90^\circ \\ \angle\lambda_{\mathbf{H}3,2}(f)\Big|_{f=0} = -90^\circ \end{cases} \tag{4.63}$$

With the calculation of the above individual phase response at frequency zero, the phase response of the open-loop transfer matrices at frequency zero is

$$\begin{cases} \angle\lambda_{\mathbf{oL},1}(f)\Big|_{f=0} = \tan^{-1}\left(\dfrac{\omega L_t}{R_t}\right) - 90^\circ \\ \angle\lambda_{\mathbf{oL},2}(f)\Big|_{f=0} = -\tan^{-1}\left(\dfrac{\omega L_t}{R_t}\right) - 90^\circ \end{cases} \tag{4.64}$$

(b) For the amplitude responses of the transfer matrices at frequency zero:
The amplitude response of $\boldsymbol{\lambda}_{H1\mathrm{i}}(\boldsymbol{s})$ can be calculated as

$$\begin{cases} |\lambda_{\mathbf{H}1,1}(s)|_{f=0} = 1/\left(\sqrt{R_t^2 + (\omega L_t)^2}\right) \\ |\lambda_{\mathbf{H}1,2}(s)|_{f=0} = 1/\left(\sqrt{R_t^2 + (\omega L_t)^2}\right) \end{cases} \tag{4.65}$$

The amplitude response of $\boldsymbol{\lambda}_{H2\mathrm{i}}(\boldsymbol{s})$ can be calculated as

$$\begin{cases} |\lambda_{\mathbf{H}2,1}(s)|_{f=0} = 1 \\ |\lambda_{\mathbf{H}2,2}(s)|_{f=0} = 1 \end{cases} \tag{4.66}$$

The amplitude response of $\boldsymbol{\lambda}_{H3\mathrm{i}}(\boldsymbol{s})$ can be calculated as

$$\begin{cases} |\lambda_{\mathbf{H}3,1}(s)|_{f=0} \to \infty \\ |\lambda_{\mathbf{H}3,2}(s)|_{f=0} \to \infty \end{cases} \tag{4.67}$$

With the calculation of the above individual amplitude response at frequency zero, the amplitude response of the open-loop transfer matrices at frequency zero is

$$\begin{cases} |\lambda_{\mathbf{oL},1}(f)|_{f=0} =\to \infty \\ |\lambda_{\mathbf{oL},2}(f)|_{f=0} =\to \infty \end{cases} \tag{4.68}$$

(c) For the phase responses of the transfer matrices at frequency positive infinity:
The phase response of $\boldsymbol{\lambda}_{H1i}(\boldsymbol{s})$ can be calculated as

$$\begin{cases} \angle\lambda_{\mathbf{H}1,1}(s)\Big|_{f\to\infty} = -90^\circ \\ \angle\lambda_{\mathbf{H}1,2}(s)\Big|_{f\to\infty} = -90^\circ \end{cases} \tag{4.69}$$

The phase response of $\boldsymbol{\lambda}_{H2i}(\boldsymbol{s})$ can be calculated as

$$\begin{cases} \angle\lambda_{\mathbf{H}2,1}(f)\Big|_{f\to\infty} = -90^\circ \\ \angle\lambda_{\mathbf{H}2,2}(f)\Big|_{f\to\infty} = -90^\circ \end{cases} \tag{4.70}$$

The phase response of $\boldsymbol{\lambda}_{H3i}(\boldsymbol{s})$ can be calculated as

$$\begin{cases} \angle\lambda_{\mathbf{H}3,1}(f)\Big|_{f\to\infty} = \tan^{-1}\left(\dfrac{\omega_0 L_t}{k_p}\right) \\ \angle\lambda_{\mathbf{H}3,2}(f)\Big|_{f\to\infty} = -\tan^{-1}\left(\dfrac{\omega_0 L_t}{k_p}\right) \end{cases} \tag{4.71}$$

With the calculation of the above individual phase response at frequency positive infinity, the phase response of the open-loop transfer matrices at frequency positive infinity is

$$\begin{cases} \angle\lambda_{\mathbf{oL},1}(f)\Big|_{f\to\infty} = \tan^{-1}\left(\dfrac{\omega_0 L_t}{k_p}\right) - 180^\circ \\ \angle\lambda_{\mathbf{oL},2}(f)\Big|_{f\to\infty} = -\tan^{-1}\left(\dfrac{\omega_0 L_t}{k_p}\right) - 180^\circ \end{cases} \tag{4.72}$$

(d) For the amplitude responses of the transfer matrices at frequency positive infinity:

The amplitude response of $\boldsymbol{\lambda}_{H1i}(\boldsymbol{s})$ can be calculated as

$$\begin{cases} \left|\lambda_{\mathbf{H}1,1}(s)\right|_{f\to\infty} = 0 \\ \left|\lambda_{\mathbf{H}1,2}(s)\right|_{f\to\infty} = 0 \end{cases} \tag{4.73}$$

The amplitude response of $\boldsymbol{\lambda}_{H2i}(\boldsymbol{s})$ can be calculated as

$$\begin{cases} \left|\lambda_{\mathbf{H}2,1}(s)\right|_{f\to\infty} = 0 \\ \left|\lambda_{\mathbf{H}2,2}(s)\right|_{f\to\infty} = 0 \end{cases} \tag{4.74}$$

The amplitude response of $\boldsymbol{\lambda}_{H3i}(\boldsymbol{s})$ can be calculated as

$$\begin{cases} \left|\lambda_{\mathbf{H}3,1}(s)\right|_{f\to\infty} \to \sqrt{k_p^2 + (\omega_0 L_t)^2} \\ \left|\lambda_{\mathbf{H}3,2}(s)\right|_{f\to\infty} \to \sqrt{k_p^2 + (\omega_0 L_t)^2} \end{cases} \tag{4.75}$$

With the calculation of the above individual amplitude response at frequency positive infinity, the amplitude response of the open-loop transfer matrices at frequency positive infinity is

$$\begin{cases} \left|\lambda_{\mathbf{oL},1}(s)\right|_{f\to\infty} = 0 \\ \left|\lambda_{\mathbf{oL},2}(s)\right|_{f\to\infty} = 0 \end{cases} \tag{4.76}$$

The boundary analysis above indicates that the magnitude of λ_{oL} decreases from a large value (infinity) to zero when the frequency increases from DC to a large value (infinity). At zero frequency, $\lambda_{oL,2}$ has a phase angle of $-\tan^{-1}(\omega L_t/R_t) - 90^0$ (larger than -180°); and with the frequency approaching infinity, its phase angle becomes $-\tan^{-1}(\omega_0 L_t/k_p) - 180^0$ (smaller than -180°). According to the mean value theorem, there must be one frequency such that the phase angle of $\lambda_{oL,2}$ crosses over -180°. If the magnitude of $\lambda_{oL,2}$ at this frequency is close to 1.0, the system will be marginally stable.

For the VSC AC system, the resistance is much smaller than the inductance. This means that the phase angle of $\boldsymbol{\lambda}_{H1,\,2}$ is approximately -90^0 at all the frequencies (ignore the dq coupling components).

$$\angle\lambda_{\mathbf{H}1,2}(f) \approx -90^0 \tag{4.77}$$

For the current measurement, the phase angle is

$$\angle\lambda_{\mathbf{H}2,2}(f) = -\tan^{-1}(2\pi f T_{im}) \tag{4.78}$$

Similarly, for the PI controller transfer matrix $\mathbf{H}_3$, if the coupling component is ignored, the phase angle of the second eigenvalue is

$$\angle\lambda_{\mathbf{H}3,2}(f)\Big| \approx \tan^{-1}\left(2\pi f\frac{k_i}{k_p}\right) - 90^\circ \tag{4.79}$$

Therefore, the phase angle of the second eigenvalue of the open-loop transfer matrix is

$$\angle\lambda_{\mathbf{oL},2}(f) \approx \tan^{-1}\left(2\pi f\frac{k_p}{k_i}\right) - \tan^{-1}(2\pi f T_{im}) - 180^\circ \tag{4.80}$$

As the monotonicity of the inverse tangent, if we can keep

$$\frac{k_p}{k_i} > T_{im} \tag{4.81}$$

the phase angle of the second eigenvalue cannot cross over -180° and the system will always be stable.

4.4.4 *Example of the parametric studies*

The closed-loop VSC system that consists of both the VSC AC dynamics and the inner-loop controls is a linear system. The stability is not related to operation conditions. Table 4.2 lists the parameters of the inner-loop controller and the VSC AC system.

With the parameter settings as shown Table 4.2, the bode plots of the eigenvalues of the open-loop transfer matrices are shown in Figure 4.13. The ratio of the inner-loop gain is 1.0/200, which is larger than 0.001. According to (4.81), the phase angle should not cross over –180°.

However, the phase angle of the bode plot (the second eigenvalue) crosses over –180° at frequency 1,300 Hz (out of the scope of the figure). This is caused by the coupling of the inner-loop controls, and the coupling of the VSC AC systems in the dq axis.

Bode plots of the two eigenvalues with different measurement time constants are shown in Figure 4.14. For the eigenvalue $\lambda_{oL,2}$, when the time constant of the current measurement is set to $T_{im} = 5$ ms, the phase angle crosses –180° at 38 Hz. The magnitude at 38 Hz is 5.0, which is out of the system stability

Table 4.2 VSC systems parameters

Variable	Value	Description
k_p	1.0	Proportional gain
k_i	200	Integral gain
T_{im}	0.001	Current measurement constant
R_t	0.01	VSC AC system resistance (p.u.)
L_t	$0.1/\omega_b$	VSC AC system inductance (p.u.)
ω_0	1.0	Nominal angular speed (p.u.)
ω_b	314.159	Angular speed base value

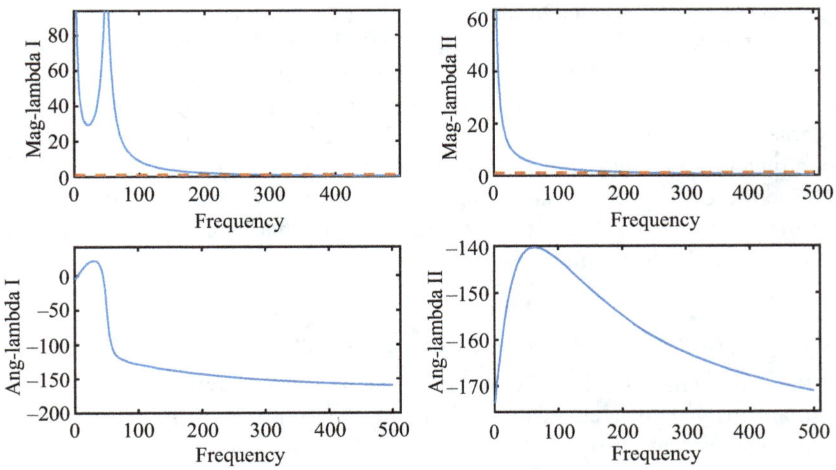

Figure 4.13 Bode plot of the eigenvalues of the VSC open-loop system

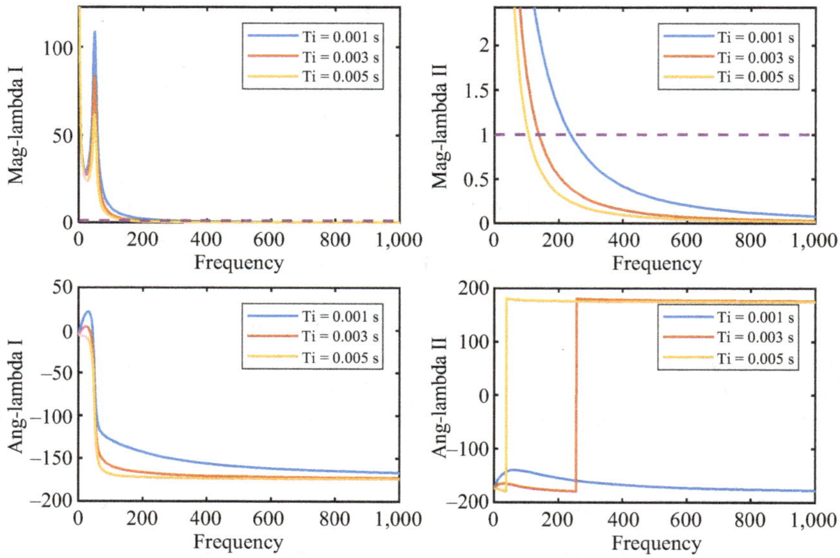

Figure 4.14 Bode plots of the eigenvalues for different T_{im} values

range (the magnitude should be less than 1.0). In addition, the magnitude crosses 1.0 at frequency 105.5 Hz. The phase is 176.75°, suggesting that the system is likely to be unstable at 105.5 Hz.

For the current measurement time constant $T_{im} = 3$ ms, the phase angle crosses −180° at frequency 255.5 Hz. Its magnitude is 0.341. The VSC system is stable with the magnitude margin of 1/0.341, which is 2.93. In the time domain, by multiplying a gain to the inner loop (including the coupling of the d-axis and q-axis), the phase angle characteristic of the bode plot does not change. The magnitude of the bode plot will increase proportionally. Figure 4.15 shows that with the step change of the multiplication factor of the PI controller from 2.8 to 3.0, the system becomes unstable. The resonant frequency is 1/0.0039 = 256.4 Hz. The time-domain simulation matches well with the above analysis.

Therefore, for VSC systems, the maximum gain of the PI in the inner-loop control should be limited. For this example, the maximum gain of inner-loop control should be less than $2.93k_p$ and $2.93k_i$.

4.4.5 Influence of the control cycle on system stability

For the above analysis, the VSC AC dynamic and inner-loop controller are solved simultaneously. However, in the real world, the controller first obtains the measured voltages and currents. Then with the designed control strategies, the controller sends out control signals to the VSC converters. This period is defined as the control cycle of the VSC system.

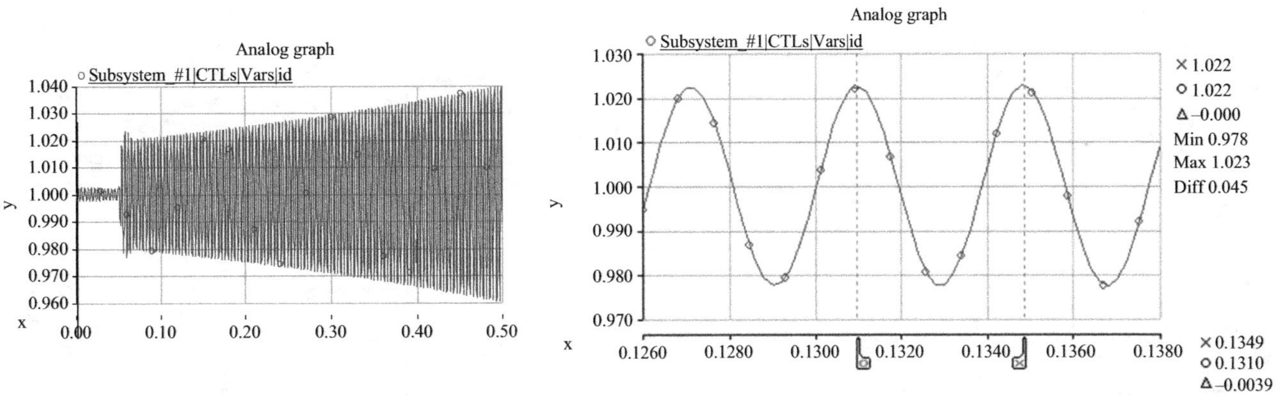

Figure 4.15 Step change of the multiplication for the inner-loop gain from 2.8 to 3.0

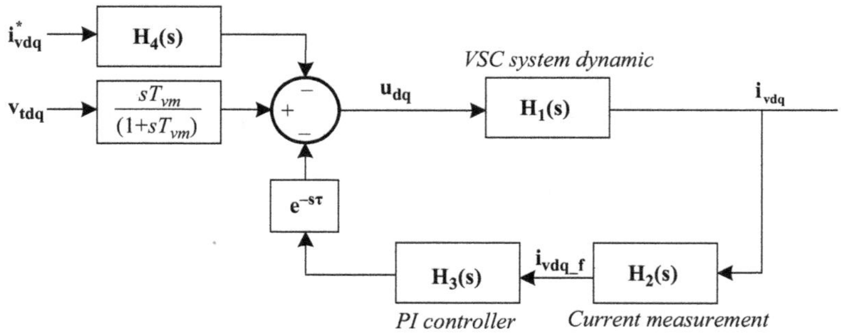

Figure 4.16 Control block representation of the simplified converter with control cycle delays

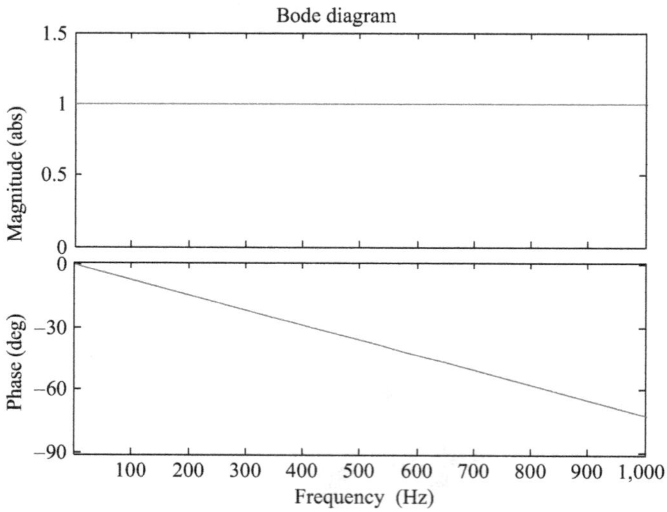

Figure 4.17 Bode plot of 200-μs delay transfer function

In order to analyze the influence of the control cycle on VSC system stability, a transfer function modeling the time delay of the control cycle, $e^{-s\tau}$ is introduced to the closed-loop transfer function.

This time delay can be considered with the solution time step in EMT simulation. As for most EMT simulation tools, the control system and power system are solved sequentially. This sequential solution will introduce one time step of delay, e.g. the solution time step is 200 μs, so the delay between the VSC AC system and its controller will be 200 μs.

The bode plot of the delay component is shown in Figure 4.17. It shows that the delay component (200 μs) will not change the magnitude characteristics. However,

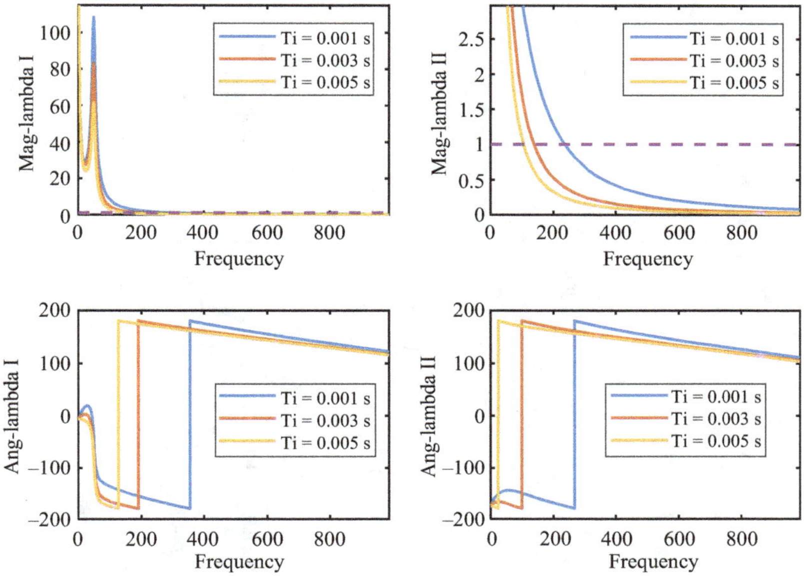

Figure 4.18 Bode plots of the eigenvalues for different T_{im} values considering the 200-μs control cycle time delay

it will linearly shift the phase angle $(-2\pi f\tau)$ in all frequency ranges. For the above system, this extra phase angle shift may reduce the system phase margin and deteriorate the VSC system performance.

Considering a 200-μs control cycle time delay, bode plots of the two eigenvalues with different measurement time constants are shown in Figure 4.18. For the second eigenvalue $\lambda_{oL,2}$ bode plot, the VSC system becomes unstable with the current measurement time constant 3 and 5 ms, as at the frequency where the phase angle crosses over −180°, the magnitude is larger than the unity value. However, as in the previous analysis, the VSC system is stable with time constant 3 ms.

For the current measurement time constant $T_{im} = 1$ ms, the VSC system is stable. The phase plot crosses −180° at frequency 267.5 Hz. Its magnitude is 0.824. The VSC system is stable with a magnitude margin of 1/0.824 = 1.214. The time-domain EMT simulation also shows that by increasing the gains of the PI control loop from 1.15 to 1.25, the system becomes unstable. The resonance frequency is 1.0/0.0038 = 263.2 Hz, which is very close to the value obtained from theoretical analysis (257.5 Hz) (Figure 4.19). Therefore, the maximum gain of inner-loop control must be less than $1.214k_p$ and $1.214k_i$.

From the above analysis and time-domain simulations, it is clearly shown that the control cycle introduces a negative phase shifting, reducing the system stability margin. The improper larger gains of the inner-loop PI parameters can further reduce the system stability margin.

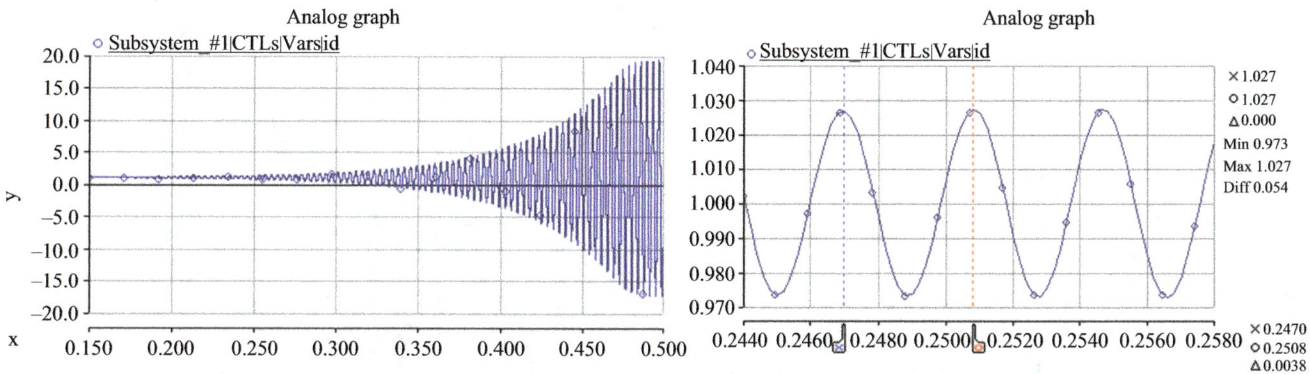

Figure 4.19 Step change of the multiplication for the inner-loop gain from 1.15 to 1.25

References

[1] U. N. Gnanarathna, A. M. Gole, and R. P. Jayasinghe, "Efficient Modeling of Modular Multilevel HVDC Converters (MMC) on Electromagnetic Transient Simulation Programs," Power Delivery, IEEE Transactions on, vol. 26, no. 1, pp. 316–324, 2011.

[2] A. Yazdani and R. Iravani. Voltage-Sourced Converters in Power Systems: Modeling, Control, and Applications. Hoboken, N.J.: IEEE Press/John Wiley, 2010.

Chapter 5
Electromagnetic modeling of DC grid

Minxiao Han[1], Guangyang Zhou[1] and Zmarrak Wali Khan[1]

5.1 Concept of multi-terminal high-voltage direct current (HVDC) and DC grid

5.1.1 Definition of multi-terminal direct current (MTDC) and DC grid

Up till now, HVDC has been predominately used for point-to-point transmission with two terminals and, in a few cases, with three terminals in radial configuration [1]. There have been quite a few multi-terminal practices for both line commutated converter (LCC)-HVDC and voltage source converter (VSC)-HVDC transmission [2]. With the fast development of large-scale remote-area renewable generation and wide-area energy transmission as in West Europe, North America, and Mainland China, the scenarios call for the need of multi-point integration, multi-point consumption, and multi-area interconnection. Studies of multi-terminal HVDC or MTDC grid have shown their potential application for the scenarios. MTDC can be defined as HVDC transmission or interconnection with more than two converter terminals. If some of the terminals in the MTDC system have more than one channel to reach each other, or there is one or more meshed structure included, it is referred to as a meshed DC grid. The MTDC can be considered as a radial-type DC grid, a typical example of DC grid. The terminology DC grid will be used in this book for both MTDC and meshed grid. The possible topology and configuration will be illustrated in the following sections.

The history of HVDC initiated from two-terminal or point-to-point configuration [3]. With more and more two-terminal HVDC links being commissioned between energy center and load center, there is a need for multi-point power integration, multi-point power consumption, and the interconnection between the links. The emerge of economic developing zone or renewable energy base in the medium reach of long-distance HVDC link needs an tapping within the HVDC channel. The concept of DC grids can also be used for low-voltage-level power supply, such as industrial park or data center with the voltage less than 1,500 V. It's also referred to as DC micro-grid. DC grids can also be configured for medium voltage such as

[1]School of Electrical and Electronic Engineering, North China Electric Power University, Changping, Beijing, China

medium-voltage direct current (MVDC) distribution system [4] with the voltage range from 2 to 50 kV. However, in this book, the discussion of DC grid will be focused on large-capacity, high-voltage, and wide-area HVDC grid.

It is likely that the HVDC grids have the potential to be used for:

1. Supply of power to urban load centers
2. Interconnection of offshore or remote generation resources
3. Connection of multiple remote generation sources to more load centers
4. Intercontinental transfer of large scale of energy
5. Overlay grid covering wide area, even globally

5.1.2 Driving force for the development of MTDC (DC grid)

The driving force for an HVDC grid is, at least, the same as that with an AC grid, which is to get increased flexibility and reliability with a lower cost. Following are some typical advantages by utilizing the DC Grid interconnection.

1. Reduce the transmission corridor: With MTDC technology, the multi-energy center can be connected to the multi-load center with one or less transmission line. Reducing of transmission lines means saving bulk of investment and tremendous work to get the right-of-way.
2. Improved flexibility and reliability: A single-point failure in the point-to-point HVDC link results in loss-of-infeed into the AC grid which might have a significant impact on the system stability depending on the amount of loss. An MTDC grid can reroute the power flow when a single-point failure occurs, thereby improving the reliability of the system. Figure 5.1 illustrates the DC grid of Zhangbei [5], China. T_1 and T_2 are the two power sources, mainly renewable generation. T_3 is connected to a pump power storage and T_4 is for the load center supply. With a meshed configuration, when, for instance, L_3 failed, the power can be taken over by lines L_1 and L_4.
3. Reduced capacity: As the peak load demand of different AC systems does not occur at the same time, the peak load demand of the MTDC system is much

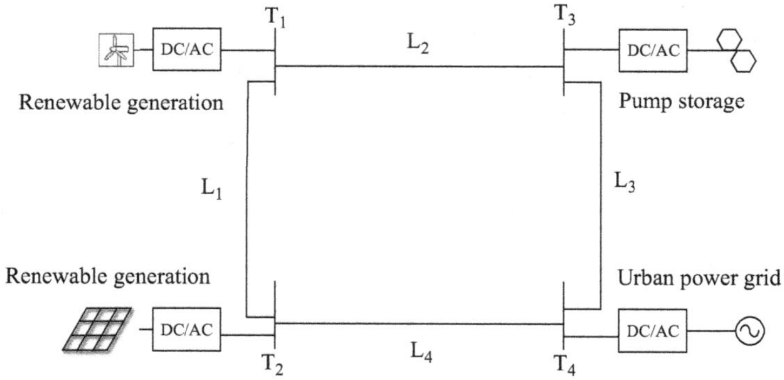

Figure 5.1 Zhangbei 4-terminal VSC-HVDC project

less than the sum of peak load demands of multiple point-to-point HVDC links. Thus, the installed capacity of power generation can be reduced as well as the ratings of the expense of the converter stations. And, the spinning reserve requirement in the AC systems can also be reduced a lot when the DC grid connects the generation to different load centers.

4. Less impact to AC system: The failure of large capacity of HVDC link can be a threat to the stability both for the sending area and the receiving area. The instability of rotor angle or voltage can happen with the imbalance of the bulk power. The impact on the AC system can be mitigated by DC grid if only one converter fails, which cause minor power imbalance and can be easily compensated for by other converters.
5. Reduced curtailment from wind farms (WFs) and photovoltaics (PVs): When multiple WFs and PVs are connected via multiple point-to-point HVDC links to different AC grids, the system operators might have to do wind curtailment from individual farms to match the load profile of the corresponding AC systems. When the WFs are integrated through the MTDC grid, the curtailment can be reduced since the power exchange can happen between different areas.
6. Smoothening of renewable generation stochastic: Since the MTDC grid can facilitate the integration of multiple WFs and PVs, it can reduce the variability in the generation profile in a large area. As shown in Figure 5.1, both T_1 and T_2 are for renewable energy integration. The stochastic numerical analysis demonstrates that the electric power deviation is much smaller when T_1 pluses T_2 as compared with a single one of them.

5.1.3 Classifications of MTDC and their properties [1,2]

1. Classified according to the topologies: parallel connection as shown in Figure 5.2 (a) and (b), series connection as shown in Figure 5.2(c), and hybrid as shown in Figure 5.2(d).

   The series and hybrid connected structure have difficulties in insulation coordination and high power loss, which make these two construction never have practical applications. All the MTDC or DC grid projects commissioned in the world are with the topologies of parallel circuits.
2. Classified according to the converters: At the early stage, the MTDC was commissioned only with LCC, which was available at that time. Because of the difficulty to inverse the power flower with LCC, MTDC almost stopped its pace with only three or four projects in operation. The planned Quebec-New England 5-terminal system was only commissioned as three terminal systems because of the huge complexity for control. HVDC technology breaks a new ground with the application of VSC. The power flow can be easily changed with VSC just by the changing of the direction of the current. DC grid can also include both LCC and VSC which is referred as hybrid DC grid [6]. A typical example in Figure 5.3 illustrates the combination of LCC and VSC for a three terminals MTDC in Southern China.

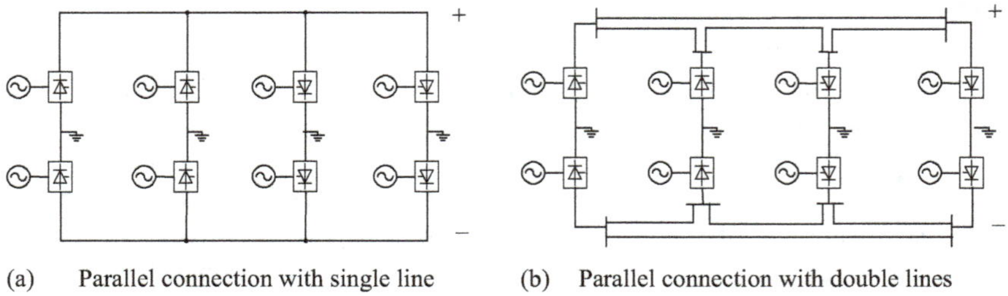

Figure 5.2 DC grid topologies

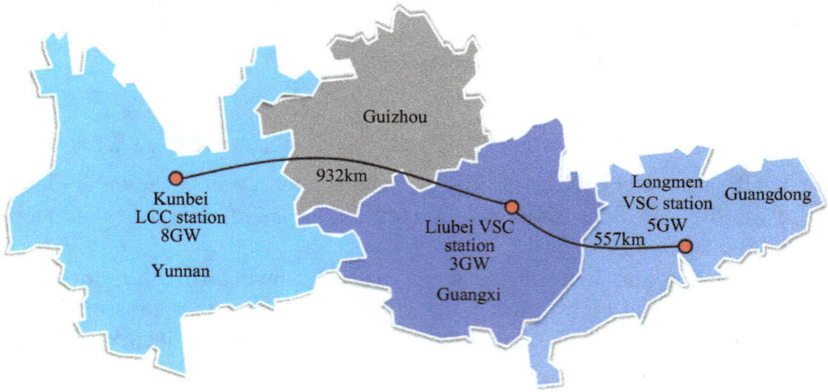

Figure 5.3 Hybrid DC grid application scenario

In many cases, the hybrid DC grid employs both LCC and VSC. VSC is used for local energy collection and integration in the sending terminal and for the inversion at the receiving end to overcome the possible commutation failure of LCC.

5.1.4 MTDC development and DC grid in the future

Multi-terminal HVDC based on LCC has been developed for many years. As described above, the commissioned projects demonstrated that the system is extremely difficult to operate with multiple terminals because of the difficulty of the coordination among the terminals. With the development of VSC HVDC, DC circuit breaker, DC/DC converter, and distributed autonomous control (DAC), a wide-area DC grid is getting more realistic with both technical and economic advantages. With the need of globalized renewable energy exploitation, transmission, and utilization, the DC grid will find a bright future in the prospective electric power system.

5.2 Modeling of the DC grid

Most devices in the main circuit of DC grid are the same as in the two-terminal HVDC, such as the AC/DC converters, the converter transformers, the filters, and the transmission lines, which have been fully discussed in the previous chapters. While devices such as the DC line circuit breaker, DC/DC converters are only found in the DC grid. The modeling of DC grid needs the sufficient studies of those devices.

5.2.1 Modeling of DC circuit breaker

1. Basic characteristics of DC circuit breaker [7]

   High-power DC circuit breaker has become one of the key factors restricting the development of multi-terminal DC systems or DC grids. Because of the natural zero-crossing point of the current in AC system, it is much easier for AC circuit breakers to crush the fault current, which could be several times of the rated value. However, the traditional AC circuit breaker cannot cut off the fault current in DC system because of no zero crossing of DC fault current. There are two options for cutting off the fault current: one is to use active or passive resonant circuit to force the fault current to zero; the other method is to use semiconductor devices to cut off the fault current, referred to as the solid-state circuit breaker.

   Due to the low damping characteristics of DC system, the fast rising speed of fault current becomes another major problem of DC current breaking. Therefore, the longer the switching operation delay, the greater amplitude of the fault current needed to be broken, which causes the arrester to absorb more energy and has poor economy. Therefore, DC circuit breakers should have a faster operating time, usually less than 5 ms.

   In the mid-1980s, a DC circuit breaker based on a resonance circuit was successfully developed. This circuit breaker realizes the fault current breaking of the LCC-HVDC system, which may also changes the current flow path. The operating time of this type of DC circuit breaker is usually 30–100 ms, in which the circuit breaker with active resonant circuit can cut off larger current (over 5 kA). In recent years, power semiconductor devices have been applied in DC circuit breakers. Solid-state circuit breakers consisting entirely of power electronic devices and hybrid circuit breakers composed of power electronic devices and mechanical switches have been successfully developed and applied to practical engineering.

   One of the main requirements of a multi-terminal DC grid is the ability to suppress fault currents within 5 ms. Only solid-state DC circuit breakers and hybrid DC circuit breakers can reach the goal. So, the two types of circuit breakers will be discussed as follows.

2. Solid-state DC circuit breaker

   In solid-state DC circuit breakers, the main circuit is equipped with semiconductor switches (such as insulated gate bipolar transistor (IGBT)) with fast switching ability. Since the current flows in both directions, a pair of bidirectional switches is used. The arresters are connected in parallel, as shown in Figure 5.4.

   High conduction losses occur when current flows through semiconductor switches during normal operation. However, when the shutdown signal is received, semiconductor devices are immediately turned off, and the voltage across them rises rapidly until the arrester clamps the voltage to a value higher than the DC voltage to demagnetize the inductance of the DC system. Solid-state DC circuit breakers do not require mechanical switches compared to other circuit breakers; so, they can suppress fault currents more quickly and are widely used in low-voltage scenarios. But the main disadvantage is that the conduction losses are significantly higher than other types of circuit breakers. In order to reduce conduction losses, integrated gate-commutated thyristor (IGCT) has become the preferred semiconductor device for solid-state DC circuit breakers.

3. Hybrid DC circuit breaker

   Two important parts of hybrid DC circuit breaker are main circuit breaker and auxiliary switch. It is also called active hybrid DC circuit breaker due to the use of a large number of power devices to realize current control and switching off. The main circuit breaker consists of a large number of submodule (SM) units. In order to ensure the bidirectional current cut-off capability, each SM unit consists of several IGBTs and their anti-parallel diodes connected back to back in series. When the main breaker is disconnected, the voltage across it increases rapidly, but the amplitude is limited by the arrester in parallel with it. There are a small number of IGBTs in the auxiliary switch. The fast mechanical switch and the auxiliary switch are connected in series on the normal current-passing branch, and the main breaker is on the bypass branch, as shown in Figure 5.5.

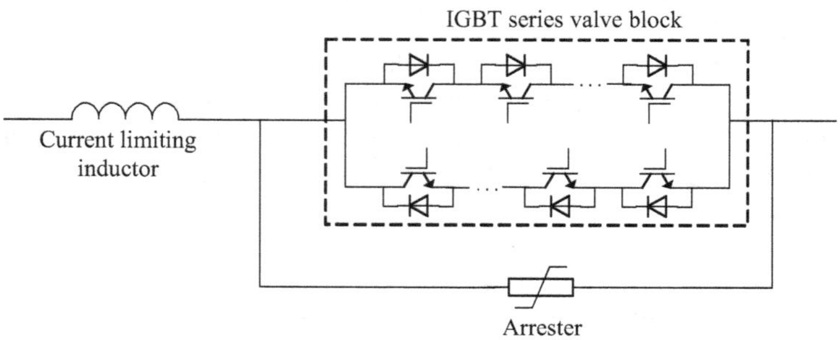

Figure 5.4 Solid-state DC circuit breaker

The function of the main circuit breaker is to break the fault current at rated voltage. In a normal operation, all load currents flow through the fast mechanical switch and auxiliary switch, not through the main circuit breaker. When the fault current exceeds the protection threshold, it transfers to the main circuit breaker by trigging off the auxiliary switch, and the fast mechanical switch can break without arc. When the fast mechanical switch is pulled open enough to withstand the DC voltage, the main breaker opens.

By controlling the main circuit breaker, this kind of circuit breaker can operate in current-limiting mode, where the voltage across the inductor is controllable, which provides a time for protection strategy to determine whether the breaker is temporarily or permanently disconnected. The main circuit breaker will remain closed until it receives the open command or the fault current rises to its maximum breaking current capacity. If the fault occurs on a line that is not directly connected to the circuit breaker, it is necessary to reclose the fast mechanical switch and open the auxiliary switch to re-transfer the current to the main current branch; otherwise, the close of the switch in the main circuit breaker will cause the voltage to rise rapidly to the clamp voltage of the arrester. The clamp voltage higher than the DC voltage produces a back electromotive force (EMF) across the reactor, forcing the fault current to drop and decrease to zero. After the fault is cleared, the disconnector isolates the fault line from the system to prevent the arrester from overheating.

Compared to solid-state DC circuit breakers, the number of IGBTs in the main current-passing branch of hybrid DC circuit breakers is significantly reduced; thus, the conduction loss is significantly reduced. Compared to resonant DC circuit breaker, semiconductor device can transfer fault current from main current branch to bypass branch quickly and fast mechanical switch can realize arc-less breaking. Compared to the circuit breaker which relies on arc voltage to realize fault current transfer, this structure further accelerates the breaking speed of fault current.

Figure 5.6 shows a hybrid circuit breaker applied to the Zhoushan 5-terminal VSC-MTDC project. Each pole of the project is equipped with one such circuit breaker. The behavior analysis during the fault is given in Section 5.4.

4. Modeling of hybrid DC circuit breaker

In this section, a mathematical model of the hybrid DC circuit breaker is established for different stages.

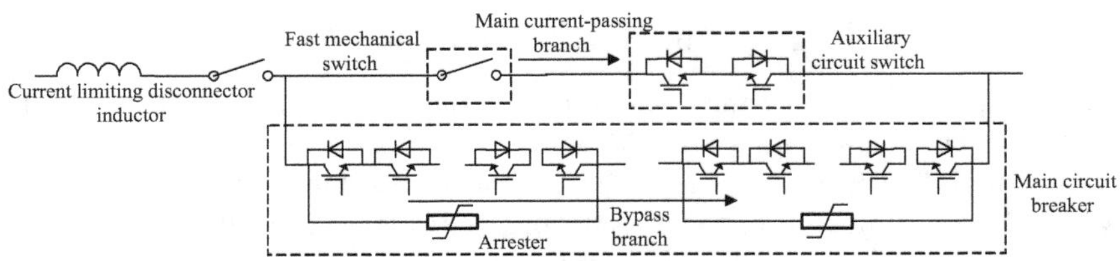

Figure 5.5 Hybrid DC circuit breaker

Figure 5.6 DC circuit breaker for Zhoushan 5-terminal VSC-MTDC project [8]

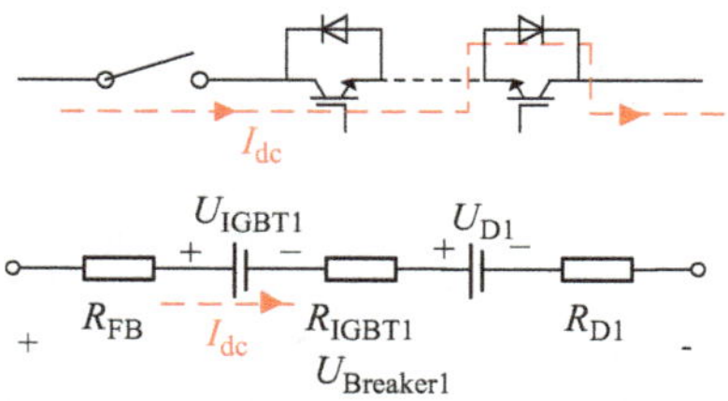

Figure 5.7 Normal operating stage of hybrid DC circuit breaker

During the normal operating state, the load current flows completely through the main branch. Taking the direction shown in Figure 5.7 as an example, the equivalent circuit of the hybrid circuit breaker at this stage is established as follows.

The circuit equation is:

$$\begin{cases} U_{IGBT1} = n_1 U_{IGBT0} \\ U_{D1} = n_1 U_{D0} \\ R_{IGBT1} = n_1 R_{IGBT0} \\ R_{D1} = n_1 R_{D0} \\ U_{breaker1} = U_{IGBT1} + U_{D1} + I_{dc}(R_{IGBT1} + R_{D1} + R_{FB}) \end{cases} \tag{5.1}$$

where U_{IGBT0} and U_{D0} are the turn-on voltage drop of a single IGBT and a single diode, respectively, and R_{IGBT0} and R_{D0} are the on-resistance of a single IGBT and a single diode, respectively. R_{FB} is the equivalent resistance of a fast mechanical switch which is almost zero, and n_1 is the number of IGBTs and diodes on the main branch (auxiliary switch). $U_{breaker1}$ is the voltage across the circuit breaker during the normal operating stage.

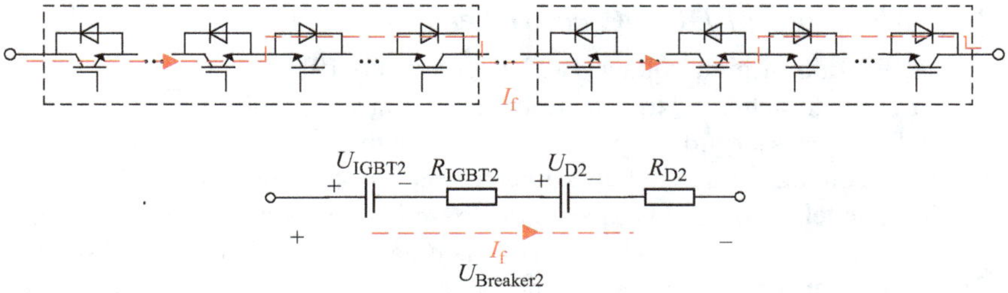

Figure 5.8 Fault current transfer to the transfer branch

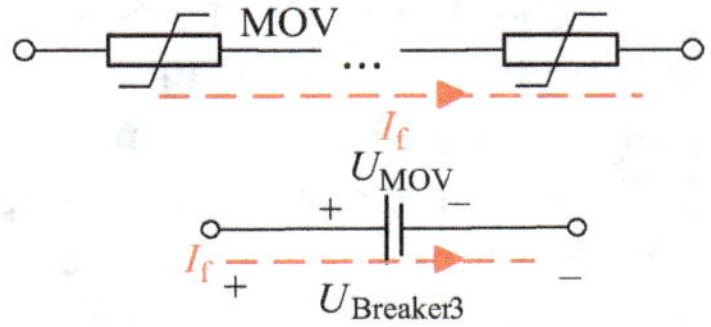

Figure 5.9 Fault current transfer to the energy-absorption branch

Following the DC-side short-circuit (SC) fault, the fault current is transferred from the main branch to the transfer branch (main circuit breaker), which consists entirely of power electronics. The real circuit and equivalent circuit of this stage are shown in Figure 5.8.

The circuit equation is:

$$\begin{cases} U_{IGBT2} = n_2 U_{IGBT0} \\ U_{D2} = n_2 U_{D0} \\ R_{IGBT2} = n_2 R_{IGBT0} \\ R_{D2} = n_2 R_{D0} \\ U_{breaker2} = U_{IGBT2} + U_{D2} + I_f(R_{IGBT2} + R_{D2}) \end{cases} \tag{5.2}$$

Here, I_f is the fault current, U_{IGBT2} and U_{D2} are the forward voltage drop of all IGBTs and diodes in the transfer branch, respectively, and R_{IGBT0} and R_{D0} are the on-resistance of all IGBTs and diodes in the transfer branch, respectively. n_2 is the number of full IGBTs and diodes in the transfer branch, and $U_{breaker2}$ is the voltage across the circuit breaker at this stage.

When all the IGBTs in the transfer branch are blocked, the fault current is transferred to the energy-absorbing branch. The equivalent circuit in this stage is shown in Figure 5.9.

In this stage, the arrester shows a nonlinear V-I characteristic, and the voltage across the arrester rises to the clamp voltage, so that the voltage across the DC circuit breaker is equal to the voltage value of the arrester.

$$U_{breaker3} = U_{MOV} \tag{5.3}$$

5.2.2 Modeling of DC/DC converter

1. Classification of high-voltage DC/DC converters [9]

   In order to achieve large-scale development and application of DC power grid, it is necessary to have a device similar to an AC transformer to realize voltage conversion and network interconnection of power grids with different voltage levels. The DC/DC converter based on power electronic conversion technology, also known as DC transformer, has the functions mentioned above, and has been paid attention to and developed in recent years. This kind of DC/DC converter can also play the role of power flow control, fault isolation, etc., and provides strong support for DC grid operation control. Due to the flexible control performance of power electronics, DC/DC converters can play a greater role in DC grid than AC transformers in AC grids.

   An important consideration for classification of high-voltage DC/DC converters is whether isolated transformers are employed for the interconnection of different DC sides. There for the converters can be divided into isolated and non-isolated converters. According to the number of DC voltage levels, the converters can be further divided into two-level converters, three-level converter and multilevel converters. According to the switching mode, there are hard switching mode converter and soft switching mode converter. The following is a brief introduction to some of the most promising DC/DC converter topologies.

2. Two-level dual-active bridge converter

   The basic structure of a two-level dual-active bridge (DAB) converter is shown in Figure 5.10. It was originally used only for low-voltage industrial DC systems. This structure connects two VSC bridges through an intermediate transformer, commonly referred to as a face-to-face connection. This design uses full-controlled devices, such as IGBT, to realize voltage source mode

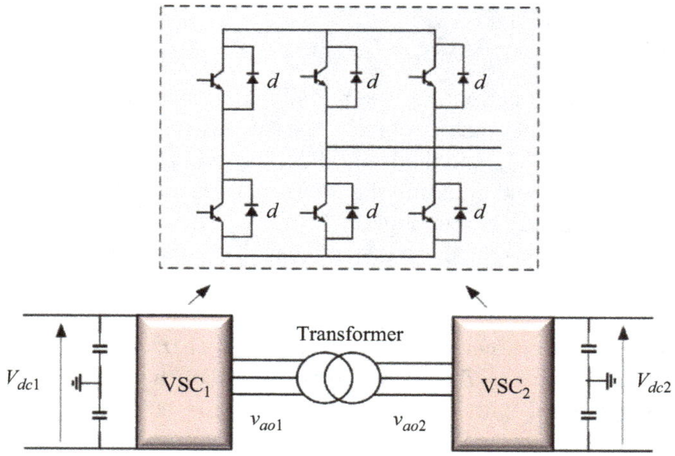

Figure 5.10 Two-level dual-active bridge converter

inversion. The typical fundamental frequency of the internal AC circuit is between 250 Hz and 1 kHz. The higher frequency reduces the weight and size of the converter. Normally, the converter operates in square wave mode at fundamental frequency, with each arm conducting 180 degrees. By controlling the phase angle of AC voltage between two converters, the power flow between two DC sides can be controlled. Under this condition, the switching process of IGBT in the bridge arm can be designed as a zero current switch (ZCS), while the anti-parallel diode is on. Switching process will not produce switching loss. Based on pulse width modulation (PWM) technology, harmonics generated by two-level DAB circuits can be reduced. But using high-frequency PWM will result in higher switching losses. By contrast, other low-frequency modulations, such as selective harmonic elimination (SHE), can control internal power while reducing harmonics.

The two-level DAB is not only simple in structure and working principle, but also has the ability to isolate faults on the two DC sides. While the topology has some drawbacks. For example, in the case of high DC voltage, the intermediate frequency transformer in it produces high *dv/dt*, which makes the transformer unable to operate normally. Therefore, two-level DAB is suitable for medium-voltage and low-power capacity applications. In addition, when converter valve consists of several IGBT devices in series, the two-level DAB requires complex strategies to balance the dynamic and static voltages of each switch.

3. Modular multilevel dual-active bridge converter

With the development of HVDC technology, the power-processing capability and DC working voltage of VSC-HVDC are rapidly improved. Conventional two-level DC-DC converters are difficult to achieve operating voltages of several thousand kV. Modular multilevel converter (MMCs) has received wide attention in recent years compared to two-level VSC. It consists of a variety of interconnected SMs, each with a small capacitor on the DC link. It has the advantages of modularity, extendibility, high efficiency, and low output harmonics. Therefore, MMC-DAB is a more pertinent choice for future HVDC applications. MMC-DAB connects two modular multilevel DC-AC converters through an internal intermediate frequency AC transformer, as shown in Figure 5.11(a). Each DC-AC converter has three phase units, and each phase unit is composed of upper and lower bridge arms. Each bridge arm of the phase unit has N SMs, which are divided into half-bridge (HB) and full-bridge (FB) structures, as shown in Figure 5.11(b) and (c), respectively. The structure inputs a required number of SMs in the conduction path as needed to generate a multilevel output voltage. This design reduces the size of passive components, converter losses and total occupied area.

There are two different modulation schemes for the operation of the converter's AC link (internal transformer), called multilevel modulation and trapezoidal modulation. The multilevel modulation scheme uses a sine wave as the modulation signal, while the trapezoidal modulation uses the trapezoidal wave as the modulation signal. Although sinusoidal modulation results in lower switching losses and lower voltage stress (*dv/dt*) in the transformer, it

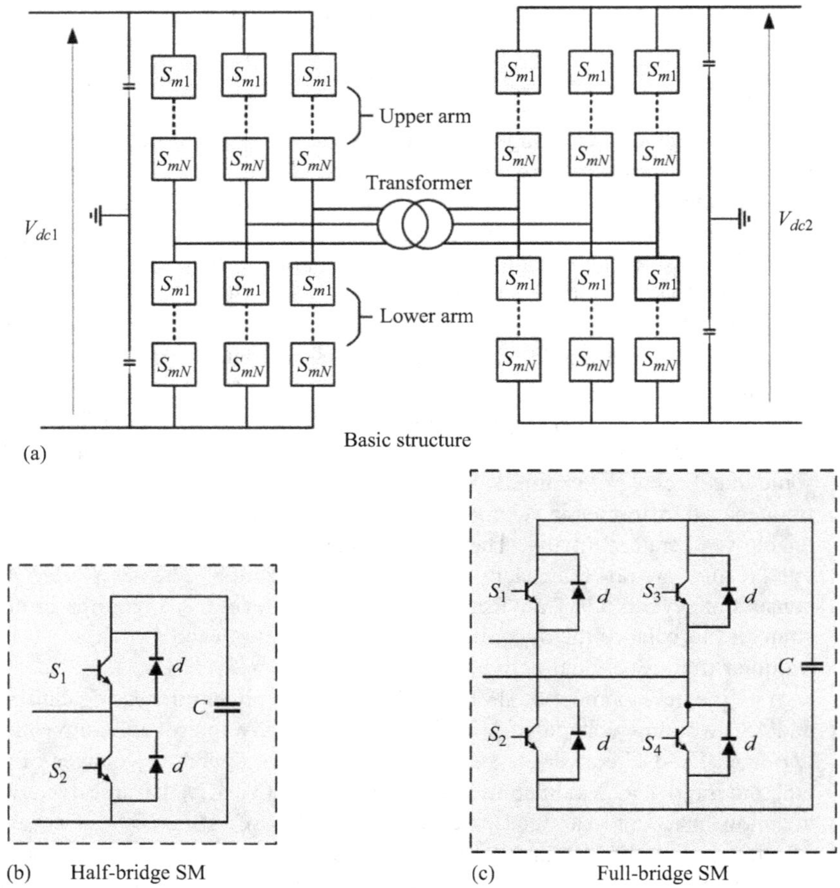

Figure 5.11 MMC-DAB converter

has a lower power density than a simple trapezoidal DAB converter. While the sinusoidal modulation does not take full advantage of zero current switching (ZCS) technology. On the other hand, trapezoidal modulation essentially reduces the energy-storage requirements, which in turn reduces the size and cost of the converter. In this case, the switching device operates at the fundamental frequency, thereby reducing power loss.

4. Switch-module hybrid DC/DC converter

The switch-module hybrid DC/DC converter topology is shown in Figure 5.12. It is characterized by a "T" structure as a whole [10]. The two arms are composed of switches S connected in series with high-voltage and low-voltage DC. The intermediate branch is composed of half-bridge or full-bridge SMs connected in series. This type of converter does not require an intermediate transformer.

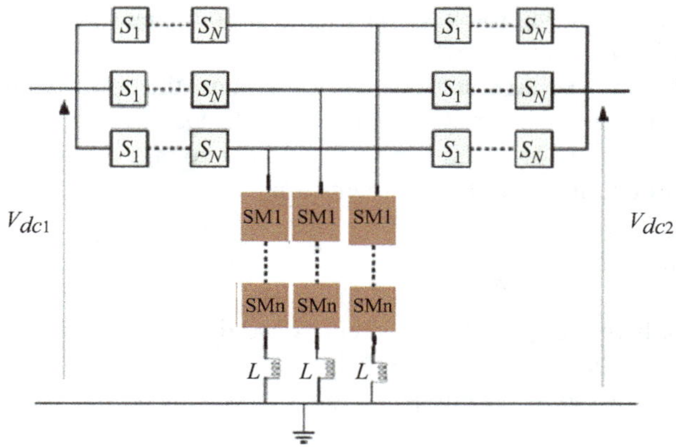

Figure 5.12 Hybrid structure-based DC-DC converter

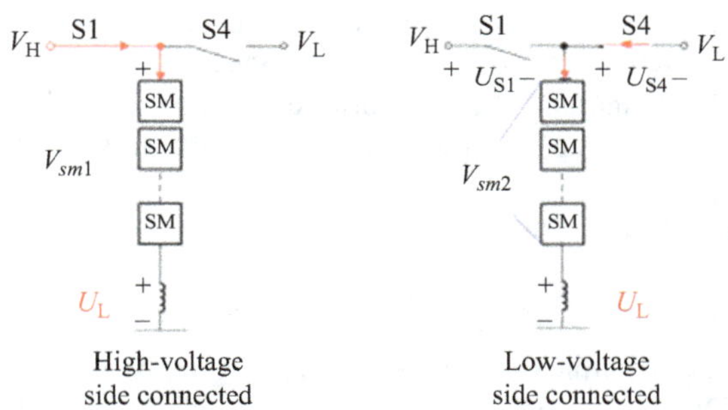

Figure 5.13 Operation process of hybrid DC/DC converter

Generally, series switches can be composed of several IGBTs and their anti-parallel diodes in series. DC on both sides and two switching arms form a two-level converter. The intermediate branch is composed of multiple half-bridge or full-bridge SMs connected in series, which interacts with power on both DC sides through control strategy. On the other hand, the cascaded SMs of the intermediate branch act as energy-storage elements, enabling the exchange of energy on both sides. The intermediate branch is also connected to an inductance L, which can limit impulse current and be used for current control. Switch-module hybrid DC/DC converters usually adopt a multi-phase interleaving mode (Figure 5.12 is a three-phase structure), in order to ensure current continuity and reduce voltage ripple.

This kind of hybrid DC/DC converter can alternately switch on and off the bridge arm switches on both sides (leaving a certain dead time in the switching process) to connect the SM series branches to the two sides of the DC system in turn, so as to realize the energy exchange. Take single-phase branch and energy transmission from the high-voltage side to the low-voltage side as an example to illustrate, as shown in Figure 5.13.

When the high-voltage side switch closes and the low-voltage side switch opens, and the suitable voltage V_{sm1} of the SM series is selected, the energy storage of the inductor of the SM series increases, and the voltage across the inductor and the current through it satisfy the following equation:

$$v_l = V_H - V_{sm1} = L\frac{di_l}{dt} \tag{5.4}$$

When the high-voltage side switch opens and the low-voltage side switch closes and the voltage of the SM series is selected as V_{sm2}, the inductor discharges through the low-voltage circuit and the energy is fed to the low-voltage side.

$$-v_l = V_{sm2} - V_H = L\frac{di_l}{dt} \tag{5.5}$$

According to the reference direction in Figure 5.13, when the inductor current is positive, the low-voltage side circuit is turned on, and actually the current flows through anti-parallel diodes. When the current is reversed, it flows through the switch tubes. When the SM adopts a half-bridge structure, the hybrid converter cannot block DC-side faults. When full-bridge structure is used, DC-side faults can be blocked, but the number of components and operation loss are increased.

5.3 Control system for DC grid

5.3.1 Introduction of DC grid control system

Both the steady and dynamic operations of DC grid greatly depend on the function of control system. DC grid is composed of the variety of resources, loads and different types of converter and with high-level complexity. DC grid has a general control objective which can only be reached by the coordination of multiple converters.

The control strategies designed for multiple converter system can be classified as the centralized control and distributed control according to the control system design philosophy and configuration.

1. Centralized control

   Centralized control strategy is the traditional way for the coordination control of multi-converter system. A top-layer unit is set for the central control with the function as follows: collection the states of converters; parameter-setting calculation for each converters, such as the DC voltage, the terminal power, the reactive power exchanging with AC, etc.; and setting information delivering to converters. The central unit has a strong communication links with all the converters, as in Figure 5.14.

   The reliable and instantaneous communication is the basic requirement to guarantee the operation of centralized system, which makes the system lack of flexibility, redundancy and difficult to expand. DC grid usually is developed with a variety of converters and different operating modes. The time scale for

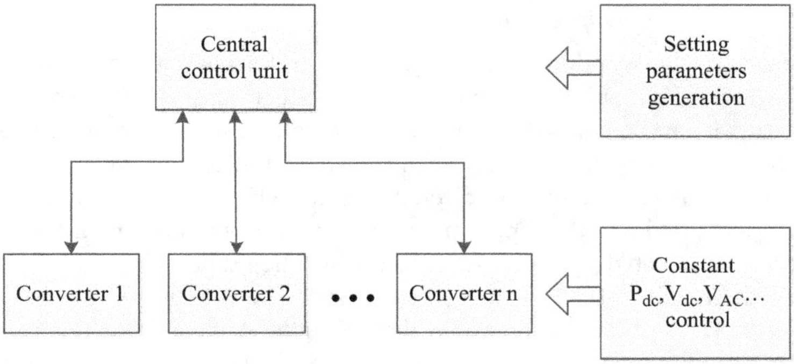

Figure 5.14 Centralized control for DC grid

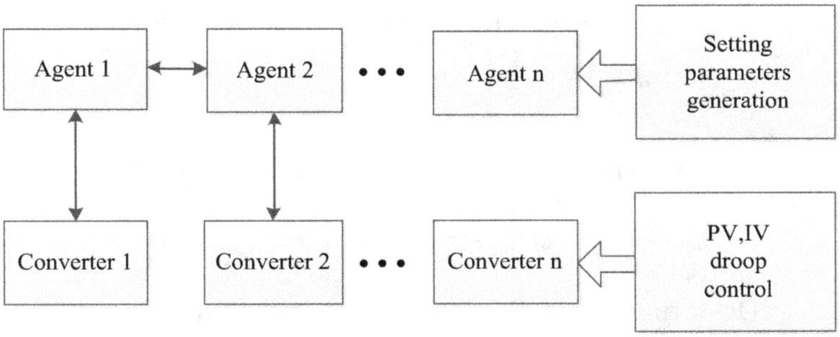

Figure 5.15 Distributed control for DC grid

different converters may range from deferent orders of time. A control system lacking flexibility cannot satisfy the need for DC grid control.

2. Distributed control

   In the past few years, the autonomous distributed system, ADS, has been well recognized and applied for many industries [11]. In the ADS, all the control policies are integrated into the agent, or converters as in DC grid and all the agent are independent, parallel and equal. Each agent can realize its function according the rule designed and the information the collect. The links built between neighbor agents are weak links and there's no need for instantaneous communication, as in Figure 5.15.

Both voltage margin control and voltage droop control have found their application in DC grid. The two control strategies can adaptively keep the system stable without the information of control center during interruptions and belong to the type of autonomous distributed control (ADC). The voltage margin control has been employed in the commissioned MTDC projects and can be found in [12] in detail. The principle of droop control will be given in the following section.

5.3.2 Voltage droop control

The primary frequency adjustment is achieved by referring to the control method of the AC system. Without considering the line voltage loss, the DC voltage of the entire network can be considered the same. Therefore, the active power output by each converter will change linearly with the change of the DC voltage. When the voltage drops, the active power injected into the DC system at the sending end should be increased and the output power of the receiving end should be reduced. Otherwise, the injected power at the sending end should be reduced and the output power of the receiving end should be increased. The purpose is to rebalance the power of the DC system. It can be seen that when the voltage droop control is used, there is no need for command communication or data exchange between the converter station or control center during the transient process. The voltage droop control follows the idea of the ADC. The characteristic of the control method is shown in Figure 5.16, in which considering the power loss and voltage drop on the line, the DC voltage of each station is slightly different during the stable operation.

Active power of a station varies linearly with its DC voltage,

$$\begin{aligned} P_{ref} &= P_{ref\_0} + \Delta P \\ &= P_{ref\_0} + K_{droop} \cdot (V_{dc} - V_{dc\_0}) \end{aligned} \tag{5.6}$$

P_{ref_0}, V_{dc_0}——Rated active power and rated DC voltage for converter;

V_{dc}——DC voltage measurement;

K_{droop}——The slope of the linear change of the active power command with the DC voltage; and

P_{ref}——Outer ring active power class control instruction value.

As shown in Figure 5.16, the solid point is its initial stable operating point of the multi-terminal DC system, and the direction in which the active power injected into the DC system is the positive direction. If the station 1 exits due to a fault, the input power in DC system will be lower than the output power, and the active power will be out of balance. At this point, the DC voltage drops due to insufficient charging of the capacitor, and the operating point of the system changes. The output active power of the converter station 2 will reduce linearly due to the drop of the DC voltage. Converter station 3, on the contrary, increases the power command

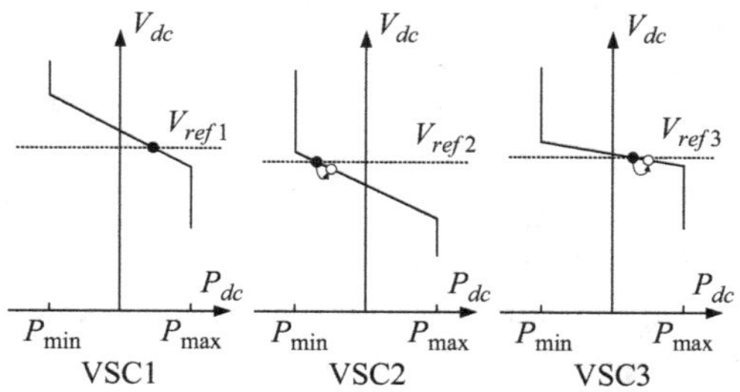

Figure 5.16 P-V characteristic curve of voltage droop control

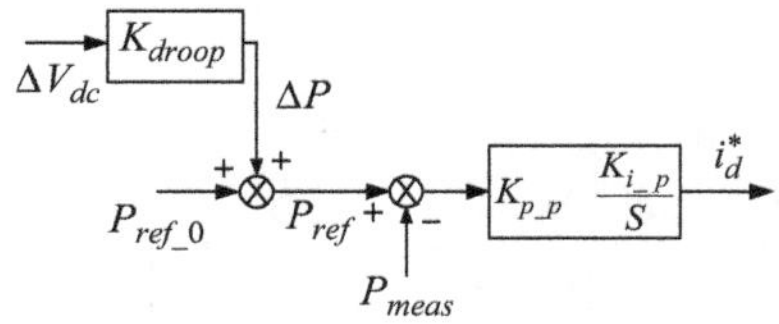

Figure 5.17 Voltage droop control system

with the decrease of DC voltage, thus injecting power that is more active until the new balance point is found between converter station 2 and converter station 3. The voltage droop control implementation is shown in Figure 5.17.

Therefore, once the power of the converter changes, the control system will automatically calculate the active power command according to the DC voltage. When the characteristics of all converters find a common operating point, the system enters a stable operation state, which is shown as follow:

$$\begin{cases} P_1 + P_2 + P_3 + P_{loss} = 0 \\ V_1 + V_{loss1} = V_2 + V_{loss2} = V_3 + V_{loss3} \end{cases} \tag{5.7}$$

In the formula, the subscript loss refers to active power loss and voltage drop on the line.

Therefore, it can be seen that the voltage droop control can automatically adjust its active power commands with the change of DC voltage without the need of inter-station communication, and realize the idea of autonomous decentralized control. The voltage droop control can effectively avoid the transient overshoot as in voltage margin control but with some limitations. When one converter station fails and exits, the input/output power of all converter stations in the DC system will change. The new system operation point can't guarantee an optimal operation, such as the minimum power loss or maximum power transmission.

5.3.3 ADC for DC grid optimized control

Although the above-mentioned autonomous decentralized control strategy can achieve stable operation of the DC grid, the control of each terminal (converter) is based on local information and its own autonomous control parameters, so the mode adjustment and optimization operation of the entire system cannot be realized.

We will use the example circuit shown in Figure 5.1 to illustrate the ADC for DC grid optimization. Figure 5.1 is a 4-terminal VSC DC grid with renewable energy access and different intensity receiving AC systems. For converters connected to renewable energy, the constant V/f control is usually adopted. The power level is determined by maximum power point tracking (MPPT) of renewable energy generation, and the DC voltage cannot be controlled. Therefore, such terminals cannot achieve autonomous decentralized control. In Figure 5.16, the two terminals T_3 and T_4 connected to the receiving AC system can implement droop control. The power obtained by the receiving AC system

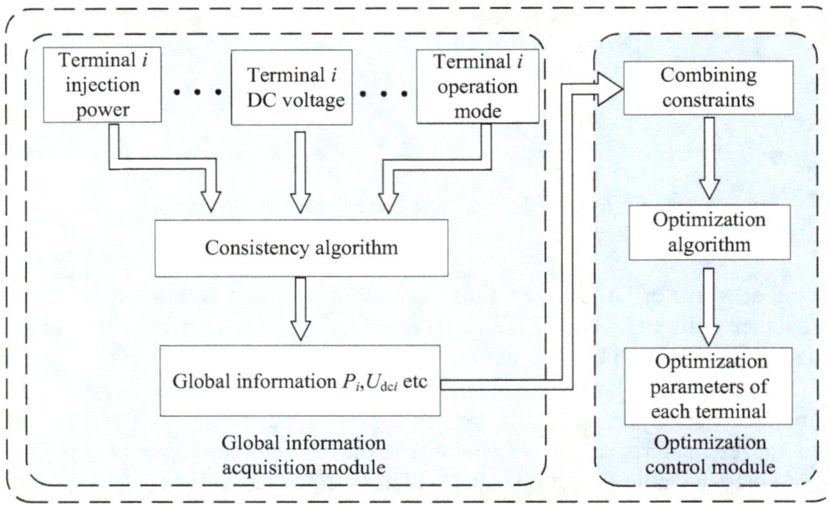

Figure 5.18 The structure of the converter agent in HVDC grid system

can be assigned by setting the droop parameters. The system may encounter the following problems during its operation:

1. According to the needs of the operation of the AC system, the power distribution relationship of T_3 and T_4 may change and needs to be adjusted by modifying control parameters.
2. With the increase of renewable energy output, the distribution of power flow is unsuitable, and a line may overload and a node voltage may exceed the limit, which needs to be adjusted by modifying control parameters.
3. When focusing on an optimization goal, such as the minimum overall network loss, it is necessary to optimize the controller parameters from the perspective of the overall network.

The above control cannot be realized only by control of the converter terminals. Therefore, it is necessary to establish upper control or secondary control based on global information to achieve the above control objectives.

The realization of secondary control can be decentralized to the control unit to each converter (or agent) to form a multi-agent system. Each agent obtains global information through consistency algorithm, and then optimizes according to optimization objective and optimization algorithm, and realizes the control of the agent terminal. In fact, each agent in the multi-agent system needs global information and optimal control, which results in a great increase in the communication and computation of the whole system. However, due to decentralized implementation, there is no need for centralized control of central control unit, which has higher reliability and is easy to expand.

The application of the consistency algorithm between the agents of the multi-agent system for information exchange can obtain the global information of the system. The consistency algorithm has low requirements on the performance and reliability of the communication system. The failure of a communication line does not necessarily affect the access of agents to global information. When a communication line fails, as long as the multi-agent system is still a connected graph, the agents can still obtain global information by exchanging information between adjacent agents.

For a piece of information x that needs to be shared in the system, the principle of the consistency algorithm used by each agent to obtain this information is as follows:

$$x_i(k+1) = x_i(k) + \sum_{j=1}^{N} \delta_{ij}[x_j(k) - x_i(k)] \tag{5.8}$$

$x_i(k)$ and $x_j(k)$ are the information obtained by agent i and j, respectively, in the k-th iteration. $x_i(k+1)$ is the updated value of $x_i(k)$. δ_{ij} is the coefficient of information exchange between agent i and j. N is the total number of agents involved in information exchange.

From the perspective of the whole system, (5.8) can be written in matrix form:

$$\boldsymbol{X}(k+1)=\boldsymbol{W}\boldsymbol{X}(k) \tag{5.9}$$

where $\boldsymbol{X}(k) = [x_1(k) \quad x_2(k) \quad \cdots \quad x_N(k)]^T$, which is the value of x in each agent in step k, and $\boldsymbol{W}$ is the information exchange coefficient matrix:

$$\boldsymbol{W} = \begin{bmatrix} 1-\sum\delta_{1j} & \cdots & \delta_{1N} \\ \vdots & \ddots & \vdots \\ \delta_{N1} & \cdots & 1-\sum\delta_{NN} \end{bmatrix} \tag{5.10}$$

$\boldsymbol{W}$ is closely related to the agents' speed of obtaining global information. The coefficient matrix of fixed weights can be designed by semi-positive convex programming to accelerate the convergence speed.

For a DC grid system consisting of multiple converter stations, a converter station agent can be set up at each converter station to form a multi-agent system [13] for the DC grid. Each agent contains two functional modules: global information acquisition and optimization control. The global information acquisition module applies a consistency algorithm to process information exchanged with neighboring agents, and obtains information such as node injection power, DC voltage, operation mode, and reactive power exchanged with the AC system. Based on the global information, the optimization-control module optimizes the system operation parameters according to the optimized objective function to achieve the optimal operation of the system.

For DC grid, the optimization effect of the autonomous decentralized control based on the consistency algorithm is also related to the completeness of the global information obtained by the agent. The higher the integrity of global information obtained by agents, the better the optimization effect. However, the higher integrity of global information also means more variables to participate in sharing and longer convergence time. If the convergence time is required to be constant, it will undoubtedly increase the requirements for the speed of the communication system. Therefore, in practical applications, the information shared by each agent should be reasonably classified according to the specific situation, and the number of variables participating in the iteration is reduced. Therefore, each agent in the system can achieve the optimal control target of the hybrid DC grid, and minimize the time required for iteration.

After obtaining the global information, each agent can adopt different optimization algorithms to optimize the parameters of the local converter controller, and realize the optimal control of the DC grid through the cooperation of terminals.

Taking the interior point method as an example, the convex optimization problem of DC grid loss optimization is explained.

The optimization model of loss is a nonlinear model and can be expressed as

$$\begin{cases} \min & f(\boldsymbol{x}) \\ \text{s.t.} & \boldsymbol{h}(\boldsymbol{x}) = \boldsymbol{0} \\ & \underline{\boldsymbol{g}} \le \boldsymbol{g}(\boldsymbol{x}) \le \bar{\boldsymbol{g}} \end{cases} \tag{5.11}$$

where $\boldsymbol{x}$ is the control variable. $f(\boldsymbol{x})$ is the objective function of the optimization. $h(\boldsymbol{x})$ is the equality constraint. $g(\boldsymbol{x})$ is the inequality constraint. The number of equality constraints is m, and the number of inequality constraints is r.

The interior point method [14] transforms the inequality constraint of the optimization model into an equality constraint by introducing a slack variable. By introducing a Lagrange multiplier, the constraints are integrated into the objective function. And the barrier function method is used to constrain the slack variable. The Lagrange function of the optimization model shown in (5.13) is as follows:

$$\begin{aligned} L &= f(\boldsymbol{x}) - \boldsymbol{y}^T h(\boldsymbol{x}) \\ &\quad - \boldsymbol{z}^T[\boldsymbol{g}(\boldsymbol{x}) - \boldsymbol{l} - \underline{\boldsymbol{g}}] - \boldsymbol{w}^T[\boldsymbol{g}(\boldsymbol{x}) + \boldsymbol{u} - \bar{\boldsymbol{g}}] - \mu\sum_{j=1}^{r}\ln(l_j) - \mu\sum_{j=1}^{r}\ln(u_j) \end{aligned} \tag{5.12}$$

where $\boldsymbol{y}^T$ is the Lagrange multiplier of the equality constraint, and $\boldsymbol{z}^T$, $\boldsymbol{w}^T$ are the Lagrange multipliers of the inequality constraint. $\boldsymbol{l}$, $\boldsymbol{u}$ are slack variables of the inequality constraint. μ is the penalty factor of the barrier function.

In order to apply the interior point method to the loss optimization control of the DC grid, it is necessary to select the control variables and establish the corresponding objective function and constraints of the system. Line losses can be expressed as a function of node voltages in the system, and converter station losses are a function of output currents. The output currents are related to the output power, which is related to node voltages of the DC grid. Therefore, the objective function is ultimately the function of node voltages in the DC grid, and they should be taken as control variables in optimization. However, since some terminals do not participate in optimal control, their node voltages are limited by the constraints and state variables are affected by control variables, but they can still be optimized together with the constraints and control variables. The DC voltage of each terminal participating in the optimal control is an independent control variable.

Corresponding to the classification of constraints by the interior point method, the constraints of hybrid MTDC can be divided into linear equality constraints, linear inequality constraints, nonlinear equality constraints, nonlinear inequality constraints, and upper and lower limit constraints. The linear inequality constraint is the current limit of each branch of the DC grid. Linear equality constraint is the relationship between voltage and current of the converter.

The nonlinear equality constraint is the relationship between the injection power P through the converter station and the DC grid voltage and grid parameters.

The nonlinear inequality constraint is the power limit of the receiving converter station.

The upper and lower limit constraints are the limits of the control variables and state variables, that is, the upper and lower limits of the voltage of each node.

By solving the above optimization problem, the node voltages of the control variables can be obtained. For converters with droop control, the voltage reference values can be taken directly from the optimization result, and the optimized voltage values can be realized by adjusting the droop parameters, including the slope and the intercept.

5.3.4 Power flow control [15]

The basic task of DC grid operation is to maintain the power (power flow) level of each converter and transmission line in the grid to the desired target, and the voltage of each node is within a reasonable range. The form, composition, and operation mode of grid may be changing constantly. Power flow control should be implemented in any feasible operation mode. The basic work of power flow control is power flow calculation. In general, the power flow calculation of the DC grid does not need to consider the phase angle, and the grid parameters are only resistors. Therefore, the calculation is much simpler than the AC grid. In the power flow calculation of the AC grid, the power and voltage are nonlinear, and the solution needs to be iterated. When the current is used to describe the DC grid power flow, it may be solved without iteration. From the perspective of the DC grid, the voltage and current characteristics of the converter are nothing more than the following control modes:

1. Constant voltage control mode: the voltage is given and the current is determined by the grid;
2. Constant current control mode: the current (power) is given and the voltage is determined by the grid; and
3. Droop control mode: voltage and current satisfy droop relationship.

Two variables are introduced into the voltage and current control modes of the converter. One is constrained and the other is determined by the DC grid. AC and

DC are interconnected in the actual power grid. However, based on the above-mentioned converter-control modes, the decoupling of AC to DC is realized, and the solution of the power flow can be performed separately.

If all of the converter terminals in Figure 5.1 adopt droop control, the converter terminal characteristics are expressed as:

$$\begin{cases} V_1 = V_{01} - k_1(I_1 + I_2) \\ V_2 = V_{02} - k_2(I_3 - I_2) \\ V_3 = V_{03} - k_3(I_4 - I_2) \\ V_4 = V_{04} + k_4(I_3 + I_4) \end{cases} \tag{5.13}$$

where V_{0i}, k_i $(i=1 \sim 4)$ are the droop parameters of terminal i. When only line resistance is considered in power flow calculation, the power flow equations of the network are as follows:

$$\begin{cases} V_1 - V_3 = R_1 I_1 \\ V_1 - V_2 = R_2 I_2 \\ V_2 - V_4 = R_3 I_3 \\ V_3 - V_4 = R_4 I_4 \end{cases} \tag{5.14}$$

Equations (5.13) and (5.14) consist of eight variables and eight independent equations. Therefore, the solution exists and is unique.

The system rated voltage shown in Figure 5.1 is 500 kV, and the current of the converter terminals and lines does not exceed 2 kA. When terminals 1 and 2 are connected to renewable energy, they do not participate in droop control. At a given moment, the current is constant. The control characteristics of the four terminals are as follows:

$$\begin{cases} I_{T1} = 1.5kA \\ I_{T2} = 1kA \\ V_{T3} = 475 - 25I_{T3} \\ V_{T4} = 475 - 12.5I_{T4} \end{cases} \tag{5.15}$$

Table 5.1 shows the power flow calculation results of the grid. It can be seen that under this mode of operation, the power flow is distributed reasonably and the voltage is also within the normal range.

Table 5.1 Power flow calculation results of the example

	No. 1	No. 2	No. 3	No. 4
Terminal current (kA)	1.5	1.0	0.85	1.65
Branch current (kA)	1.186	0.314	1.314	0.336
Terminal voltage (kV)	502.18	501.54	496.25	495.63

The solvability of the power flow does not guarantee controllability. In fact, for a n-node grid, there can theoretically be $(n^2-n)/2$ current channels. If the converter adopts droop control, the power flow value will be the result of the balance of the electric quantities. Power flow cannot be completely controlled, and it is even more difficult to achieve power flow optimization. The reality is that the power flow does not require strict control and only needs to be controlled within a reasonable range. The control of the power flow can be realized by the parameter setting of the converter terminals, or by connecting power flow control devices in the line. In addition, some scholars have proposed that DC/DC converters can be used to achieve power flow control.

5.4 DC grid fault protection and control

5.4.1 Properties of DC grid faults [16]

DC grid faults including the DC line (overhead line or cable) faults, converter faults, and AC-side faults. The converters faults or AC line faults normally only affect one terminal and the power and voltage of the DC system can regain the proper balance by the ADC as describe above. While the DC line faults will have an impact on the overall DC grid and will be discussed in this section. The transient behavior of DC line fault is greatly dependent on the system topology and configuration. The pole-to-pole SC fault or the pole-to-ground SC fault in the low-resistance grounding system have the same properties and are the most challenging scenarios, which need to be investigated carefully.

The major components in DC grids include VSC converters, DC capacitors, DC cables/lines, possibly DC/DC converters, and fault-protection equipment, i.e., DC circuit breakers. The DC-side components deserve special attention in fault studies since they may both contribute to the SC fault current and experience fault current. LCC-HVDC has no contribution to the uncontrolled SC current.

A complex DC grid may have numerous VSC converters, numerous DC lines and possibly some DC/DC converters in the future. However, a DC grid will have only three energy sources that can feed DC fault: the AC grids connecting through VSC converters, the DC/DC converters connected to other pieces of DC grid, the stored energy in DC capacitors and the DC lines. The initial SC level greatly decided by the capacitor connected with the DC/AC or DC/DC converters. The capacitors will significantly increase the derivative and magnitude of fault current because of the small circuit impedance. For a typical half-bridge MMC DC/AC converter, the SC current can be divided into two steps, which are before and after the blocking of IGBTs. The first step is that the capacitors in SMs discharge through the SC point, as show in Figure 5.19 and the equivalent circuit is show as Figure 5.20. After IGBTs blocking, the SC current goes to the second step and is mainly supplied by AC source through anti-parallel diodes as illustrated in Figure 5.21, and its equivalent circuit is shown in Figure 5.22.

The impedance of the DC line is relatively small since only the resistance part applies (reactance has no effect on the steady-state DC current). This implies higher

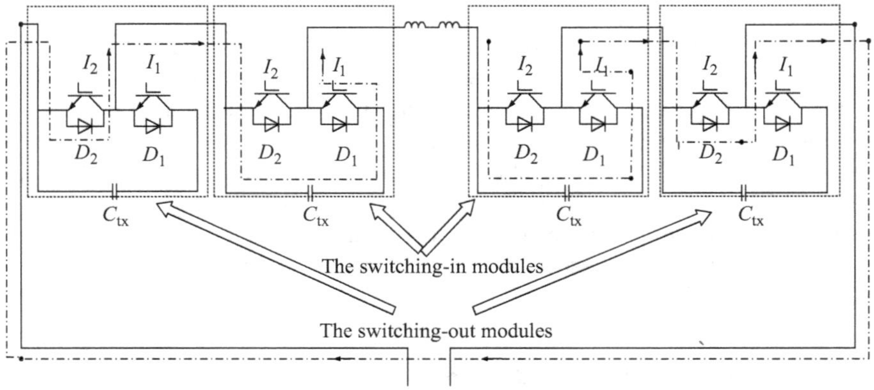

Figure 5.19 The discharging route of capacitors in the SM

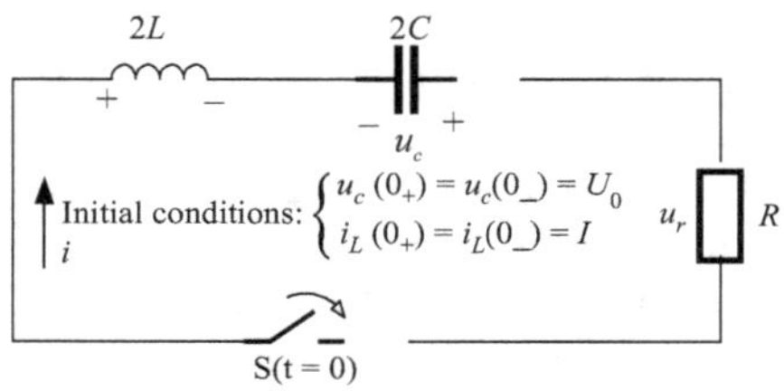

Figure 5.20 The equivalent discharging circuit of capacitors in the SM

fault currents compared to the AC voltage source. The small time constant for the capacitor-discharge circuit means the high derivative of the initial current. The fast transient process brings a challenge for the fault detection and protection, typically within 5 ms. Besides, different from the counterpart AC system, there is no zero-crossing point in the DC fault current, which makes the fault-cleaning process much more difficult.

As compared to the traditional AC system, the DC grid system, mostly composed of power electronic equipment, is with a low inertia. After the occurrence of a SC fault, both the converters and the energy-storage components on the DC side are discharged through the fault point, resulting in the fault response speed at least one order of magnitude faster than its counterpart AC system, and both the amplitude and the rising speed of the fault current are greater. However, the overcurrent capability of power electronic devices is relatively low, and the SC faults can easily cause overcurrent damage to the devices. Besides, the DC grid system has to remove the fault segment after DC-side SC fault, and may lose some transmission lines or even converters, which will affect the continuous and reliable operation of the DC grid. Furthermore, the SC fault of the DC system not only affects the DC system itself, but also affects the connected AC system. One DC SC fault is equivalent to a three-phase

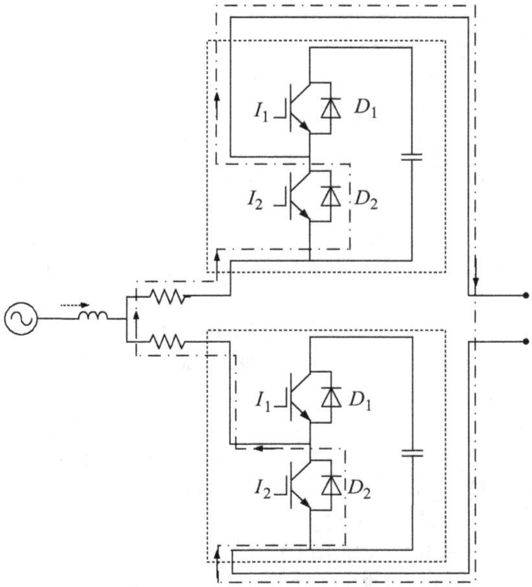

Figure 5.21 Current supplied by AC source through anti-parallel diodes

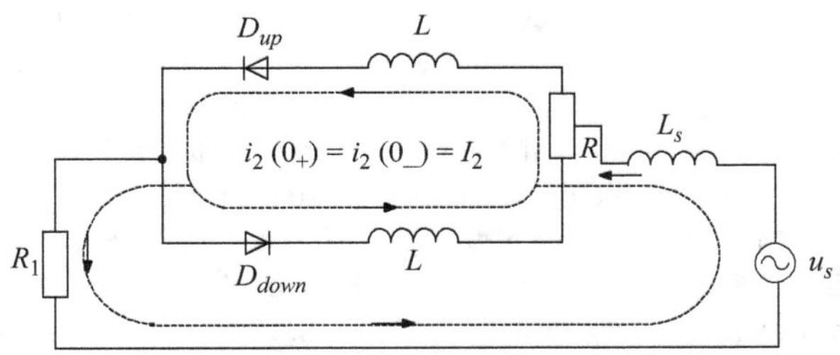

Figure 5.22 Equivalent discharging circuit after IGBTs blocking

SC fault from the perspective of the AC system, which will seriously endanger the safe and stable operation of the power system.

5.4.2 Fault clearance and fault current control

One method of DC fault isolation is to trip the AC circuit breaker. The logic of this method is very simple. When the DC-protection system detects the occurrence of a fault, the protection system will immediately send trip signals to the corresponding AC circuit breakers, thereby preventing the AC side from supplying SC current to the fault point. Thus, equipment in the DC system together with the converters can be protected. When the fault current on the DC line has decayed to zero, the fault line is then cut out from the system through the DC switch, so as to realize the isolation of the DC fault. Although such protection scheme is convenient, and is easy to implement in real engineering practices, its shortcomings are also obvious. Firstly, due to the large amplitude of the DC fault current and the rapid rise speed, the DC protection is supposed to complete a full set of actions (including the fault detection, the trip of circuit breakers, etc.) within a few milliseconds, in order to prevent the DC fault from causing damage to the DC system and converter. However, the operation of AC circuit breakers generally requires two to three cycles. Therefore, the DC fault-isolation method based on AC circuit breaker tripping cannot meet the requirement of the operating time. Besides, when a SC fault occurs at a certain point in the DC grid, all the transmission lines in the DC system may suffer from overcurrent. Therefore, the fault-isolation method based on the AC circuit breaker may cause the entire system to stop operating, which cannot guarantee the power supply reliability and anti-interference capability of the DC grid. Although the fault-isolation method based on AC circuit breaker has been applied in the current engineering practice, this method is only a stopgap measure because there is no other ideal isolation method at present. The development of DC grid is still inseparable from in-depth research and exploration of DC fault-isolation technology.

Another technical solution is to use DC circuit breakers to achieve DC fault isolation [17]. The DC circuit breaker has a fast action speed, which enables it to only disconnect the faulty line without affecting the operation of other parts of the system. However, since the fault current has no natural zero-crossing point during a DC fault, the arc-extinguishing process for the DC circuit breaker to realize during the fault isolation becomes difficult. At present, high-voltage DC circuit breakers with the ability to quickly break large fault currents are still in the research stage and have not yet achieved large-scale engineering applications.

The third type of the technical solution for fault isolation is that the converter is composed of SMs with fault blocking capability, and the typical structure is a full-bridge submodule (FB-SM). The principle of the FB-SM to isolate the SC fault is illustrated in Figure 5.23: after detecting the fault, all IGBTs in the converter are immediately blocked. At this time, the capacitor in the SM is connected into the fault current flowing path with reverse polarity, so the fault current charges the capacitor and quickly decays to zero. Once the fault current decays to zero, the unidirectional conductivity of the diode can ensure that it cannot increase negatively after zero crossing, but is always clamped to zero. Therefore, the DC fault isolation is achieved by the FB-SM.

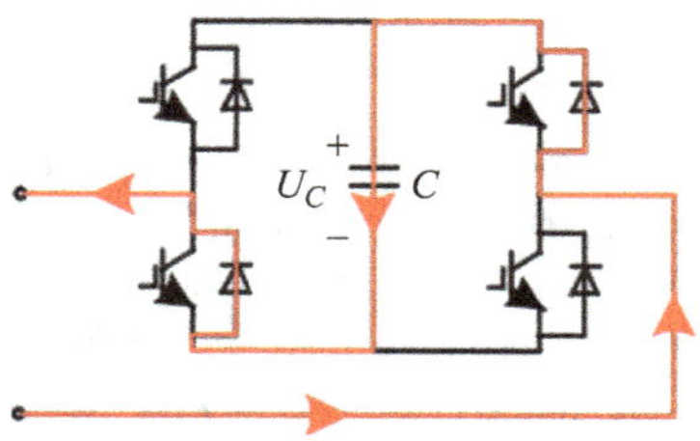

Figure 5.23 Structure of FB-SM

5.4.3 Post fault recovery

To ensure overall power system stability, the HVDC systems are required to be able to recover power transmission quickly. Generally, the post-fault recovery strategy for the DC grid system embedding DC circuit breakers (DCCBs) is originated from the AC system. Namely, after the arc deionization of faulted transmission lines, the DCCBs will be automatically re-closed within a predetermined time. However, in the case of reclosing to permanent faults, the system will subject to DC faults again and therefore suffer from second strikes. Unlike the AC circuit breaker that breaks the fault current at the zero-crossing point of the AC current, the DCCB needs to break the current without the zero-crossing point of the fault. Therefore, multiple breaking will bring great challenges to the selection of arresters and the design of the main circuit of the DCCB.

5.4.4 DC fault simulation example

Based on the pre-mentioned four-terminal HVDC system in Figure 5.1, the protective measures for fault isolation by tripping the DC circuit breaker after the fault inception are simulated. The topology of all four converters is half-bridge MMC. Here, assuming that all DC lines are equipped with DC circuit breakers at both ends. A transient pole-to-ground fault is occurred on the transmission line between T_1 and T_3 at 1.5 s and lasts for 50 ms. The simulation verifies the effectiveness of the protection of the DC circuit breaker on the system. Simulation results are shown as follow.

As shown in Figure 5.24, if the MTDC system is not equipped with a DC circuit breaker, the AC-side circuit breaker will not be able to operate in such a tiny time span. After the occurrence of the fault at 1.5 s, voltages of all converter

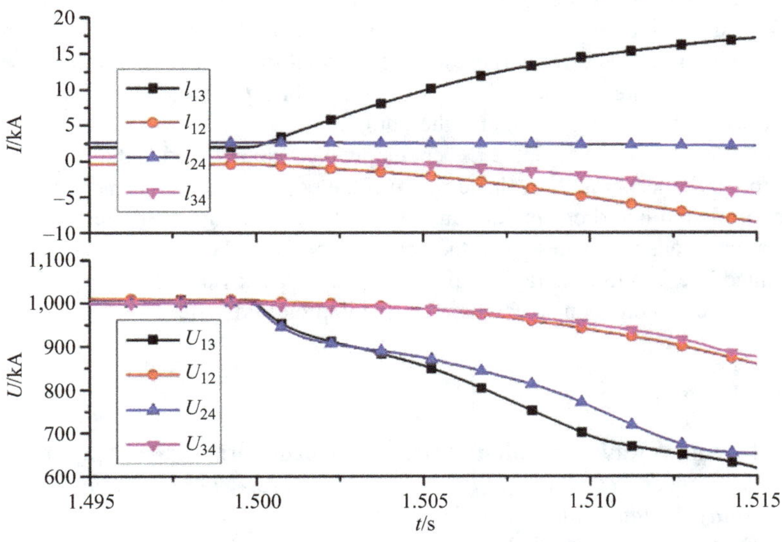

Figure 5.24 Simulation result after fault without DC circuit breaker

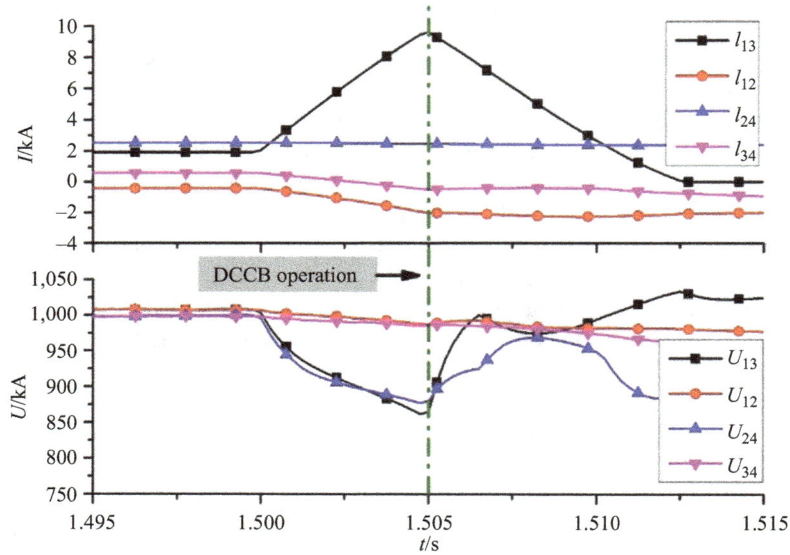

Figure 5.25 Simulation result after fault with DC circuit breaker

dropped rapidly. Terminals on both sides of the fault line, T_1 and T_3, suffer the most significant voltage drops. The minimum values are 613 and 649 kV, respectively, which are only about 60% of the rated value. After the SMs are blocked due to overcurrent, its equivalent circuit of the converter is a six-pulse rectifier circuit as mentioned above, which has no fault-isolation capability, so the fault current on the DC side will keep rising. Among them, the current on the fault line l_{13} has the fastest rising speed and the largest amplitude, and the maximum value can exceed eight times the rated value, which is bound to bring a great threat to the safety operation of the entire system and the equipment.

As shown in Figure 5.25, the DC circuit breakers are installed in the MTDC system. After the fault inception, the voltage between the positive and the negative poles of the DC transmission line drops rapidly due to the SC fault, but gradually restores the rated value after a transient process with the operation of the DC circuit breakers. Current on the faulted line l_{13} still rises rapidly after fault, but its peak value is only 9.63 kA, which can see a great depression as compared to the unprotected system.

References

[1] Nilanjan Ray Chaudhuri, Balarko Chaudhuri, Rajat Majumder, and Amirnaser Yazdani, *Multi-Terminal Direct-Current Grids Modeling, Analysis, and Control. IEEE, Inc.*, 2014

[2] Dragan Jovcic and Khaled Ahmed, *High Voltage Direct Current Transmission Converters, Systems and DC Grids. John Wiley & Sons, Ltd*, 2015

[3] Graeme Bathurst, George Hwang, and Lalit Tejwani. 'MVDC - The New Technology for Distribution Networks', *11th IET International Conference on AC and DC Power Transmission*, 10–12 Feb. 2015
[4] Siemens Reports. *Planned HVDC projects in China.* [Online] Available from: www.siemens.com /energy/hvdc
[5] Gen Li, Ting An, Jun Liang, *et al.* 'Power reversal strategies for hybrid LCC/ MMC HVDC systems'. *CSEE Journal of Power and Energy Systems*, Vol. 6, No. 1, 2020
[6] Nadew Adisu Belda, Cornelis Arie Plet, and Rene Peter Paul Smeets. 'Analysis of faults in multiterminal HVDC grid for definition of test requirements of HVDC circuit breakers'. *IEEE Transactions on Power Delivery* , Vol. 33 , No. 1, 2018
[7] Christian M. Franck, 'HVDC circuit breakers: A review identifying future research needs'. *IEEE Transactions on Power Delivery*, Vol. 26, No. 2, 2011
[8] Wandi Zhou , Xiaoguang Wei , Sheng Zhang , *et al.* 'Development and test of a 200 kV full-bridge based hybrid HVDC breaker'. EPE'15 ECCE-Europe, 2015
[9] Grain Philip Adam, Islam Azmy Gowaid, Stephen Jon Finney, Derrick Holliday, and Barry W. Williams, 'Review of dc-dc converters for multi-terminal HVDC transmission networks'. *IET Power Electronics*, Vol. 9, No. 2, pp. 281–296, 2016
[10] Jie Yang, Zhiyuan He, Hui Pang, and Guangfu Tang, 'The hybrid-cascaded DC-DC converters suitable for HVDC applications'. *IEEE Transactions on Power Electronics*, Vol. 30, No. 10, pp. 5358–5363, 2015
[11] Dong Xu and Lei Wan, 'Hierarchical optimal power flow control for loss minimization in hybrid multi-terminal HVDC transmission system'. *CSEE Journal of Power and Energy Systems*, Vol. 2, No. 1, 2016
[12] Hong Rao, Yuebin Zhou , Shukai Xu, *et al.*, 'Key technologies of ultra-high voltage hybrid LCC-VSC MTDC systems'. *CSEE Journal of Power and Energy Systems*, Vol. 5, No. 3, 2019
[13] Haojie Wang, Minxiao Han, Josep M. Guerrero, Juan C. Vasquez, and Bitew G. Teshager, 'Distributed secondary and tertiary controls for I–V droop-controlled-paralleled DC–DC converters'. *IET Generation, Transmission & Distribution*, Vol. 12, No. 7, 2018
[14] José Luis Martinez Ramos, Antonio Gómez Exposito, and Victor H. Quintana, 'Transmission power loss reduction by interior-point methods: Implementation issues and practical experience'. *IEE Proceedings - Generation, Transmission and Distribution*, Vol. 152, No. 1, 2005
[15] Kumars Rouzbehi, Seyed Saeid Heidary Yazdi, and Negin Shariati Moghadam, 'Power flow control in multi-terminal HVDC grids using a serial-parallel DC power flow controller'. *IEEE Access*, Vol. 6, 2018
[16] Marius Langwasser, Giovanni De Carne, Marco Liserre, and Matthias Biskoping, 'Fault current estimation in multi-terminal HVDC grids considering MMC control'. *IEEE Transactions on Power Systems*, Vol. 34, No. 3, pp. 2179–2189, 2019
[17] Oliver Cwikowski, Alan Wood, Allan Miller, Mike Barnes, and Roger Shuttleworth, 'Operating DC circuit breakers with MMC'. *IEEE Transactions on Power Delivery*, Vol. 33, No. 1, 2018

Chapter 6

Electromagnetic simulation of HVDC transmission

Shengtao Fan[1]

6.1 Introduction

Electromagnetic simulation is an effective approach to study the dynamic behaviors of high-voltage direct current (HVDC) transmission system. Transient characteristics of HVDC transmission system can be described with a set of differential equations, which are time varying and nonlinear. To include the distributed parameter effects of HVDC transmission line, partial differential equations are also needed. Special programs can be developed for the given type of electromagnetic-transient study. However, it is usually easier to setup the simulation of HVDC transmission system with the help of some general-purpose electromagnetic-transient programs. The general-purpose electromagnetic-transient simulation programs fall into two categories.

Simulation programs of the first category such as SimPowerSystems [1] use state equation approach. For this type of programs, the state equations are explicitly obtained. Different numerical methods can be used to give the time-domain solutions of the obtained state equations. With the state equations, the well-established linear and nonlinear analyses methods can also be directly used. To establish the state equation of one circuit, the graph theory-based method can be used [2,3]. In [4], based on their branch-node representation, an algorithmic method of developing the state equations of complex power circuits and systems including power electronics has been set forth.

For the second category of simulation programs, the state equations are never explicitly established. Instead, certain numerical method (Euler, trapezoidal, Gear, etc.) is directly applied to the dynamic components in the system. The companion models of diverse components are obtained at the branch level. A significant advantage of this approach is that the formulation of state equations of complex circuits is avoided, which can be a complex and tedious task. This category of simulation programs includes SPICE type programs [5], electromagnetic transients program (EMTP) [6], and PSCAD/EMTDC [7]. Currently, this type of programs are widely used for the modeling and simulation of HVDC systems, which is also the main topic of this

[1]Shengtao Fan Consulting, Winnipeg, Canada

chapter. From Sections 6.2 to 6.6, the discussion covers several fundamental aspects, which are important to construct electromagnetic-transient-simulation programs. In Section 6.7, some power electronics simulation-related problems are discussed, which is important to the simulation of HVDC system. A benchmark HVDC system is used to show the simulation results in Section 6.8.

6.2 Principles of EMT simulation

To simulate the transient phenomena, an EMT-type program provides a sequence of snapshot pictures of the network quantities at discrete intervals, achieved through a step-by-step procedure that proceeds along the time axis with a variable or fixed time step. In each step of the procedure, it computes the quantities of the network from their known values from previous steps [3,6], which is shown in Figure 6.1.

In each pass, the companion circuits (Norton equivalents) of network components are first calculated through applying numerical method (e.g. trapezoidal rule)

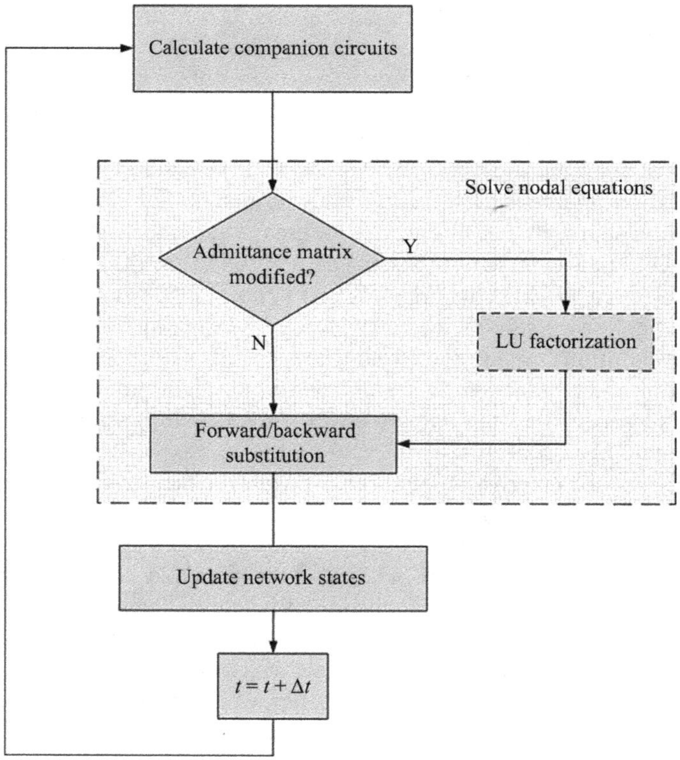

Figure 6.1 Flowchart of EMT simulation

to their describing differential equations. For the duration of the current time step, the entire network then is represented by a purely resistive network with Norton current sources. The nodal analysis (NA) or modified nodal analysis (MNA) approach can then be used to obtain the system equations of this network. The obtained system equations are sets of linear equations with node voltages as unknown variables in NA. If MNA is used, currents of some branches (e.g. ideal voltage sources) are also treated as unknown variables.

To solve this system of linear equations (nodal equations), a widely used procedure is first factorizing the admittance matrix into lower and upper triangular matrices. The solutions are then obtained through one forward and one backward substitutions [8]. If the nodal admittance matrix has been modified due to switching events, etc., a re-factorization of the admittance matrix into one lower and upper triangular matrix is also required. Then, based on the known node voltages, the network states (branch currents and voltages, etc.) can be updated and then be used to calculate the companion circuits in the next loop or output the simulation results.

6.3 Numerical methods of EMT simulation

In the EMT-type program, numerical methods are applied at the branch level to obtain the equivalent companion circuits validated in one time step. Among the numerical methods, the trapezoidal rule (TR) is the most commonly used. Backward Euler method is another commonly used approach to suppress the possible numerical oscillations generated by switching actions of power electronics.

Suppose we are seeking the solutions of the ordinary differential equation (ODE) as:

$$\frac{dx}{dt} = f(t, x(t)) \tag{6.1}$$

6.3.1 Discretization formula

The TR gives the discretized solution of this ODE with the following iteration formula:

$$x(t_0) = x(t_0 - h) + \frac{h}{2}[f(t_0 - h, x(t_0 - h)) + f(t_0, x(t_0))] \tag{6.2}$$

where h is the time step. The TR can be obtained in different ways [9–13]. Integrating both sides of (6.1) from $t_0 - h$ to t_0, we have

$$x(t_0) = x(t_0 - h) + \int_{t_0 - h}^{t_0} f(t, x(t))dt \tag{6.3}$$

The integration term $\int_{t_0-h}^{t_0} f(t, x(t))dt$ is equal to the shadow area as shown in Figure 6.2.

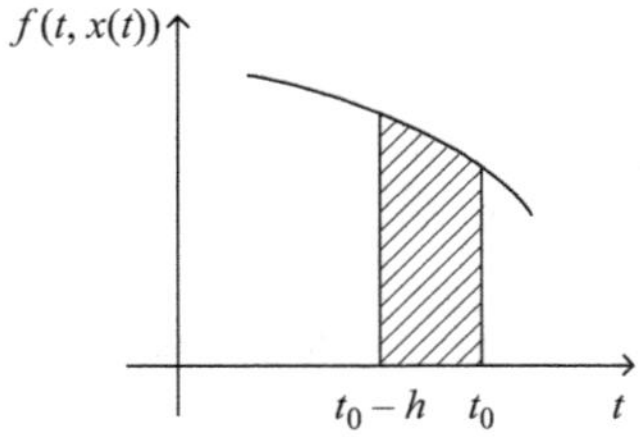

Figure 6.2 Integration of $f(t, x(t))$

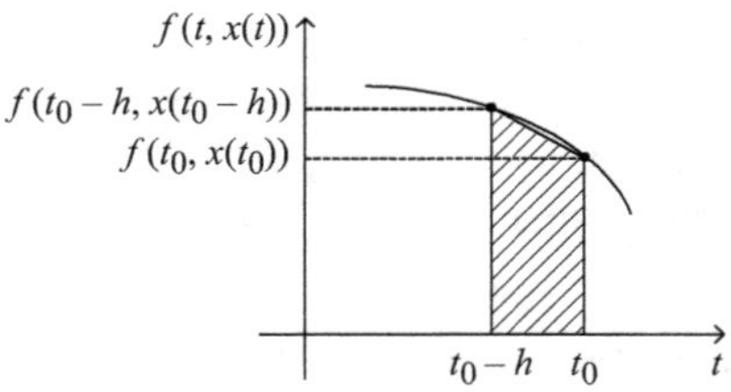

Figure 6.3 The trapezoidal rule

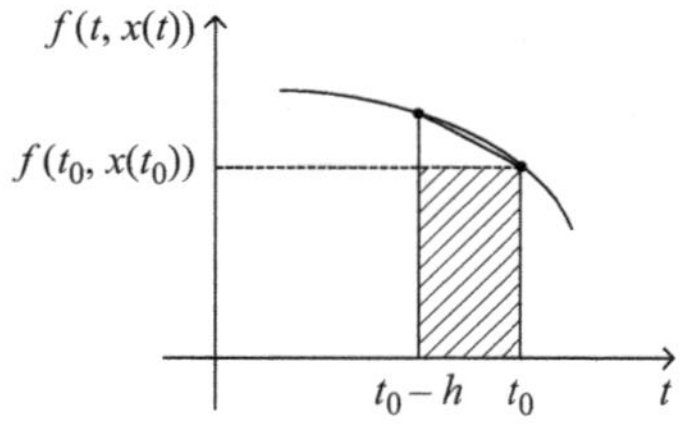

Figure 6.4 The backward Euler

We have different ways to approximate this area, among which the TR is widely used.

As illustrated in Figure 6.3, if the trapezoid rule is used to approximate the value of the area, then we have

$$\int_{t_0-\Delta t}^{t_0} f(t, x(t))dt \approx \frac{h}{2}[f(t_0 - h, x(t_0 - h)) + f(t_0, x(t_0))] \tag{6.4}$$

The backward Euler method estimates the shadow area with rectangle, which is shown as in Figure 6.4.

$$\int_{t_0-\Delta t}^{t_0} f(t, x(t))dt \approx f(t_0, x(t_0))h \tag{6.5}$$

And the discretization formula is

$$x(t_0) = x(t_0 - h) + f(t_0, x(t_0))h \tag{6.6}$$

The forward Euler method also uses the rectangle rule as shown in Figure 6.5 to estimate the shadow area.

$$\int_{t_0-h}^{t_0} f(t, x(t))dt \approx f(t_0 - h, x(t_0 - h))h \tag{6.7}$$

The discretization formula is

$$x(t_0) = x(t_0 - h) + f(t_0 - h, x(t_0 - h))h \tag{6.8}$$

Because the appearance of differential term at t_0, the backward Euler and the TR are implicit methods. The implicit property of these methods generates a set of simultaneous equations to be solved. However, the advantage of these methods is that their numerical stability is better.

6.3.2 Accuracy

It is important to study the errors of the numerical methods and how these errors are accumulated. Local truncation error (LTE) is one important measures of the error made on each step. If the obtained solutions x_{n-1}, x_{n-2}, and x_0 are accurate, the errors of solution x_n at t_n obtained with the numerical method is called the LTE [9–12].

$$\varepsilon(t_n) = \|x_n - x(t_n)\| \tag{6.9}$$

where $x(t_n)$is the exact solution of (6.1). The LTE reflects how closely the discretization formula approximates the integration.

The Taylor expansion at $t_0 - h$ yields

$$x(t_0) = x(t_0 - h) + h\frac{d}{dt}x(t_0 - h) + \frac{h^2}{2}\frac{d^2}{dt^2}x(\lambda) \tag{6.10}$$

or

$$x(t_0) = x(t_0 - h) + f(t_0 - h, x(t_0 - h))h + \text{o}(h^2) \tag{6.11}$$

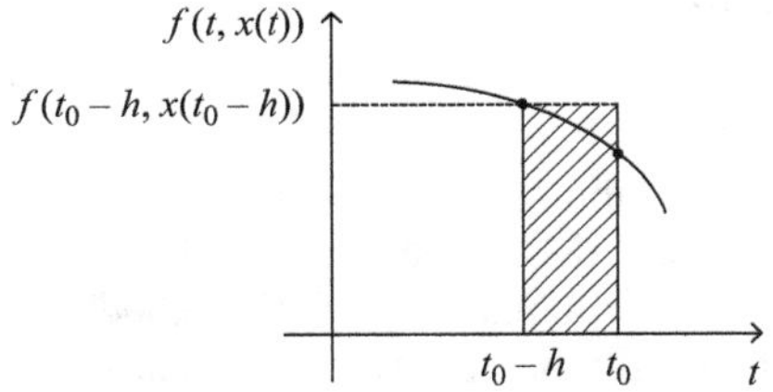

Figure 6.5 The forward Euler

where $\lambda \in [t_0 - h, \quad t_0]$. Therefore, the LTE of the forward Euler method is

$$\varepsilon_{FE} = \mathrm{o}(h^2) \tag{6.12}$$

Similarly, for the backward Euler method, we have

$$x(t_0) = x(t_0 - h) + f(t_0, x(t_0))h + \mathrm{o}(h^2) \tag{6.13}$$

and

$$\varepsilon_{BE} = \mathrm{o}(h^2) \tag{6.14}$$

Since the shadow areas are approximated with rectangle rule for both the forward Euler and the backward Euler methods, the orders of the LTE are also the same.

To obtain the LTE of the TR, expand $x(t)$ and $\frac{dx(t)}{dt}$ at $t_0 - h$ with Taylor series.

$$x(t_0) = x(t_0 - h) + h\frac{d}{dt}x(t_0 - h) + \frac{h^2}{2}\frac{d^2}{dt^2}x(t_0 - h) + \frac{h^3}{6}\frac{d^3}{dt^3}x(\delta) \tag{6.15}$$

$$\frac{d}{dt}x(t_0) = \frac{d}{dt}x(t_0 - h) + h\frac{d^2}{dt^2}x(t_0 - h) + \frac{h^2}{2}\frac{d^3}{dt^3}x(\xi)h^2 \tag{6.16}$$

where $\delta \in [t_0 - h, \quad t_0]$, $\quad \xi \in [t_0 - h, \quad t_0]$.

With (6.16), we have

$$\frac{d^2}{dt^2}x(t_0 - h) = \frac{1}{h}\left[\frac{d}{dt}x(t_0) - \frac{d}{dt}x(t_0 - h) - \frac{h^2}{2}\frac{d^3}{dt^3}x(\xi)h^2\right] \tag{6.17}$$

Substituting (6.17) into (6.15) yields:

$$x(t_0) = x(t_0 - h) + \frac{h}{2}\left[\frac{d}{dt}x(t_0 - h) + \frac{d}{dt}x(t_0)\right] + h^3\left[\frac{1}{6}\frac{d^3}{dt^3}x(\delta) - \frac{1}{4}\frac{d^3}{dt^3}x(\xi)\right] \tag{6.18}$$

The LTE of the TR is

$$\varepsilon_{TR} = h^3\left[\frac{1}{6}\frac{d^3}{dt^3}x(\delta) - \frac{1}{4}\frac{d^3}{dt^3}x(\xi)\right] = \mathrm{o}(h^3) \tag{6.19}$$

The LTE of the TR is one order higher than the orders of the forward and backward Euler methods.

6.3.3 Stability

Compared to accuracy, the stability of numerical methods is a more important aspect to the EMT simulation. The EMT simulation needs to solve differential equations with boundary values and excitations. Numerical methods give approximations of the solution at a series of discrete points. The LTEs can be generated at any points. Furthermore, this error can be accumulated in solutions of subsequent points. The error generated at one point may increase as simulation time proceeds, which will lead to a diverse simulation.

However, it is difficult to measure the stability of certain numerical method with general ODEs. A widely used approach is to use a simple but typical test equation as in [9–12,14]

$$\frac{dx(t)}{dt} = \lambda x(t) \tag{6.20}$$

where λ is a scalar complex constant.

If $\text{Re}(\lambda) < 0$, then $|x(t)|$ decays exponentially. The numerical solution of (6.20) gives a series of discrete values:

$$x_0, \; x_1, \; x_n, \; \cdots \tag{6.21}$$

A reasonable solution requires:

$$|x_k| < |x_{k-1}|, \quad k = 1, \; 2, \; \cdots \tag{6.22}$$

which leads to the definition of absolute stability.

For a given numerical method, the region of absolute stability is the region of the complex z-plane with $z = h\lambda$, where the numerical solution satisfying the absolute stability requirement (6.22), when the method is applied to the test equation (6.20) with step h.

To check the stability of the forward Euler method with the test equation, applying (6.8) to (6.20) yields:

$$x(t_0) = x(t_0 - h) + h\lambda x(t_0 - h) = (1 + h\lambda)x(t_0 - h) \tag{6.23}$$

Assuming a constant step, we have

$$x(t_0) = (1 + h\lambda)x(t_0 - h) = (1 + h\lambda)^2 x(t_0 - 2h) = \cdots = (1 + h\lambda)^n x(0) \tag{6.24}$$

We can obtain the condition of absolute ability as

$$|1 + h\lambda| < 1 \tag{6.25}$$

The region of the absolute stability of the forward Euler method is the shaded area as shown in Figure 6.6.

For the TR method, we can similarly obtain the condition of absolute ability as

$$\left|\frac{2 + h\lambda}{2 - h\lambda}\right| < 1 \tag{6.26}$$

The region of the absolute stability of the TR is the whole left plane as shown in Figure 6.7.

For the backward Euler method, we can obtain the condition of absolute ability as

$$|1 - h\lambda| > 1 \tag{6.27}$$

The region of the absolute stability of the backward Euler method is the whole plane excluding the shaded area as shown in Figure 6.8.

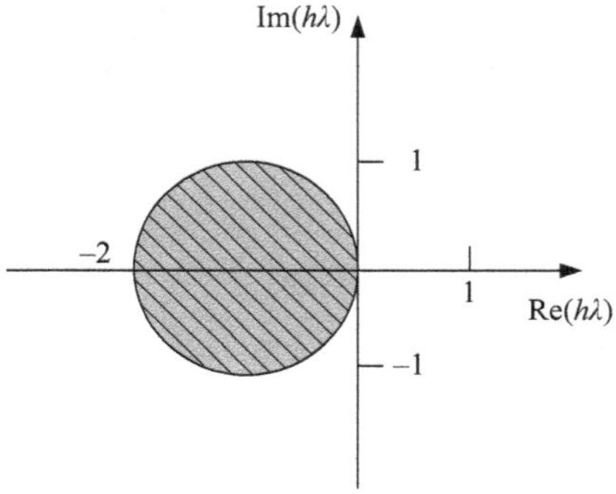

Figure 6.6 The region of the absolute stability of forward Euler method

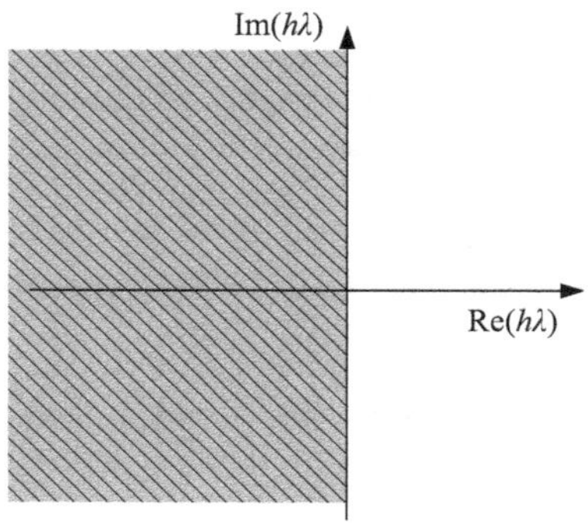

Figure 6.7 The region of the absolute stability of the trapezoidal rule

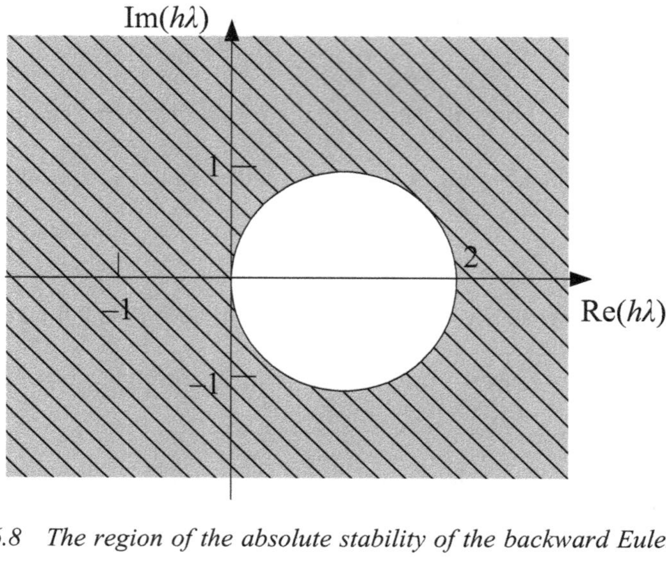

Figure 6.8 The region of the absolute stability of the backward Euler method

From Figures 6.7 and 6.8, we can find that the regions of the TR and the backward Euler method include the left complex plane. Such a type of numerical method is called A-stable method. For A-stable methods, the numerical solution of the test equation (6.20) is always stable, as long as the real solution is stable ($\mathrm{Re}(\lambda) < 0$).

As discussed in appendix of [14], the stability of numerical methods is a more important property to electromagnetic-transient simulation of electrical networks. It guarantees an acceptable simulation result when the system has both fast and slow transients (stiff system). The time step doesn't need to be reduced to very small values to maintain numerical stability due the presence of fast transient phenomenon. Although the errors of fast transient phenomenon may be large, the slow transient can be accurately captured, which is often of the interests.

Although the concept of A-stability introduced base on the test equation (6.20) is a good metric of the stability of numerical methods, it should be noted that A-stable methods do not always guarantee stable numerical solutions for general differential equations. However, for the linear ODEs as

$$\frac{d}{dt}\mathbf{x}(t) = \mathbf{A}\mathbf{x}(t) \tag{6.28}$$

where $\mathbf{A}$ is a constant matrix, whose eigenvalues are all in left complex plane, A-stable method can always give stable numerical solutions [10,13].

For the test equation (6.20), if $\mathrm{Re}(\lambda) < 0$, which implies the real system is stable, both the TR and the backward Euler method give stable numerical solutions. This is due to the A-stability of these two methods. If the real system itself is unstable ($\mathrm{Re}(\lambda) > 0$), the TR will also give an unstable numerical solution, which agrees with the real solution. However, Figure 6.8 shows that the backward Euler method may generate a stable solution. The backward Euler method may provide misleading information regarding the stability of the original system. However, this property also gives the backward Euler method the capability of suppressing the numerical oscillation generated by the TR after switching events, which will be discussed later.

The capability of suppressing the numerical oscillation comes from the property of stiff decay [10]. For a test equation as (6.29),

$$\frac{dx(t)}{dt} = \lambda[x(t) - g(t)] \tag{6.29}$$

where $g(t)$ is an arbitrary but bounded function.

A numerical method is stiff decay, if

$$|x_n - g(t_n)| \to 0 \quad \text{as} \quad h_n\mathrm{Re}(\lambda) \to -\infty \tag{6.30}$$

Applying the backward Euler method to (6.29) yields:

$$x_n - g(t_n) = (1 - h_n\lambda)^{-1}[x_{n-1} - g(t_n)] \tag{6.31}$$

Therefore, the backward Euler method is stiff decay.

The numerical method with the stiff decay property is suitable for solving very stiff systems. For one-step methods, stiff decay is equivalent to the L-stability property [15].

6.4 Companion circuits

Applying the numerical method to different components, the equivalent circuits of current time step can be obtained at the branch level, which is called the companion circuits of the components. The companion circuits can be represented in the Norton equivalent with conductance and current source.

6.4.1 Inductor

The branch characteristic of the inductor (Figure 6.9) is described by

$$L\frac{di_L}{dt} = v_L \tag{6.32}$$

With the discretization formula (6.2) of the TR, we have

$$i_L(t_0) = i_L(t_0 - \Delta t) + \frac{\Delta t}{2}\left(\frac{v_L(t_0)}{L} + \frac{v_L(t_0 - \Delta t)}{L}\right) \tag{6.33}$$

Let

$$g_L = \frac{\Delta t}{2L} \tag{6.34}$$

$$i_{hL}(t_0 - \Delta t) = g_L v_L(t_0 - \Delta t) + i_L(t_0 - \Delta t) \tag{6.35}$$

where $g_L$and I_{hL} are the admittance and history current, respectively. Then, we can rewrite (6.33) as

$$i_L(t_0) = g_L v_L(t_0) + I_{hL}(t_0 - \Delta t) \tag{6.36}$$

According to (6.36), we can obtain the inductor companion circuit as in Figure 6.10.

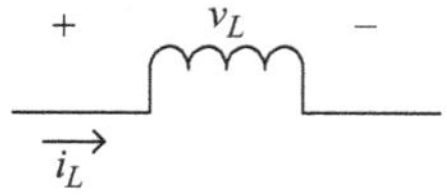

Figure 6.9 The inductor

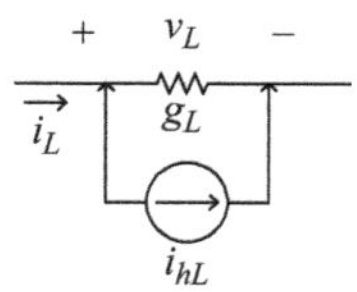

Figure 6.10 Companion circuit of inductor

6.4.2 Capacitor

As for capacitor (Figure 6.11), we have

$$C\frac{dv_C}{dt} = i_C \tag{6.37}$$

Applying the TR (6.2) to (6.37) yields:

$$v_C(t_0) = v_C(t_0 - \Delta t) + \frac{\Delta t}{2}\left(\frac{i_C(t_0)}{C} + \frac{i_C(t_0 - \Delta t)}{C}\right) \tag{6.38}$$

Let

$$g_C = \frac{2C}{\Delta t} \tag{6.39}$$

$$i_{hC}(t_0 - \Delta t) = -g_C v_C(t_0 - \Delta t) - i_C(t_0 - \Delta t) \tag{6.40}$$

where g_C and I_{hC} are the conductance and history current of the companion circuit, respectively. And (6.38) can be rewritten as

$$i_C(t_0) = g_C v_C(t_0) + i_{hC}(t_0 - \Delta t) \tag{6.41}$$

Comparing (6.39) with (6.34), we can find time step Δt has reverse effects on the value of conductance for capacitors. With the decrease of time step, the conductance will increase quickly. This may lead to ill-conditional admittance matrix. To overcome this problem, the Norton equivalent form as in Figure 6.12 can be transformed into the Thevenin equivalent form as shown in Figure 6.13, where the resistor and history voltage are

$$r_C = \frac{\Delta t}{2C} \tag{6.42}$$

$$v_{hC}(t_0 - \Delta t) = v_C(t_0 - \Delta t) + r_C i_C(t_0 - \Delta t) \tag{6.43}$$

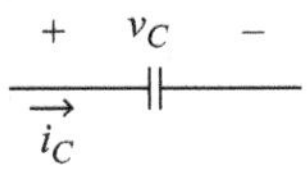

Figure 6.11 The capacitor

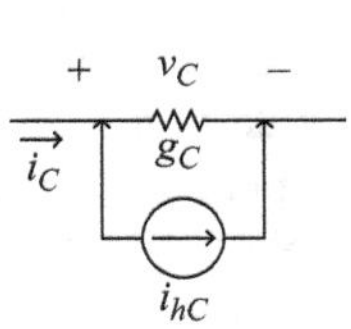

Figure 6.12 Capacitor companion circuit in Norton equivalent form

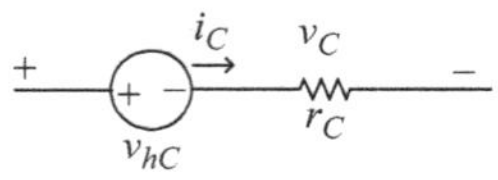

Figure 6.13 Capacitor companion circuit in Thevenin equivalent form

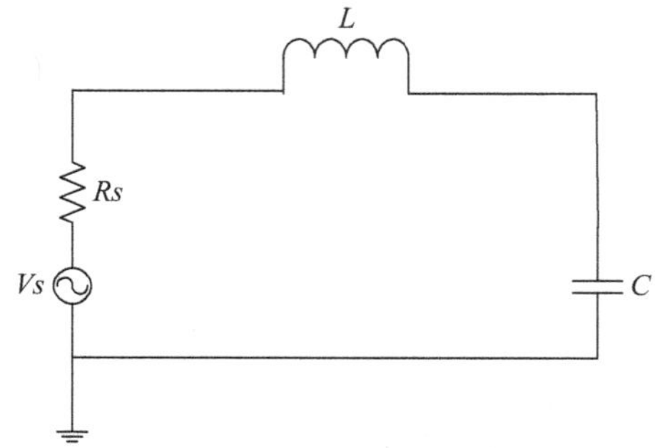

Figure 6.14 Simple RLC circuit

The shortcoming of Thevenin equivalent form is that the formulation of system equations needs to handle ideal voltage source. In this situation, instead of NA, some methods such as MNA can be used to generate the system equations.

6.5 Formulation of system equations

6.5.1 Nodal analysis

The NA can be used to formulate the networks consists of only current-defined branches. A branch is called current-defined branch, if the current of this branch can be explicitly expressed as the functions of the branch voltages or other control variables [16–18]. The current source and conductance of Nortorn equivalent are current-defined branches. The obtained equations of the system has the form as (6.44).

$$\mathbf{YV} = \mathbf{J} \tag{6.44}$$

where **Y** is the admittance matrix. **J** is the injected currents of the nodes. Both **Y** and **J** are variables with known values. However, the node voltages **V** are unknown variables to be solved. Each row of (6.44) represents the Kirchhoff's Current Law (KCL) equation of one node, where the left side is the sum of the currents flow out from this node whereas the right side is the sum of the injected currents into the same node. Due to the linear independence, the generation of (6.44) is actually quite straightforward and is suitable for computers. It can be obtained through summing all the contributions of each branch [15].

Consider a simple RLC circuit as in Figure 6.14. The voltage source *Vs* in series with resistor R_s can be transformed to a Norton equivalent as shown in

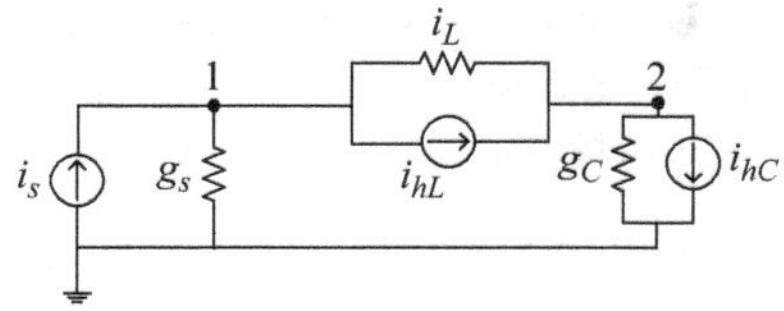

Figure 6.15 Companion circuits

Figure 6.15, where the conductance is $g_s = \frac{1}{R_s}$, the current source is $i_s = g_s v_s$. The companion models of the capacitor and the inductor in each time step are also obtained at the branch level, as in Section 6.5.1. The numbers of node are generated continuously from 1 as shown in Figure 6.15. The node of ground is numbered as 0.

We can form the NA matrix equation as follows:

$$\begin{bmatrix} g_s + g_L & -g_L \\ -g_L & g_L + g_C \end{bmatrix} \begin{bmatrix} v_1 \\ v_2 \end{bmatrix} = \begin{bmatrix} i_s(t) - I_{hL}(t - \Delta t) \\ I_{hL}(t - \Delta t) - I_{hC}(t - \Delta t) \end{bmatrix} \tag{6.45}$$

For each time step, the history currents for inductor and capacitor are calculated with the previous time-step value and the source current is also known, so the current matrix **J** is known. Thus, we can calculate the node voltages so that the system values can be solved step by step from time zero.

6.5.2 Modified nodal analysis

The NA can formulate the equations of current-defined branches. To handle other type of components, the branch must be transformed into a current-defined branch (e.g. the way to process the voltage source in Figure 6.14). However, for some branches, it is not possible to be converted into a current-defined one (e.g. ideal voltage source). To overcome the limitations of NA, the MNA is introduced by C. W. Ho [19]. The MNA can process the networks consisting of both current-defined branches and voltage-defined branches. Similar to a current-defined branch, a branch is called the voltage-defined branch, if the voltage of this branch can be explicitly expressed as the functions of the branch currents or other control variables [16–18] (e.g. resistor and voltage source). Besides the node voltages, the currents of the voltage-defined branches are also selected as unknown variables in the MNA. The system equations of the MNA have the following form:

$$\begin{bmatrix} \mathbf{Y} & \mathbf{B} \\ \mathbf{C} & \mathbf{D} \end{bmatrix} \begin{bmatrix} \mathbf{V} \\ \mathbf{I} \end{bmatrix} = \begin{bmatrix} \mathbf{J} \\ \mathbf{E} \end{bmatrix} \tag{6.46}$$

where **V** are unknown node voltages. **I** is the unknown branch currents. For the first set of equations,

$$\mathbf{YV} + \mathbf{BI} = \mathbf{J} \tag{6.47}$$

Each row represents one KCL equation of one node. The left side of one row is the sum of the currents flowing out this node. The right-side **J** is the sum of injected currents. The elements of **B** can be 1, –1, or 0, which reflects the contribution of branch currents to the sum of currents flowing out the node.

For the second set of equations,

$$\mathbf{CV} + \mathbf{DI} = \mathbf{E} \tag{6.48}$$

Each row represents one branch constitutive equation.

Due to the superposition property, (6.46) can be obtained through summing all the contributions of each branch, which is suitable for implementation on computers [15].

For the previous case as Figure 6.14, we need to convert the Thevenin voltage source into its Norton equivalent to obtain a network consisting of only current-defined branches. However, if the internal resistor R_s of the voltage source is small or even zero, the obtained conductance g_s will be large numbers or infinity. It will cause numerical problems in the subsequent solution of the system equation (6.44).

Figure 6.16 shows a simple LC circuit, where the internal resistor of the voltage sources is zero. Also, when small time step is used, the Thevenin equivalent form of capacitor as in Figure 6.13 can be used to prevent the problem of ill-conditional matrix.

The obtained companion circuit is shown in Figure 6.17. The numbers of node and voltage-defined branch are generated continuously from 1 as shown in Figure 6.17. The node of ground is numbered as 0.

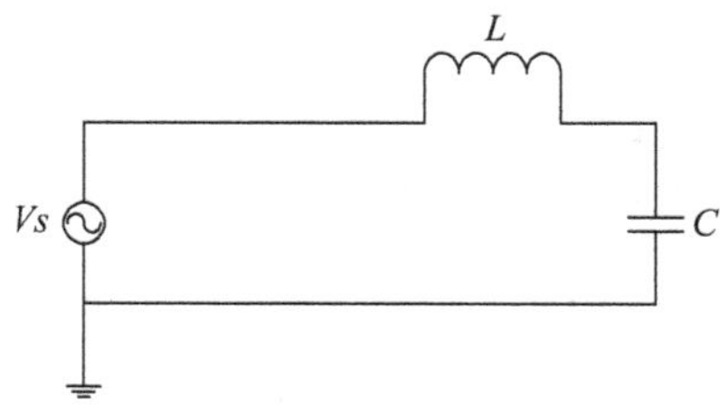

Figure 6.16 A simple LC circuit

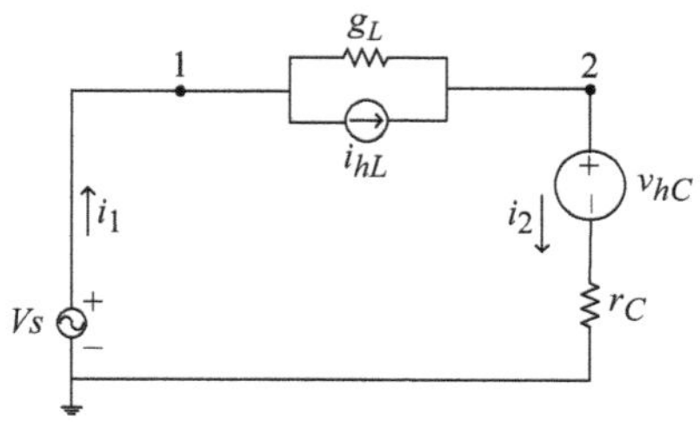

Figure 6.17 Companion circuits

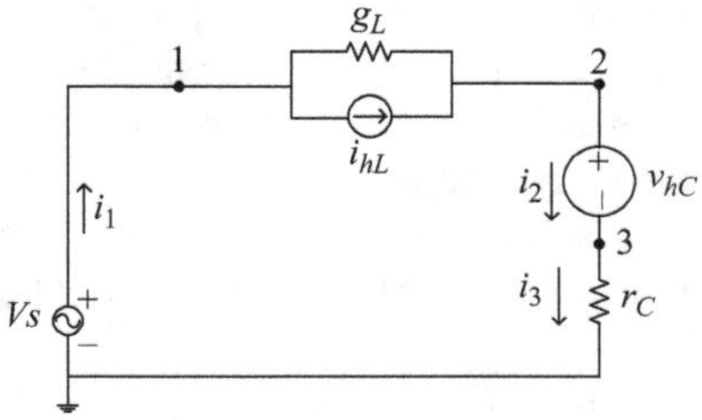

Figure 6.18 Companion circuits

Using the MNA, we can obtain the system equation as (6.49)

$$\begin{bmatrix} g_L & -g_L & -1 & 0 \\ -g_L & g_L & 0 & 1 \\ 1 & 0 & 0 & 0 \\ 0 & 1 & 0 & r_C \end{bmatrix} \begin{bmatrix} v_1 \\ v_2 \\ i_1 \\ i_2 \end{bmatrix} = \begin{bmatrix} -i_{hL} \\ i_{hL} \\ V_s \\ v_{hC} \end{bmatrix} \tag{6.49}$$

With the MNA, we can also treat the Thevenin equivalent of capacitor as two voltage-defined branches connected in series as shown in Figure 6.18. The new system equation is shown in (6.50).

$$\begin{bmatrix} g_L & -g_L & 0 & -1 & 0 & 0 \\ -g_L & g_L & 0 & 0 & 1 & 0 \\ 0 & 0 & 0 & 0 & -1 & 1 \\ 1 & 0 & 0 & 0 & 0 & 0 \\ 0 & 1 & -1 & 0 & 0 & 0 \\ 0 & 0 & 1 & 0 & 0 & -r_C \end{bmatrix} \begin{bmatrix} v_1 \\ v_2 \\ v_3 \\ i_1 \\ i_2 \\ i_3 \end{bmatrix} = \begin{bmatrix} -i_{hL} \\ i_{hL} \\ 0 \\ V_s \\ v_{hC} \\ 0 \end{bmatrix} \tag{6.50}$$

It should be noted that, because of the added additional node voltage v_3 and currents of voltage-defined branch i_3, the dimension of the obtained linear equations is higher than the previous one. This will increase the computation burden when solving the linear equations. However, comparing (6.49) with (6.50), only 4 additional nonzero elements are added in the coefficient matrix. If the sparse matrix techniques are employed to obtain the solution, the computation burden does not increase as much as in dense case.

6.6 Solving linear equations

In every time step, the NA or the MNA will give a system of linear equations with node voltages or branch currents as unknown variables. Suppose, the linear equations to be solved are as follows:

$$\mathbf{Ax} = \mathbf{b} \tag{6.51}$$

where the $n \times n$ matrix $\mathbf{A}$ has a nonzero determinant and then the system has a unique solution.

Besides the EMT simulation, finding the solution of system of linear equations is also of fundamental importance to many applications. As discussed by G. W. Stewart in [20], "Suppose that all matrix algorithms save one were to disappear, which one would you choose to survive? A group of experts would quickly agree that they could not do without the ability to solve linear equations. Their algorithm of choices would naturally be Gaussian elimination—the most versatile of all matrix algorithms."

6.6.1 Gaussian elimination

We can calculate the solutions of (6.51) by

$$x_i = \frac{\det(\mathbf{A}_i)}{\det(\mathbf{A})} \quad i = 1, \cdots, n \tag{6.52}$$

where $\mathbf{A}_i$ is the matrix formed by replacing the ith column of $\mathbf{A}$ by $\mathbf{b}$, det( ) means determinant of a matrix. This method is called Cramer's rule. Cramer's rule is theoretically correct but computationally inefficient, because it needs $O((n+1)!)$ operations to obtain the solution for n-order equation. Therefore, it is primarily used for low-order systems with two or three unknowns [11,12]. For power system electromagnetic simulations, Gaussian elimination is more widely used.

Consider the system equation below

$$\begin{bmatrix} a_{11} & \cdots & a_{1n} \\ \vdots & \ddots & \vdots \\ a_{n1} & \cdots & a_{nn} \end{bmatrix} \begin{bmatrix} x_1 \\ \vdots \\ x_n \end{bmatrix} = \begin{bmatrix} b_1 \\ \vdots \\ b_n \end{bmatrix} \tag{6.53}$$

Augmented matrix is firstly constructed as (6.54).

$$\left[\begin{array}{ccc|c} a_{11} & \cdots & a_{1n} & b_1 \\ \vdots & \ddots & \vdots & \vdots \\ a_{n1} & \cdots & a_{nn} & b_n \end{array}\right] \tag{6.54}$$

Solving a system of linear equation with Gaussian elimination consists of two stages: the forward elimination and the backward substitution. The idea is to transform the matrix in (6.54) into an upper triangular matrix form through a finite number of elementary row operations. In each round, one variable is eliminated. After $k-1$ rounds, the obtained matrix is shown as in (6.55).

$$\left[\begin{array}{ccccccc|c} a_{11}^{(1)} & & & \cdots & & & a_{1n}^{(1)} & b_1^{(1)} \\ & \ddots & & \vdots & & & \vdots & \vdots \\ & & a_{k-1k-1}^{(k-1)} & \cdots & & & a_{k-1n}^{(k-1)} & b_{k-1}^{(k-1)} \\ & & & a_{kk}^{(k)} & & \cdots & a_{kn}^{(k)} & b_k^{(k)} \\ & \mathbf{0} & & \vdots & & \ddots & & \vdots \\ & & & a_{nk}^{(k)} & & \cdots & a_{nn}^{(k)} & b_n^{(k)} \end{array}\right] \tag{6.55}$$

To eliminate the kth variables, the kth row is multiplied by $-a_{ik}^{(k)}/a_{kk}^{(k)}$ and added to the ith row, where $i = k+1, \cdots n$. After $n-1$ rounds, we will obtain the upper triangular matrix as (6.56), which has the same solution as (6.53). The process of variable elimination is usually called the forward elimination.

$$\left[\begin{array}{cccc|c} a_{11}^{(1)} & & \cdots & a_{1n}^{(1)} & b_1^{(1)} \\ & a_{22}^{(2)} & \cdots & a_{2n}^{(2)} & b_2^{(2)} \\ & & \ddots & \vdots & \vdots \\ \mathbf{0} & & & a_{nn}^{(n)} & b_n^{(n)} \end{array}\right] \tag{6.56}$$

The solution of (6.56) can be obtained with a back substitution process, since it has the upper triangular form. The forward elimination and the backward substitution described with the C code are as follows:

```
double A[n][n + 1];
int i, j, k;
double x[n];
double s;
// forward elimination
void GaussFE()
{
  for (k = 0; k<n-1; k++)
  {
    for (i = k+1; i<n; i++)
    {
      s = A[i][k] / A[k][k];
      for (j = k; j<n + 1; j++) A[i][j] -= s*A[k][j];
    }
  }
}
// backward substitution
void GaussBS()
{
  for (i = n-1; i >= 0; i-)
  {
    x[i] = A[i][n];
    for (j = i + 1; j<n; j++)
      x[i] -= A[i][j] * x[j];
    x[i] /= A[i][i];
  }
}
```

We can find that the needed arithmetic operations in forward elimination is $O(n^3)$, while the back substitution process only needs $O(n^2)$operations. Compared to Cramer's rule, the Gaussian elimination is much more efficient.

One should notice that in the kth round, the diagonal element $a_{kk}^{(k)}$ must not be zero. If it happens to be zero, as long as the original matrix $\mathbf{A}$ is non-singular, we can find a nonzero element in the vector $\left[a_{kk}^{(k)} \quad \cdots \quad a_{nk}^{(k)}\right]^T$. Furthermore, through row permutation, we can then make the diagonal element to be nonzero and continue the process of variable elimination. The process of finding nonzero diagonal elements is called pivoting.

6.6.2 LU factorization

In electromagnetic-transient simulations, it often happens that the obtained coefficient matrix $\mathbf{A}$ in (6.51) is the same as in previous time step, only the values of the right-hand vector $\mathbf{b}$ are changed. In this situation, we can first factorize matrix $\mathbf{A}$ into a lower triangular matrix $\mathbf{L}$ and an upper triangular matrix $\mathbf{U}$ as shown in (6.57), which is usually called LU factorization.

$$\mathbf{A} = \mathbf{LU} \tag{6.57}$$

Then, the solution of (6.51) can be obtained only through solving two systems of linear equations in the triangular form as in (6.58).

$$\begin{cases} \mathbf{Ly} = \mathbf{b} \\ \mathbf{Ux} = \mathbf{y} \end{cases} \tag{6.58}$$

The Gaussian elimination process itself can actually give the LU factorization of one matrix, which means that the computation complexity of LU factorization is also $O(n^3)$. There are different of variations of the classical Gaussian elimination, such as Pickett's charge east method (also called the left-looking method), Crout's method, etc. Although the required operations are fixed, these variations have advantages when writing to cache is more expensive than reading from it [20].

If matrix $\mathbf{A}$ is a positive definite (e.g. the admittance matrix obtained from a purely resistive network with NA), then $\mathbf{A}$ can be uniquely factorized as

$$\mathbf{A} = \mathbf{U}^T\mathbf{U} \tag{6.59}$$

where $\mathbf{U}$ is an upper triangular matrix with positive diagonal element. This factorization is called the Cholesky decomposition, which only needs roughly half of the operations of the LU factorization [20,21].

The advantage of factorizing $\mathbf{A}$ into lower triangular and upper triangular matrices is obvious. If the coefficient matrix does not change, then we can obtain the solution with two substitution processes as in (6.58), which only needs $O(n^2)$operations. It is much more efficient than the original Gaussian elimination process (6.54–6.56) based on row operations of the augmented matrix.

It is frequent that only a small amount of elements in matrix $\mathbf{A}$ has been changed. This is common when simulating the HVDC systems. In this situation, to avoid a full factorization, some compensation or partial factorization methods can be used to further improve the efficiency [22].

6.6.3 Sparse matrix techniques

The coefficient matrices generated with the NA or the MNA have a lot of zero elements, which are usually called sparse matrices. Sparsity comes from the loose coupling of systems. In power systems, one node usually only connects with a few other nodes. For a network, the elements y_{ij} and y_{ji} of the admittance matrix are zero, if there are no branches between nodes i and j. We say that a matrix is sparse, if it can bring an advantage to exploit its zeros, which relates to the nonzero patterns, the architecture of the computer, etc. Sparse matrix also appears in many other applications. For such a sparse matrix populated primarily with zeros, the required memory will be greatly reduced, if proper data structure is used to avoid storing of the zero elements. The number of operations is also greatly reduced, if the operations involving zero elements are avoided. Without sparse techniques, it is impractical to solve some very large systems with the Gaussian elimination method.

6.6.3.1 Storage of sparse matrix

For sparse matrices, to save the required memory space and facilitate the operations, various static or dynamic structures can be used to store sparse matrix. Among them, the triplet form and compressed-column form / compressed-row form are widely used [21,23]. The triplet form uses two integer arrays i and j and one real array x of length equals to the number of nonzeros to represent the matrix. As example, consider the matrix shown in (6.60).

$$\mathbf{A} = \begin{bmatrix} 1.0 & 0 & 5.0 & 7.0 \\ 0 & 3.0 & 0 & 0 \\ 2.0 & 0 & 6.0 & 8.0 \\ 0 & 4.0 & 0 & 9.0 \end{bmatrix} \tag{6.60}$$

The sparse matrix as shown in (6.60) can be stored with the triplet form as (6.61).

$$\begin{aligned} \text{int } n &= 4; \\ \text{int } nnz &= 9; \\ \text{int } i[] &= \{0,2,1,3,0,2,0,2,3\}; \\ \text{int } j[] &= \{0,0,1,1,2,2,3,3,3\}; \\ \text{int } x[] &= \{1.0,2.0,3.0,4.0,5.0,6.0,7.0,8.0,9.0\}. \end{aligned} \tag{6.61}$$

The array i is used to store the row numbers, while j the column numbers. The numerical values are stored in array x.

The triplet form is simple to create but difficult to access the elements. However, the compressed-column form is easier to access the elements with indirect addressing. As the triplet form, compressed-column form also puts the subsequent nonzeros of the matrix columns in contiguous locations. In the compressed-column form, the matrix is described with four parameters:

n: The matrix demission.
p: An integer array of size $n+1$. The first entry $p[0] = 0$, and the last entry $p[n] = nnz$ is number of the nonzero entries in the matrix.
i: An integer array of size nnz. The row indices of entries in column j of A are located in $i[p[j]] \cdots i[p[j+1]-1]$.
x: A real array of size nnz. The numerical values in column j of a real matrix are located in $x[p[j]] \cdots x[p[j+1]-1]$.

The same sparse matrix as in (6.60) can be stored using compressed-column form as in (6.62).

$$\begin{aligned} \text{int n} &= 4; \\ \text{int p} &= \{0, 2, 4, 6, 9\}; \\ \text{int i []} &= \{0, 2, 1, 3, 0, 2, 0, 2, 3\}; \\ \text{double x []} &= \{1.0, 2.0, 3.0, 4.0, 5.0, 6.0, 7.0, 8.0, 9.0\}. \end{aligned} \tag{6.62}$$

Similarly, the compressed-row form puts the subsequent nonzeros of the matrix rows in contiguous locations, which is essentially the compressed-column form of the transformed matrix.

6.6.3.2 LU factorization considering sparsity

(a) None-zero fill-in

Because only nonzero elements are stored in a sparse storage structure, the operations involving zero elements can be detected and avoided, when solving the linear equations with the Gaussian elimination. The main step of the Gaussian elimination or LU factorization is multiplying one row with a number and then adding it to another row. This step may create new nonzeros, which are called fill-ins. The process of creating fill-in is shown in Figure 6.19.

The fill-in will greatly affect the number of operations needed to perform Gaussian elimination and the forward/backward substitution. And the number of generated fill-ins is greatly affected by ordering of the matrix. As an example, consider the nonzero structure of the sparse matrix shown in Figure 6.20. After one step of the Gaussian elimination, the obtained matrix is shown in Figure 6.21. It becomes a dense matrix due to the generated fill-ins, which are represented with red color.

However, if the matrix is reordered with proper row and column permutations, we can obtain a matrix as shown in Figure 6.22. Assuming the matrix is a positive definite so that no pivoting is required in the process of elimination. In the whole process of the Gaussian elimination, no fill-ins are generated as shown in Figure 6.23.

(b) Graph representation

The nonzero pattern of a square sparse matrix can be represented by a graph. Often, it is possible to gain an insight into sparse matrix techniques by working with the graph associated with the matrix. Some problems of sparse matrix can

nonzero		nonzero	
		↓	
nonzero	→	fill-in	

Figure 6.19 The creation of a fill-in

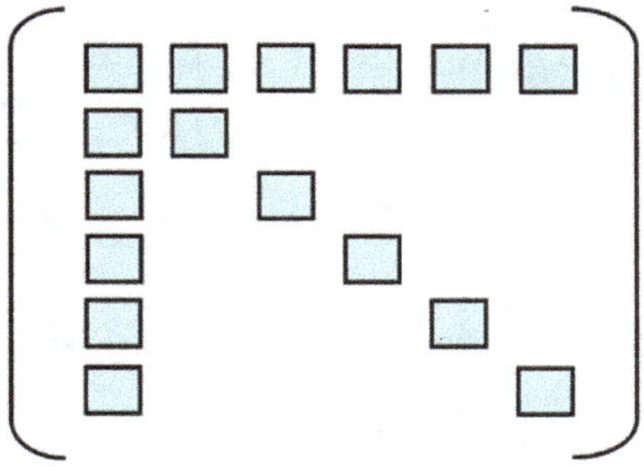

Figure 6.20 The nonzero structure

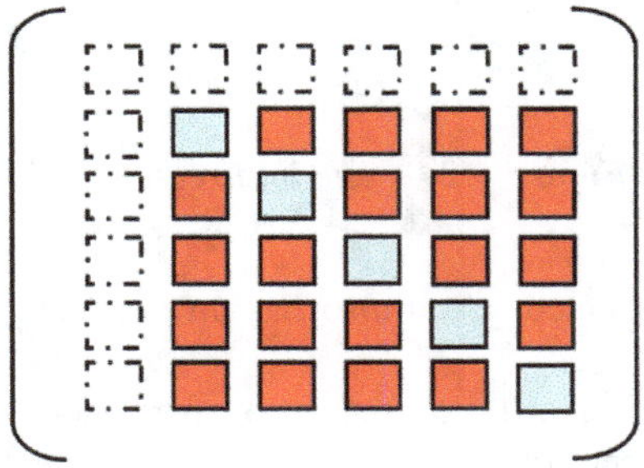

Figure 6.21 The nonzero structure after one step of the Gaussian elimination

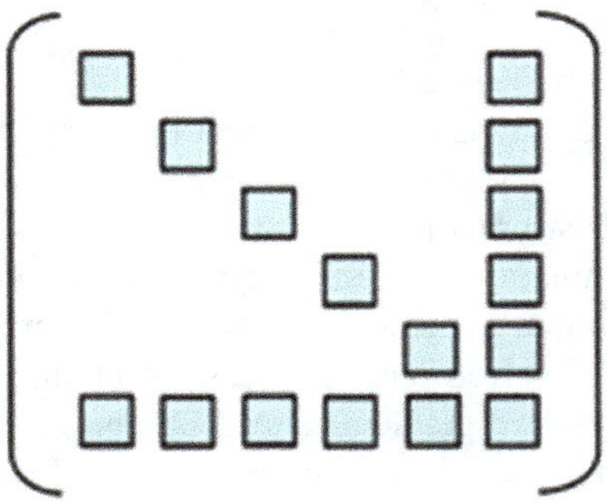

Figure 6.22 The nonzero structure after permutations

find answers in the graph theory. It is also a good tool to visualize what is happening in sparse matrix computation.

A directed graph consists of a set of nodes (vertices) and directed edges between nodes. The directed graph connects with the sparse matrix as follows: For any square sparse matrix **A**, the number of vertices in the graph equals to the order of the matrix. If a_{ij} is a nonzero entry, there is an edge from node i to node j in the directed graph. Any square sparse matrix has an associated directed

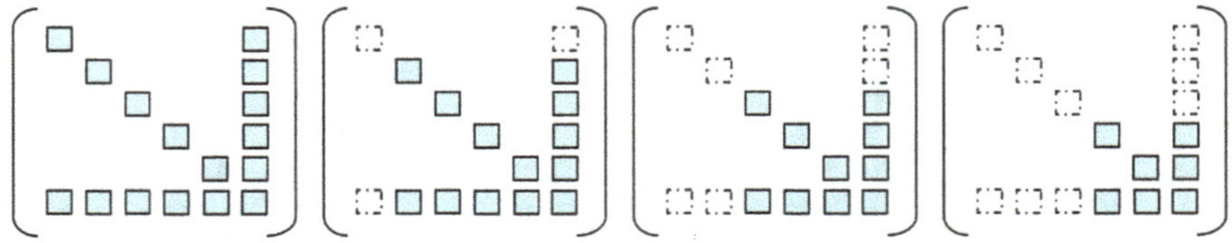

Figure 6.23 The Gaussian elimination without a fill-in

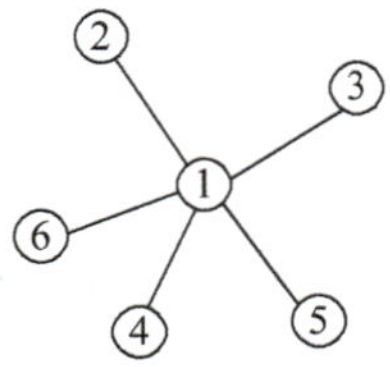

Figure 6.24 The undirected graph

graph, and any directed graph also has an associated square matrix nonzero pattern.

For a symmetric matrix, a connection from nodes i to j implies the existence of the connection from nodes j to i. Therefore, the arrows may be dropped and we obtain an undirected graph. For example, the sparse matrix in Figure 6.20 can be represented with the undirected graph in Figure 6.24. The elimination of node 1 will make any pair of the remaining nodes have connections, which is shown in Figure 6.25.

(c) Ordering: Minimum Degree Reordering

The fill-in will greatly affect the number of operations needed to perform the Guassian elimination and the forward/the backward substitution. Therefore, we need to find the best ordering to generate the least fill-ins. However, the bad news is that determining the best ordering in the elimination process, which results in the minimum number of fill-ins, is NP-Hard. This implies that minimizing the work of performing the Gaussian elimination is more costly than itself, which is a P-Hard problem. The good news is that we can approximate this minimum using graph-based heuristics. One of such heuristic is to always select the vertex with minimum degree [21,24].

If we renumber the nodes of the graph associated with the matrix in Figure 6.20 with the minimum degree rule, which is shown Figure 6.26. This ordered graph also relates to the nonzero pattern of the matrix in Figure 6.22, which will generate totally no fill-in.

6.6.3.3 Main steps of typical sparse solver

To solve a sparse linear equation, one typical sparse solver consists of the following steps:

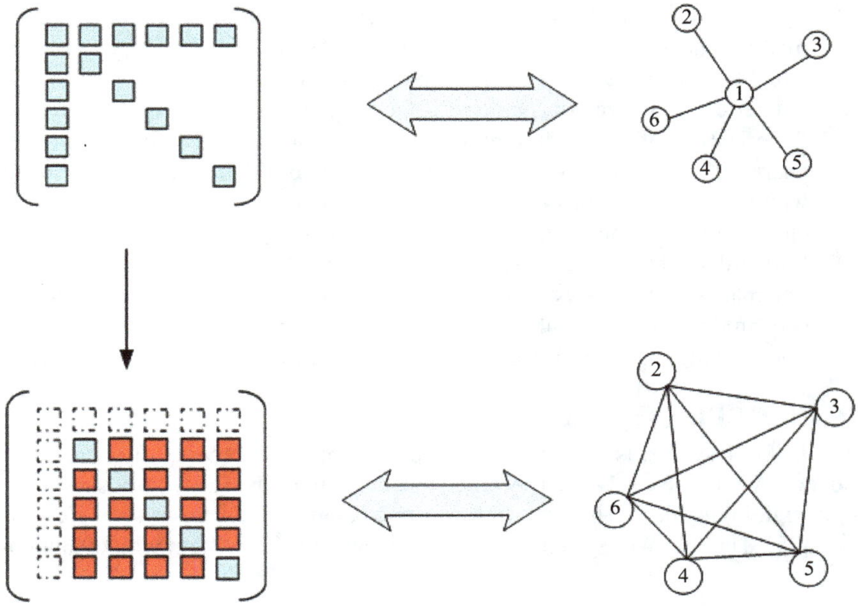

Figure 6.25 The undirected graph after one step of the Gaussian elimination

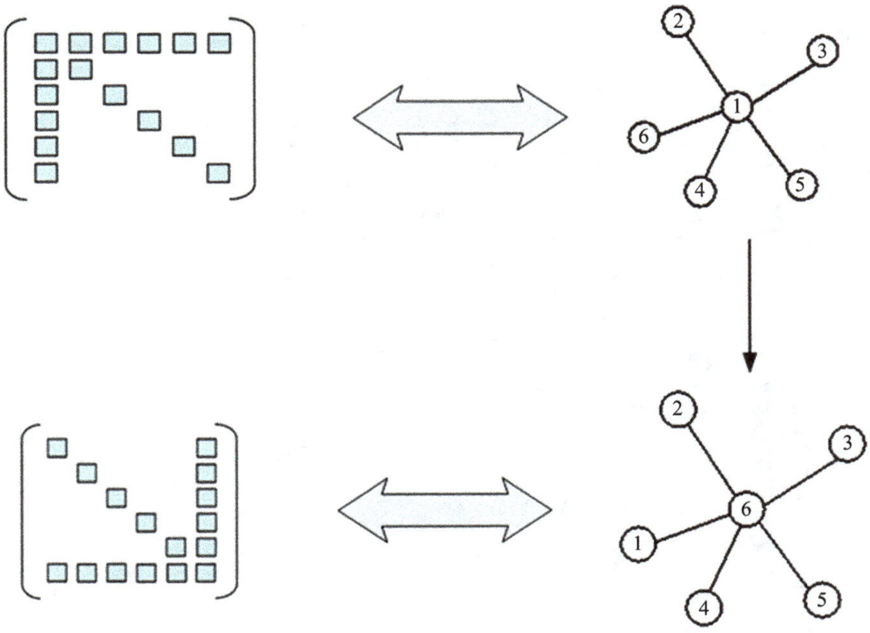

Figure 6.26 The undirected graph after ordering

(a) Symbolic analysis. Symbolic analysis usually orders the matrix with certain method and identify the nonzero element positions in the LU factorization, which is sometime also called symbolic factorization.
(b) Numerical factorization. Numerical factorization performs the numerical LU factorization of the matrix based on the results outputted by the symbolic analysis. For a matrix whose numerical values of the matrix elements change while nonzero pattern keeps unchanged, numerical factorization can be executed multiple times with only one symbolic analysis.
(c) Forward and backward substitution. Once we obtained the LU factorization of one matrix, we can solve the linear equations with the forward and backward substitution. Also, if only the right-hand vector changes, we can solve the linear equations without LU refactorization of the coefficient matrix.

6.6.3.4 IEEE-14 system case

The IEEE-14 system is used as an example to show the process of solving linear equations with sparse solvers. The single line diagram of the IEEE 14-bus test system is given in Figure 6.27. The static and dynamic parameters for this system are chosen from [25]. The IEEE-14 system generates a linear equation system with 60 unknown

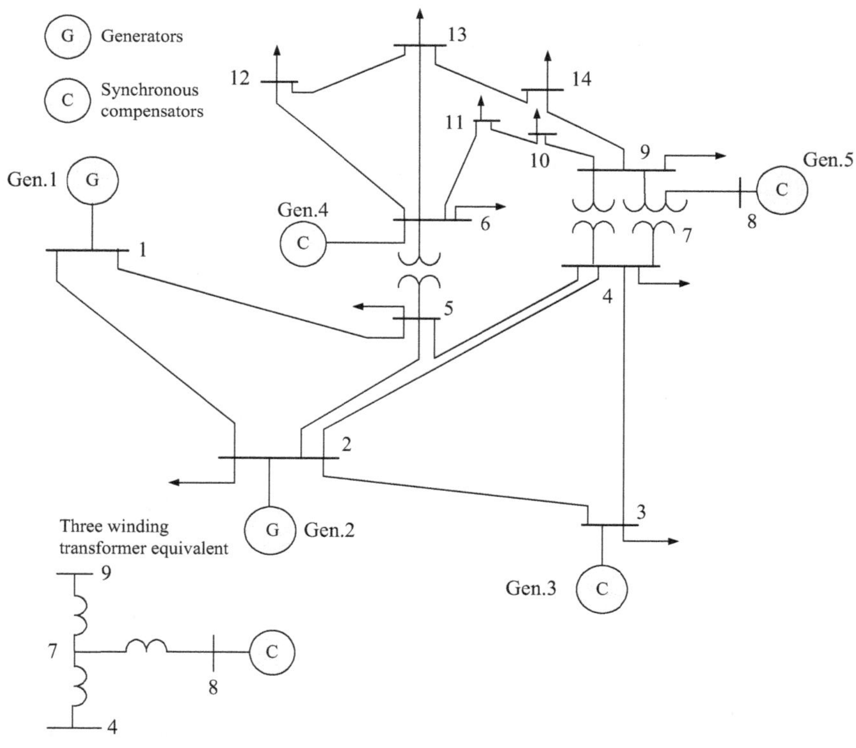

Figure 6.27 IEEE-14 system

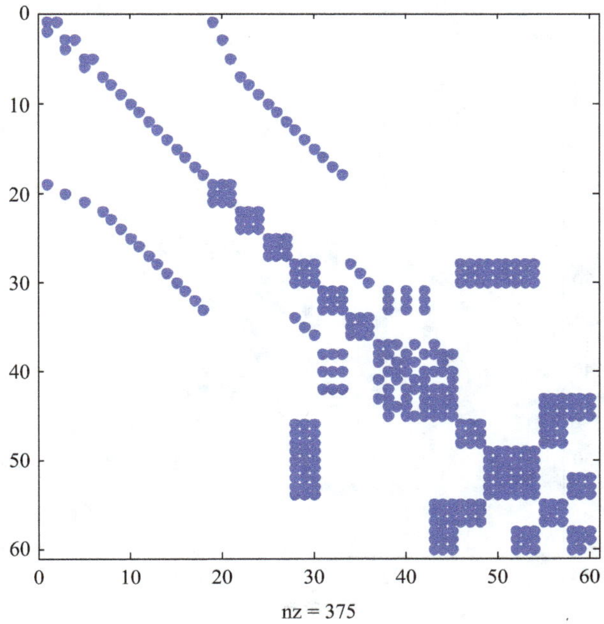

Figure 6.28 Nonzero pattern of coefficient matrix with natural order

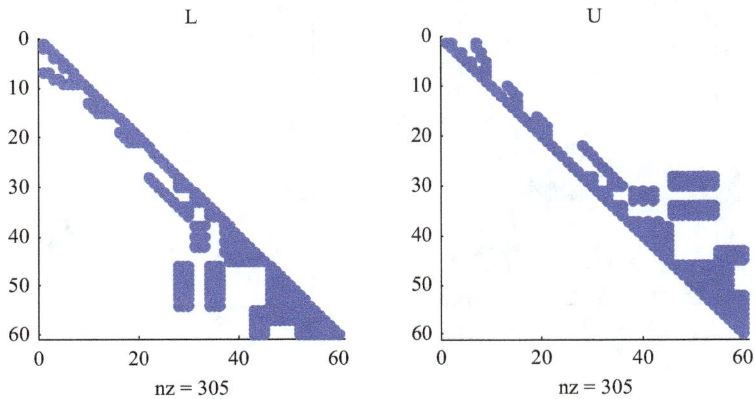

Figure 6.29 Nonzero pattern of LU matrix, natural order

variables. The nonzero pattern of the original coefficient matrix of the linear equation system (before ordering) is shown in Figure 6.28. After the LU factorization, the nonzero patterns of the obtained L and U matrices are shown in Figure 6.29.

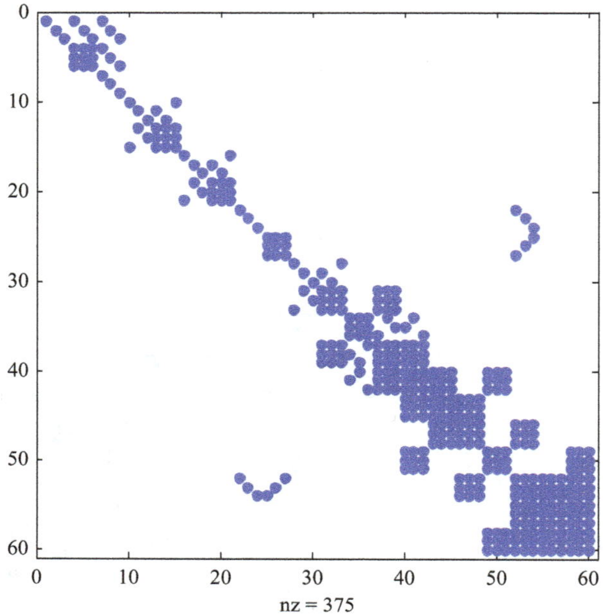

Figure 6.30 Nonzero pattern of coefficient matrix with AMD ordering

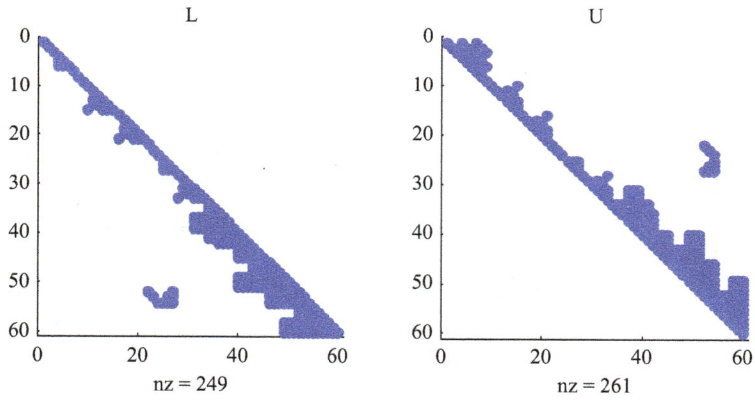

Figure 6.31 Nonzero pattern of LU matrix, AMD ordering

With the approximate minimum ordering method (AMD) [26], which is a similar method as minimum ordering but more efficient, the nonzero pattern of the ordered matrix is shown in Figure 6.30. After the LU factorization, the nonzero patterns of the obtained L and U matrices are shown in Figure 6.31. While comparing Figure 6.29 with Figure 6.31, it can be found that the number of fill-ins is reduced by the reordering.

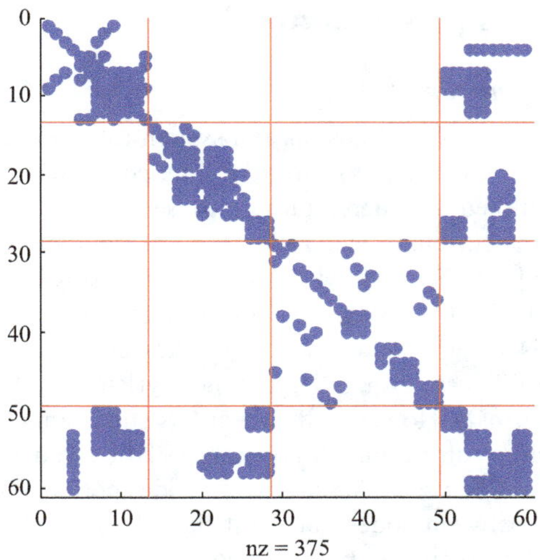

Figure 6.32 The nonzero pattern of the BBD form matrix

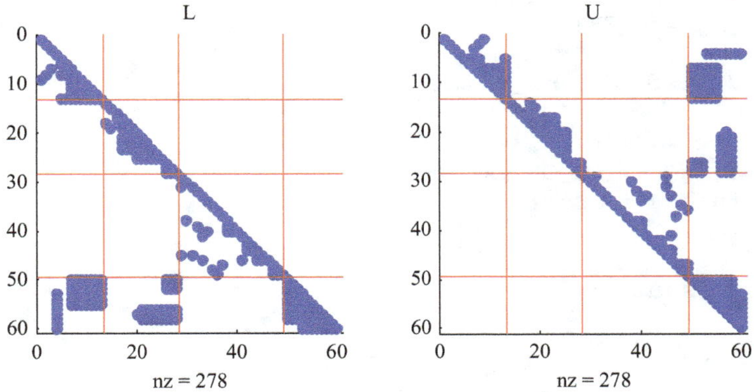

Figure 6.33 Nonzero pattern of LU matrix, BBD form

We can also order the original coefficient matrix into a blocked border diagonal (BBD) form, which is shown in Figure 6.32. The obtained L and U matrices after factorization also have the BBD form, which are shown in Figure 6.33. Compared with the AMD ordering, the BBD form may generate more fill-ins and, therefore, is less efficient. However, due the independence of each block, the obtained BBD form is more suitable for parallel implementation, which can accelerate the simulation through exploiting the hardware resources [27].

6.7 Simulation of power electronics

6.7.1 Model of power electronics

The implementation of HVDC transmission system relies on the underlying power electronic element – the thyristor for the line-commutated converter HVDC. A thyristor can be turned on when a turn on pulse is issued and is forward biased. And, it turns off if the current tries to reverse. To model the on and off states, it can usually be modeled as resistors with values that change as a function of the switching event [2]. For OFF state, the thyristor can be represented with a large resistor. With the NA, the conductance of resistor can be set as zero, which can simulate the ideal OFF state. For ON state, the thyristor can be represented with a small resistor (e.g. 1 mΩ). However, the ideal ON state cannot be reached with the NA, which will lead to an infinite conductance. To overcome this limitation, the MNA formulation can be used. The following codes could be used to implement a thyristor connected between nodes i and j. It should be noted that the modification of the elements in the matrix can be achieved in a more efficient incremental way, because multiple devices can contribute to the same elements in the admittance matrix.

```
int state = OFF; //Initial state is set as OFF
bool refactor = false;
if (state == OFF)
{
  if (Vij > 0 && fire_pulse == ON)
  {
    state == ON;
    G[i][j] += (G_ON - G_OFF);
    refactor = true;
  }
}
else if (state == ON)
{
  if (Iij < 0)
  {
    state == OFF;
    G[i][j] += (G_OFF - G_ON);
    refactor = true;
  }
}
if (refactor)
{
  //Trigger refactorization of the admittance matrix
}
```

The other types of power electronics (e.g. diode, insulated gate bipolar transistor (IGBT), etc.) can also be similarly modeled.

6.7.2 Interpolation

For real power electronic elements, the switching actions often happen at the instants when the branch voltage is forward biased or the current is inversed. However, electromagnetic-transient simulator usually uses a fixed time step strategy. Therefore, the real instant of event (switching actions) may lie between time steps. And, the action has to be postponed to the end of time step. This may cause spurious voltage spikes or incorrect harmonics. To overcome this limitation, the variable time step can be used. When events happened, go back to the time grid before events and then use a much smaller time step to catch more accurate instant of events. However, this method will cause the reformulations of dynamic elements, because the conductance of the Norton equivalent companion model relates to the used time step. Also, the variable time step may lead to other problems, such as the less efficient implementation of distributed transmission lines. A more practical and popular approach is to use interpolation. The real instant of event is firstly estimated with interpolation. And, the state of the whole system is approximated to the instant of event with interpolation. The linear interpolation is commonly used for the approximation, some nonlinear interpolation approaches can also be used to obtain better accuracy [28]. Then, the simulation can be restarted from the more accurate instant of event. Another interpolation can also be applied to bring the simulation back to the normal time grid and give the states of the network on time grid. This information is required by the control system, when the control system and the network exchange information on time grids. The steps of interpolation for handling events are shown in Figure 6.34.

6.7.3 Numerical oscillation suppression

Numerical oscillation is another phenomenon in power electronics simulation. Numerical oscillation can be caused by fast dynamics (e.g. turn off an inductive

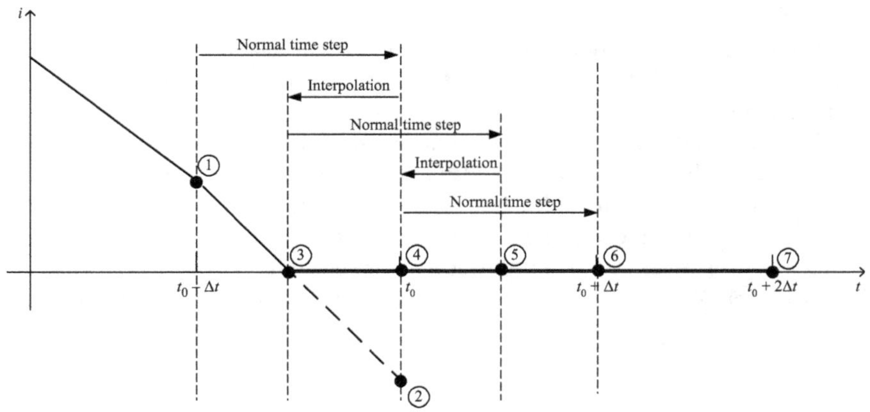

Figure 6.34 Interpolation of events

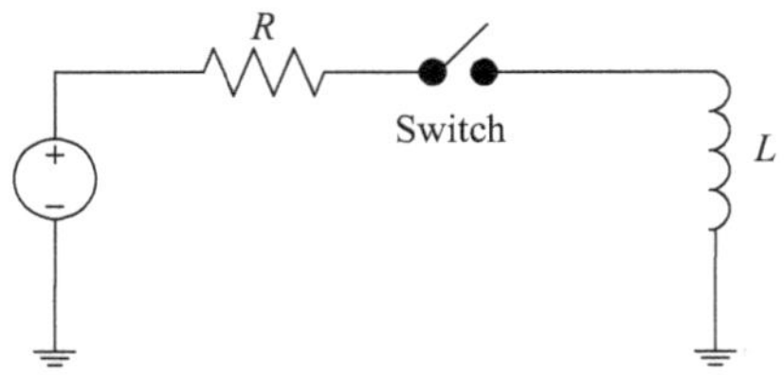

Figure 6.35 Switched RL branch

branch), whose time constant is usually much smaller than the used time steps. For the simple case as shown in Figure 6.35, the oscillation can be easily inferred.

With (6.35) and (6.36), it can be found that, after switching off, the voltages of the inductance branch satisfy (6.63)

$$u(t) = -u(t - \Delta t) \tag{6.63}$$

If the voltage of the inductance before switching off is not zero, the voltages will have oscillations according to (6.63). Actually, as pointed by Adam Semlyen in the discussion of [29], numerical oscillations can be caused by any fast dynamics whose time constant is much smaller than the used time step, which includes not only switching events but also the abrupt change of excitation. Numerical oscillation does not decay rapidly because the TR is not stiff decay. To see this, applying the TR to the test equation (6.20) with a fixed time step h, we have:

$$x_n = \frac{2 + h\lambda}{2 - h\lambda} x_{n-1} \tag{6.64}$$

It follows that

$$\frac{2 + h\lambda}{2 - h\lambda} \to -1 \text{ as } h_n\text{Re}(\lambda) \to -\infty \tag{6.65}$$

The artificial oscillations may lead to false interpretation of the real response and cause incorrect behavior of the control system. There are some remedies to fix the problem of numerical oscillations.

To suppress the numerical oscillations caused by the switching of power electronics, the snubber circuits can be added in the model of power electronics. The RC series circuits are often included to model the snubber circuit of real power electronics devices, which can provide extra damping to the high-frequency numerical oscillations. However, if the real power electronics do not have snubber circuits or the researchers want to study the ideal switch model, this method will have limitations. Furthermore, the snubber circuits may not provide enough damping effects to damp the oscillations rapidly.

Since the numerical oscillation is caused by the TR numerical method, maybe the best way to solve this problem is from the approaches of numerical method.

The critical damping adjustment (CDA) technique [29,30] changes the TR to the backward Euler method to suppress the oscillations, when the conditions which may cause numerical oscillations are detected. The step size of the backward Euler is reduced to half of the normal step, which can keep the conductance of companion circuit unchanged and avoid additional construction and factorization of the admittance matrices. As illustrated in Section 3.4, the stiff decay property gives the backward Euler method the capability of damp the oscillations. Usually, two steps of the backward Euler method will suppress the possible oscillations to an acceptable level.

The half time step interpolation is another method used to suppress numerical oscillations [31–33], which has been used in PSCAD/EMTDC [34] for many years. The idea of using half step interpolation to suppress oscillations probably comes from the intuition that the midpoint (the average value) of the line connecting the adjacent two points lies on the correct solution. It is obvious that the adjacent two points with numerical oscillations evenly distributed on the two sides of the real solution. The half step interpolation method has another advantage which may not be obvious at the first glance. As stated in Section 6.7.2, interpolation is often used to capture the accurate instant of switching events. In this situation, the facilities of interpolation have already been built, which makes the implementation of half step interpolation for the suppression of numerical oscillations quite straightforward. Zhang Yi of RTDS Technologies, Inc., pointed out that, for dynamic elements (L, C, etc.), the effect of half step interpolation is equivalent to the half step backward Euler method. For example, for the inductor, we have the relation of the branch voltages and currents between two adjacent points (t_0 and $t_0 + \Delta t$) as (6.66), when the TR is used.

$$i(t_0 + \Delta t) = \frac{\Delta t}{2L} v(t_0 + \Delta t) + \left[\frac{\Delta t}{2L} v(t_0) + i(t_0)\right] \tag{6.66}$$

With the linear interpolation assumption, we have

$$\begin{cases} i\left(t_0 + \dfrac{\Delta t}{2}\right) = \dfrac{1}{2}[i(t_0) + i(t_0 + \Delta t)] \\ v\left(t_0 + \dfrac{\Delta t}{2}\right) = \dfrac{1}{2}[v(t_0) + v(t_0 + \Delta t)] \end{cases} \tag{6.67}$$

Substituting (6.67) into (6.66) gives

$$i\left(t_0 + \frac{\Delta t}{2}\right) = \frac{\Delta t/2}{L} v\left(t_0 + \frac{\Delta t}{2}\right) + i(t_0) \tag{6.68}$$

If the backward Euler method is used to obtain the companion model at $t_0 + \Delta t/2$, we will get the same formula as (6.68), which means that these two methods used to suppress the numerical oscillations are equivalent. Therefore, we can also use the stiff decay property to explain the capability of damping the oscillations for half step interpolation method.

A similar method was proposed in [35], which is named as trapezoidal history term averaging (THTA). Instead of using the average value of the voltages and currents, THTA modifies the history current source as the average value of two subsequent history currents calculated with half time step TR. THTA is suspected to be equivalent to the CDA method [35].

6.8 Simulation case

In this section, the benchmark HVDC system presented in [36] is used for the test. The configuration of the system is shown as in Figure 6.36 and the parameters can be found in [36]. The time step of the simulation is 50 μs.

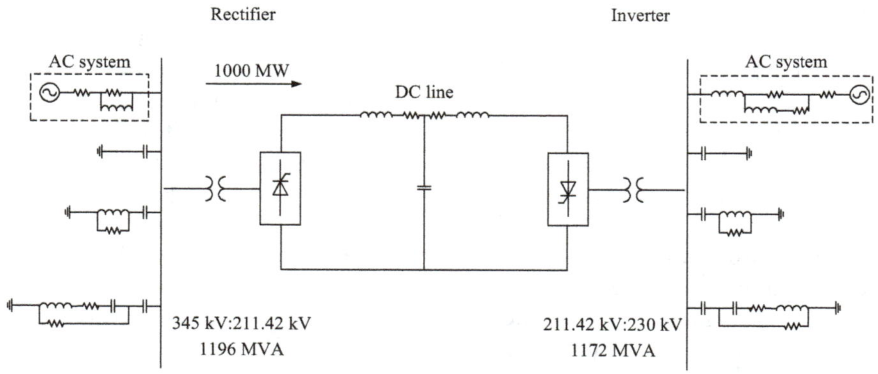

Figure 6.36 Configuration of HVDC benchmark system

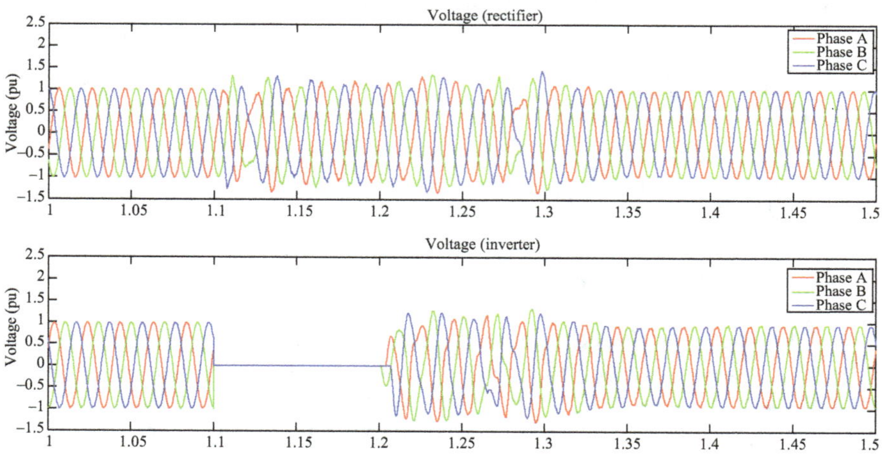

Figure 6.37 AC voltage of rectifier and inverter commutation buses

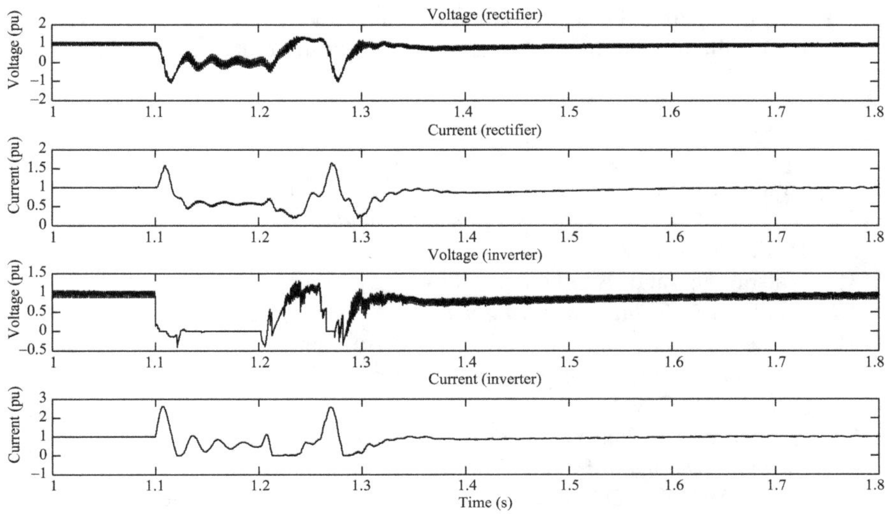

Figure 6.38 DC voltage and current

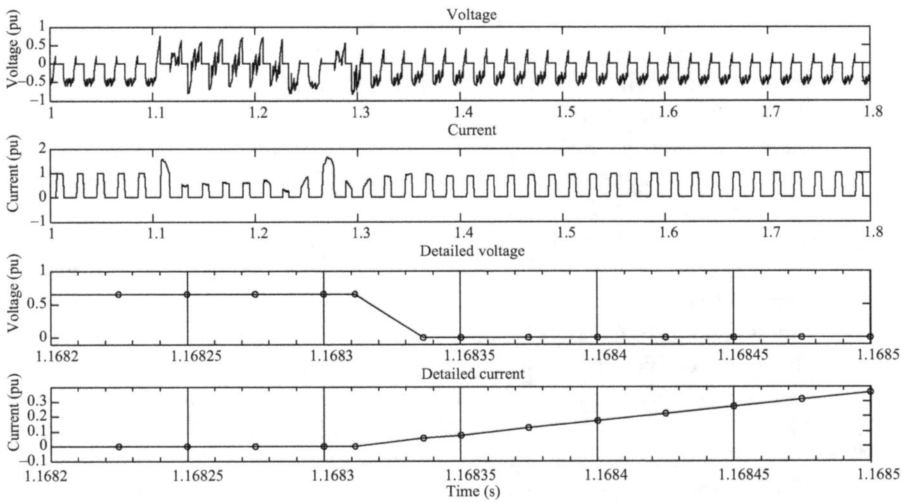

Figure 6.39 Voltage and current of one thyristor

After the system arrives at a steady state, a three-phase solid fault is applied at 1.1 s. The fault is applied on the inverter commutation bus for 5 cycles (0.1 s). Figure 6.37 shows the three-phase voltages at the rectifier and inverter commutation buses. The voltages and currents of the DC line at both the rectifier and inverter sides are shown in Figure 6.38. The voltage and current of one thyristor are also shown in Figure 6.39, where the time steps of interpolation and suppression of numerical oscillation can be clearly identified from the more detailed sub-figures.

References

[1] SimPowerSystems. *User's Guide, Version 5*. The MathWorks Inc.; 2009.

[2] Gole A. M. *Power Systems Transient Simulation, Course Notes*. University of Manitoba; 2018.

[3] Watson N., and Arrillaga J. 'Power systems electromagnetic transients simulation'. *Institution of Engineering & Technology*. 2003, p. 448.

[4] Wasynczuk O., and Sudhoff S. D. 'Automated state model generation algorithm for power circuits and systems'. *IEEE Transactions on Power Systems*. 1996, vol. 11(4), pp. 1951–1956.

[5] Ramshaw R. S., and Schuurman D. *PSpice Simulation of Power Electronics Circuits. Chapman & Hall*. 1996

[6] Dommel H. W. 'Digital computer solution of electromagnetic transients in single- and multiphase networks'. *IEEE Transactions on Power Apparatus and Systems*. 1969, vol. PAS-88(4), pp. 388–399.

[7] Woodford D. A., Gole A. M., and Menzies R. W. 'Digital simulation of dc links and ac machines'. *IEEE Transactions on Power Apparatus and Systems*. 1983, vol. PAS-102(6), pp. 1616–1623.

[8] Tinney W. F., and Walker J. W. 'Direct solutions of sparse network equations by optimally ordered triangular factorization'. *Proceedings of the IEEE*. 1967, vol. 55(11), pp. 1801–1809.

[9] Hairer E., Nørsett S. P., and Wanner G. *Solving Ordinary Differential Equations I. Non-stiff Problems*. Springer Series in Computational Mathematics 8. Berlin: Springer; 1993.

[10] Ascher U., and Petzold L. *Computer Methods for Ordinary Differential Equations and Differential-Algebraic Equations*. Society for Industrial & Applied Mathematics; 1998.

[11] Yun C., and Jiang B. *Computer Aided Circuit Analysis (in Chinese)*. Shandong University Press; 2015.

[12] Yan Q. *Numerical Analysis, 4th ed (Chinese)*. Beihang University Press; 2012.

[13] Ali D. *Improved Models of Electric Machines for Real Time Digital Simulation*. University of Manitoba; 2010. Appendix B: Numerical Stability of a Discretized System Using Rectangular and Trapezoidal Integration.

[14] Dommel H. W. *EMTP Theory Book, 2nd ed*. Vancouver: Microtran Power System Analysis; 1992.

[15] Butcher J. C. *Numerical Methods for Ordinary Differential Equations*. New York: John Wiley and Sons; 2008.

[16] Ogrodzki J. *Circuit Simulation Methods and Algorithms*. USA: CRC Press; 1994.

[17] Acary V., Bonnefon O., and Brogliato B. 'Nonsmooth modeling and simulation for switched circuits'. *Lecture Notes in Electrical Engineering*. 2011, vol. 69.

[18] Chua L. O., and Lin P. *Computer-Aided Analysis of Electronic Circuits: Algorithms & Computational Techniques*. Englewood Cliffs, NJ: Prentice-Hall; 1975.

[19] Ho C. W., Ruehli A. E., and Brennan P. A. 'The modified nodal approach to network analysis'. *IEEE Transactions on Circuits and Systems*. 1975, vol. CAS-22, pp. 504–509.

[20] Stewart G. W. *Matrix Algorithms. Volume I: Basic Decompositions*. Philadelphia: Society for Industrial and Applied Mathematics; 1998.

[21] Timothy A. D. *Direct Methods for Sparse Linear Systems*. Philadelphia: Society for Industrial and Applied Mathematics; 2006.

[22] Chan S. M., and Brandwajn V. 'Partial matrix refactorization'. *IEEE Transactions on Power Systems*. 1986, vol. 1(1), pp. 193–199.

[23] Iain S. D., Albert M. E., and John K. R. *Direct Methods for Sparse Matrices*. New York: Oxford University Press, Inc.; 1986.

[24] Tinney W. F., and Walker J. W. 'Direct solutions of sparse network equations by optimally ordered triangular factorization'. *Proceedings of the IEEE*. 1967, vol. 55(11), pp. 1801–1809.

[25] Kodsi S. K. M., and Cañizares C. A. *Modeling and Simulation of IEEE 14-bus System with FACTS Controllers*. Waterloo; 2003.

[26] Amestoy P. R., Davis T. A., and Du I. S. 'Algorithm 837: AMD, an approximate minimum degree ordering algorithm'. *ACM Transactions on Mathematical Software*. 2004, vol. 30(3), pp. 381–388.

[27] Fan S., Ding H., Kariyawasam A., and Gole A. M. 'Parallel electromagnetic transients simulation with shared memory architecture computers'. *IEEE Transactions on Power Delivery*. 2018, vol. 33(1), pp. 239–247.

[28] Tant J., and Driesen J. 'On the numerical accuracy of electromagnetic transient simulation with power electronics'. *IEEE Transactions on Power Delivery*. 2018, vol. 33(5), pp. 2492–2501.

[29] Marti J. R., and Lin J. 'Suppression of numerical oscillations in the EMTP power systems'. *IEEE Transactions on Power Systems*. 1989, vol. 4(2), pp. 739–747.

[30] Lin J., and Marti J. R. 'Implementation of the CDA procedure in the EMTP'. *IEEE Transactions on Power Systems*. 1990, vol. 5(2), pp. 394–402.

[31] Kuffel P., Kent K., and Irwin G. 'The implementation and effectiveness of linear interpolation within digital simulation'. *Proceedings of the International Conference on Power Systems Transients (IPST 1995)*. 1995, pp. 449–504.

[32] Krueger K. H., and Lasseter R. H. 'HVDC Simulation Using NETOMAC'. IEEE Montech '86, Conference on HVDC Power Transmission; Montreal, Canada, 1986.

[33] Gole A. M., Woodford S. A., Nordstrom J. E., and Irwin G. D. 'A Fully Interpolated Controls Library for Electromagnetic Transients Simulation of Power Electronic Systems'. IPST 2001; Rio de Janeiro, Brazil, June 2001, p. 681.

[34] Woodford D. A., Gole A. M., and Menzies R. W. 'Digital simulation of dc links and ac machines'. *IEEE Transactions on Power Apparatus & Systems*. 1983, vol. PAS-102(6), pp. 1616–1623.

[35] Ferreira L. F. R., Bonatto B. D., Cogo J. R., de Jesus N. C., Dommel H. W., and Martí J. R. 'Comparative solutions of numerical oscillations in the trapezoidal method used by EMTP-based programs'. IPST; Cavtat, Croatia, June 2015, pp. 147–153.

[36] Szechman M., Wess T., and Thio C.V. 'First benchmark model for HVDC control studies'. *CIGRE WG 14.02 Electra*. 1991, vol. 135, pp. 54–73.

Chapter 7
Electromechanical transient simulation of LCC HVDC

Junxian Hou[1], Lei Wan[1] and Jian Zhang[1]

7.1 General introduction of electromechanical transient simulation of LCC HVDC

The electromagnetic simulation and electromechanical simulation are two group methods in power system simulation. The electromagnetic simulation [1,2] usually resorts to the simulation of electric and magnetic reactions in a very short time. The detail voltage and current change process can be simulated as in Chapter 6. The time step may be several microseconds or even smaller, and the dynamic process may be observed within several milliseconds.

The power system is always in three phases and symmetric in steady state. When the disturbance has happened, the electromagnetic process is usually finished in short term and the focus is changed to the power system stability, then the electromechanical simulation is employed. The electromechanical transient simulation [3] is focused on the reaction between electrics and mechanics, and is usually used for large-scale power system simulation, which is based on the hypothesis that the electric power system is symmetric, and the harmonics are not considered. The time step may be several milliseconds and the dynamic process may be extended to several seconds or minutes.

Another important reason to use the electromechanical simulation is the computation amount. The electromagnetic simulation needs much longer computation time than electromechanical simulation, which is unendurable for the large scale of power system.

The line commutated converter based high-voltage direct current (LCC HVDC) is widely used in power system to transmit a large amount of power for a long distance, which may have a great effect on the whole power system's stability. The LCC HVDC model has been modeled and simulated in detail in the electromagnetic simulation software [4–8] to ensure the accuracy of LCC HVDC itself. However, because of the great impact on the large scale of power system stability, the LCC HVDC should also be involved in the simulation of the whole system with an electromechanical model.

[1]Electric Power Research Institute, Beijing, China

When used in the electromechanical simulation, the LCC HVDC model should be simplified considering the transient stability (TS)–simulation characteristics [9–11], and the simulation accuracy should be ensured as much as possible. The LCC HVDC model includes the main circuit and control system, the main circuit is the same as the steady-state model and the control system is a dynamic model, so the total model is usually called quasi-steady-state model. The reaction between LCC HVDC and power system usually is simulated as the variable power or load, which is shown in Figure 7.1.

How the detailed LCC HVDC model is being simplified for electromechanical simulation is the main concern and is discussed in detail in the present chapter. The commutation failure and model parameters' identification are also key topics for HVDC electromechanical simulation and are treated in this chapter.

7.2 Electromechanical transient simulation of LCC HVDC

To establish an electromechanical transient model or the TS model of HVDC, it is necessary to consider the main requirements and characteristics of the electromechanical transient of the power system based on detailed electromagnetic transient models. Functions and modules that have an important influence on the DC-transient characteristics will be adopted in the system model and simulation. Based on a thorough study of the manufacturer's pole-control logic, simplification is necessary to ensure that the main characteristics of the TS model are basically the same as the actual control and protection characteristics of the DC project.

Simplification process mainly includes the following:

1. Simplification of pole-control function

   The actual controller is complex with many functions. It is the pole-control system that plays a major role in the DC-transient characteristics. Therefore,

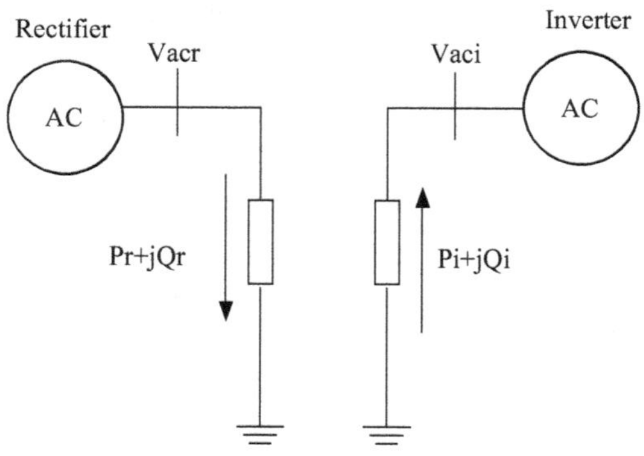

Figure 7.1 The reaction between LCC HVDC and power system

the electromechanical transient modeling mainly deals with the control functions in the pole-control system.

- Focus on the modules that play a major role in DC regulation performance in the polar control system, such as pole power control (PPC), current control, low-voltage current-limiting control, and extinction angle control. Ignore the engineering-related modules, such as DC start–stop, no-load voltage test, etc.
- Emphasize on the mainline of control logic, ignoring other numerous controls.
- The control function with a longer time scale and longer adjustment period, such as tap changer control, reactive voltage control, etc., is treated with different sections.

2. Simplification of models involving data acquisition, measurement, and instantaneous value processing

   Physical quantities of the AC system in the electromechanical transient simulation are generally the fundamental frequency phasor. Therefore, for the module related to the AC instantaneous quantity in the actual controller, the model is converted into an approximate module for processing root mean square value (RMS), such as minimum trigger angle control on the rectifier side and commutation failure prediction.
3. Simplification of models involving trigger pulse control

   The trigger control of the valve group is not considered in converter model in electromechanical transient.

7.2.1 Simulation model of the main circuit

In the electromechanical transient simulation model of power system, the quasi-steady model is commonly used for DC converter stations. Assuming that the commutating bus voltage is a three-phase symmetrical sine wave, the AC positive sequence voltage is generally used. Converters and the DC lines are the main circuit models and interact with the control system. The model structure is shown in Figure 7.2.

From Figure 7.2, the main circuit model is equivalent to a controlled voltage source on the DC side. The rectifier and inverter are combined with the DC line to solve the problem. On the AC side, the equivalent active and reactive power injections are considered for the integration of the converter to the grid.

7.2.1.1 Normal operation

The relationship between the ideal no-load DC voltage and the AC voltage of the converter is as follows:

$$U_{di0} = \frac{3\sqrt{2}}{\pi}U \approx 1.35U \tag{7.1}$$

In which, U is the RMS of the line voltage of the converter transformer value side, and $Udi0$ is the ideal no-load DC voltage.

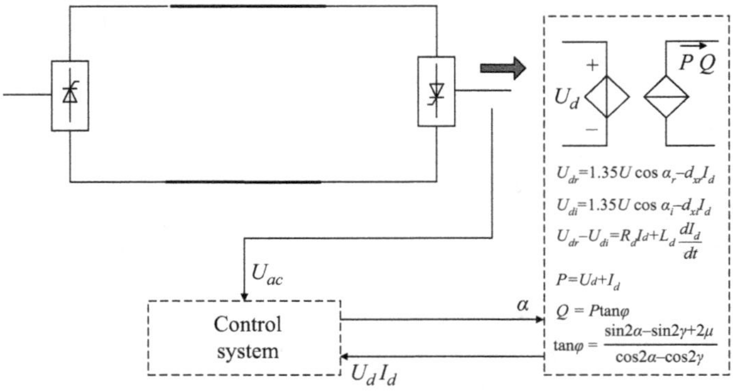

Figure 7.2 Structure principle of electromechanical transient model for DC transmission

The DC voltage at the outlet of the rectifier-side converter can be expressed as:

$$U_{dr} = U_{di0r} \cos \alpha - d_{xr}I_{dr} \tag{7.2}$$

In which, α is the trigger angle of the rectifier-side converter. I_{dr} is the DC current of the rectifier side, U_{di0r} is the ideal no-load DC voltage of the rectifier side, and d_{xr} is the equivalent reactance of the rectifier side, which can be derived as follows:

$$d_{xr} = \frac{3}{\pi}X_{cr} \tag{7.3}$$

In which, X_{cr} is the leakage reactance of the converter transformer.

Similarly, the DC voltage at the output of the inverter-side converter can be expressed as:

$$U_{di} = U_{di0i} \cos \beta + d_{xi}I_{di} \tag{7.4}$$

In which, β is the trigger angle of the inverter side. I_{di} is the DC current of the inverter side, U_{di0i} is the ideal no-load DC voltage of the inverter side, and d_{xi} is the equivalent reactance of the inverter side.

Therefore, the DC voltage is the difference between the voltage source associated with the trigger angle and the voltage drop of commutation reactance. It can be represented by a serialized version of the Thevenin circuit, as shown in Figure 7.3.

Replace the DC line with a resistor and inductor series circuit. The rectifier and inverter are, respectively, represented as Thevenin circuit, and the both ends of DC transmission line are connected to the Thevenin circuits, then a complete DC primary circuit model is formed, as shown in Figure 7.4.

In Figure 7.4, the resistance and inductance of the line are divided into two halves, which form a symmetrical form with the smoothing reactor and converter

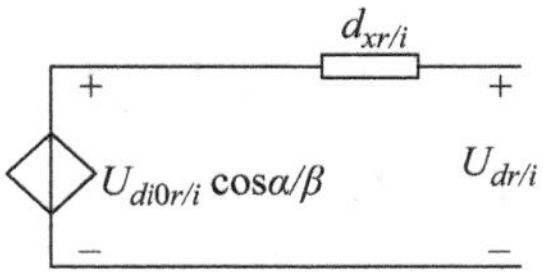

Figure 7.3 Inverter equivalent circuit

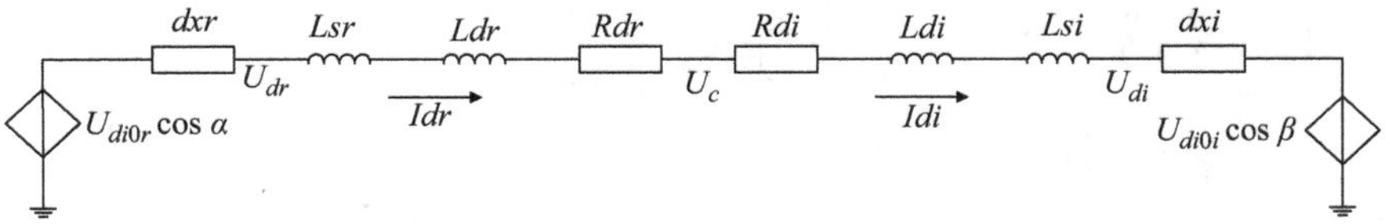

Figure 7.4 DC-side circuit model during normal operation

reactor. Then, (7.5) can be obtained:

$$\begin{cases} U_c = U_{d0ir}cos\alpha - I_{dr}(dxr + Rdr) - (Lsr + Ldr)\dfrac{dI_{dr}}{dt} \\ U_c = U_{di0i}cos\beta + I_{di}(dxi + Rdi) + (Lsi + Ldi)\dfrac{dI_{di}}{dt} \\ I_{dr} = I_{di} \end{cases} \tag{7.5}$$

where U_c is the midpoint voltage.

On the AC side, according to the conservation of power, the active power injected into the AC grid is obtained:

$$P = U_{di}I_{di} \tag{7.6}$$

Reactive power is related to the power factor, and the latter is related to the operating angle of the trigger angle, commutation angle, and extinction angle, which can be expressed as follows:

$$\begin{cases} Q = P\tan\varphi \\ \tan\varphi = \dfrac{\sin 2\alpha - \sin 2\gamma + 2\mu}{\cos 2\alpha - \cos 2\gamma} \end{cases} \tag{7.7}$$

In which, φ is the power factor angle, γ is the extinguish, and μ is the commutation angle.

7.2.1.2 Commutation failure

When the commutation failure occurs, the valve group on the bridge arm of the same phase in the inverter-side converter is simultaneously turned on, that is, the DC bus of the inverter is short-circuited. The DC-side circuit status is shown in Figure 7.5.

Comparing Figures 7.4 and 7.3, it can be seen that when the commutation failure occurs, the inverter-side loses the adjustment capability, and the current

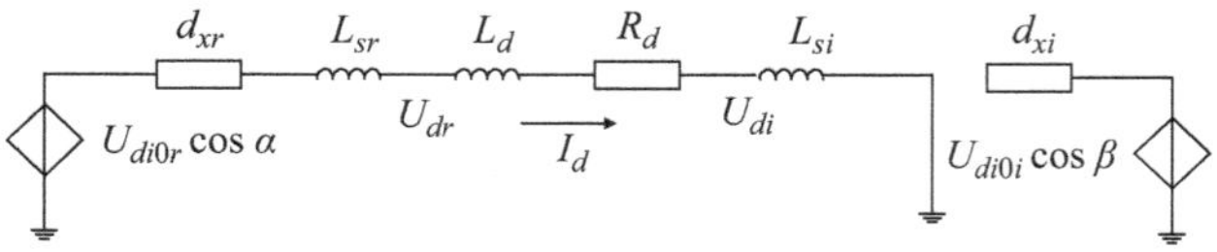

Figure 7.5 DC-side circuit model in commutation failure state

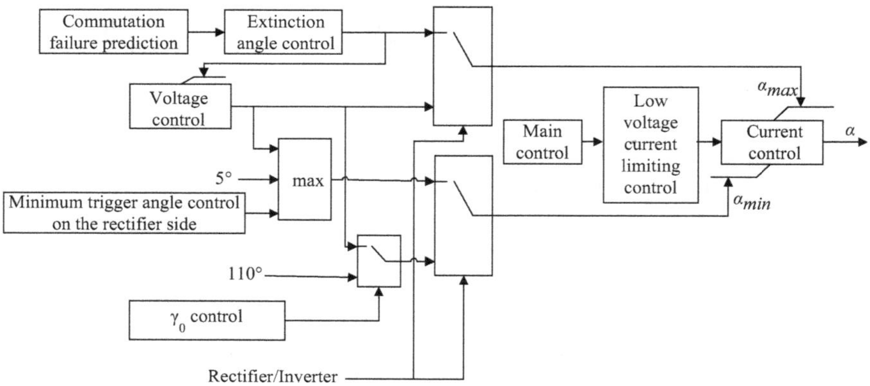

Figure 7.6 Electromechanical transient model structure of control system

change on the DC line is completely dependent on the regulation characteristics of the rectifier-side converter. The circuit equation is as follows:

$$U_{d0ir}\cos\alpha - I_d(d_{xr} + R_d) - (L_{sr} + L_d + L_{si})\frac{dI_d}{dt} = 0 \tag{7.8}$$

7.2.2 Simulation model of the control system [12,13]

The simplified HVDC control system model is shown in Figure 7.6.

The control system model includes the following control modules: main control, low-voltage current-limiting control, current control, voltage control, extinction angle control, rectifier-side minimum trigger angle control, commutation failure prediction, and γ_0 control.

The main control blocks of the control system are

- The main control link calculates the current command I_o according to the set transmission power reference value P_{ref} and the current DC voltage. If the constant current mode is selected, the current reference value is directly given.
- The low-voltage current-limiting control link limits the I_o according to the I_o and the current DC voltage and sends it into the current control link.

- In the current control link, the current command is deviated from the measured value. Considering the current margin on the inverter side, calculate the trigger angle α of the system.

The functions of other control modules are realized by the dynamic limiting of current control. The specific cooperation relationship is:

- On the rectifier side, the lower limit of the α angle is the maximum of the voltage control, the minimum trigger angle control on the rectifier side, and 5° (minimum constant trigger angle).
- On the inverter side, the upper limit of the α angle is the minimum value of the extinction angle control and the voltage control, which means that the α angle of the inverter side is actually the minimum value of the extinction angle control, voltage control, and current control.

Each module is as follows:

7.2.2.1 Main control module

The main control realizes the conversion of the power command into the current command. It integrates the main control functions of the bipolar power control and PPC in the actual controller. The control logic is shown in Figure 7.7.

In Figure 7.7:

U_d—— DC voltage;
I_d—— DC current;
P_{ref}—— Power reference value;
I_{ref}—— Current reference value;
I_o—— Current command;
U_{dfilt}—— The filtered DC voltage;
Mode—— Control mode = 0 means constant current control, while = 0 means constant power control.

The main control includes the power / current control mode selection, DC voltage filtering, current command calculation, and current command compensation function. It is the integration and simplification of the bipolar power control and the PPC part of the actual project, and only the main line of the current command calculation is retained. The power/current control mode is selected by the

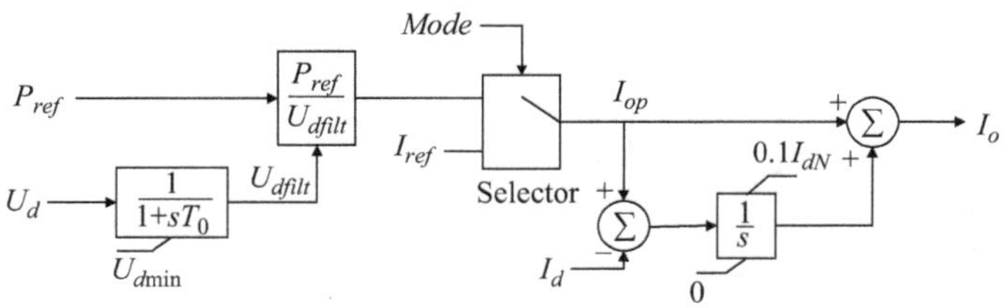

Figure 7.7 Main control logic diagram

selector to determine whether the system operates in constant power control mode or constant current control mode.

7.2.2.2 Low-voltage current-limiting control

The low-voltage current-limiting control limits the current command of the PPC according to the DC voltage level and uses the linear interpolation method to calculate the limit value of the current command, as shown in Figure 7.8.

When the DC voltage is higher than *Udhigh*, the low-voltage current-limiting control exits, that is, the current command is not limited; when the DC voltage is lower than *Udlow*, the low-voltage current-limiting control outputs *Iomin_vdcl*.

7.2.2.3 Current control

The current control mainly obtains the trigger angle according to the current command, and the logic is as shown in Figure 7.9.

In Figure 7.9:

Gain—— Current gain;
K_{p}_I—— Proportional gain;
T_{i}_I—— Integration time constant.

7.2.2.4 Voltage control

The control logic of the voltage control module is shown in Figure 7.10.

In Figure 7.10:

K_{p}_V—— Proportional gain;
T_{i}_V—— Integration time constant.

7.2.2.5 Minimum trigger angle control on the rectifier side

The logic diagram of the minimum trigger angle control module on the rectifier side is shown in Figure 7.11.

In Figure 7.11:

K1_ra—— First-stage AC bus voltage threshold;

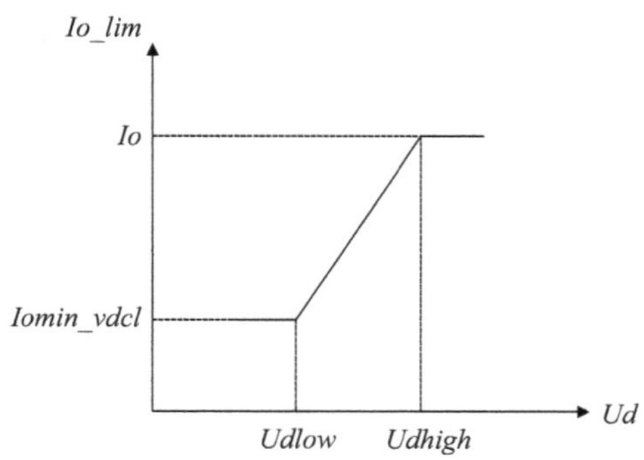

Figure 7.8 Low-voltage current-limiting link

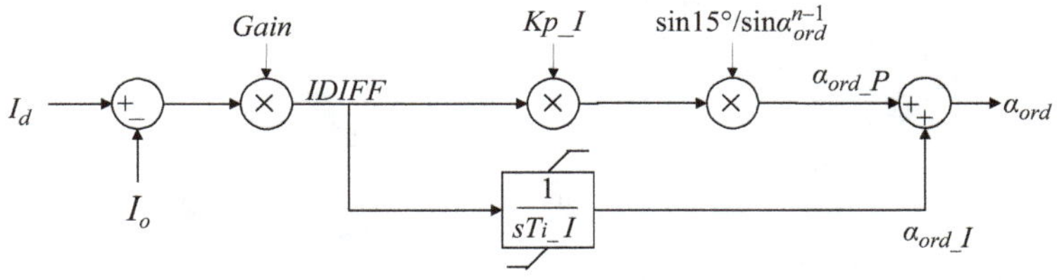

Figure 7.9 Current control module logic diagram

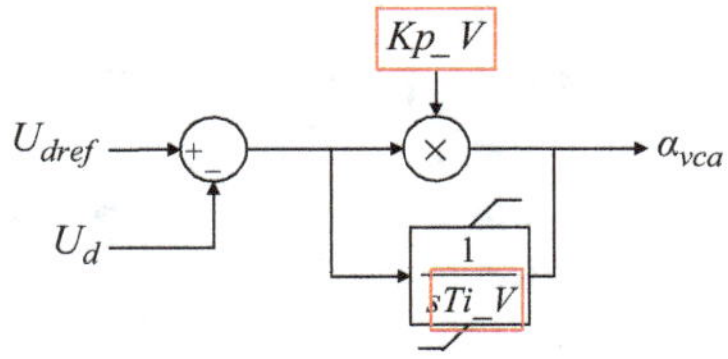

Figure 7.10 Voltage control module logic diagram

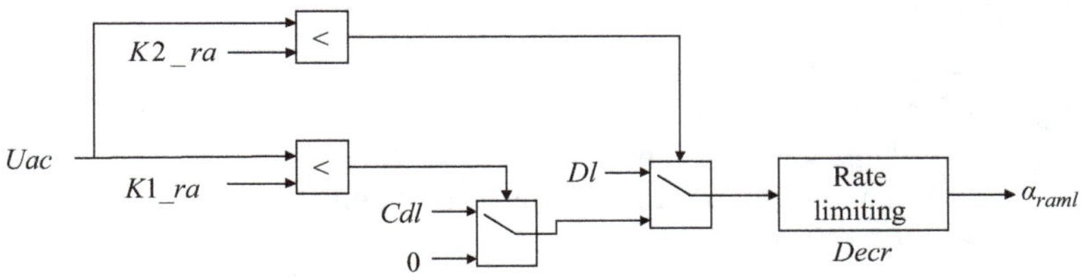

Figure 7.11 Minimum trigger angle control module logic diagram on the rectifier side

K2_ra—— Second-stage AC bus voltage threshold;
Cdl—— First-stage set angle on the rectifier side;
Dl—— Second-stage set angle on the rectifier side;
Decr—— Drop rate of output angle after the link exits.

When the AC fault occurs on the rectifier side, the output of α_{raml} is decided by the voltage amplitude. The output angle of α_{raml} is 0 in normal case. If the the voltage amplitude U_{ac} of the commutating bus is lower than the first-stage threshold *K1_ra*, the output angle of α_{raml} is *Cdl*. If the bus voltage continues to decrease to less than the second-stage threshold *K1_ra*, the output angle of α_{raml} is *Dl*. *Dl* is usually greater than *Cdl*. When the fault is cleared and the bus voltage is restored to above *K1_ra*, the output angle gradually decreases to zero according to a certain slope (related to *Decr*).

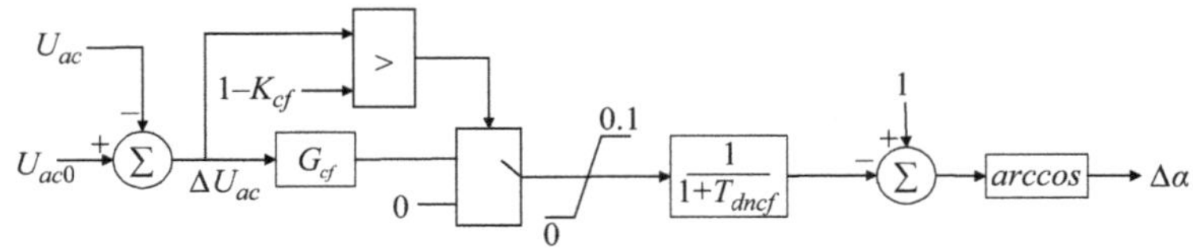

Figure 7.12 Commutation failure prediction control block diagram

7.2.2.6 Commutation failure prediction module

The purpose of the commutation failure prediction is to judge whether the commutation failure may occur according to the AC bus voltage level of the commutation bus and to realize the function of shortening the commutation failure time and reducing the risk of commutation failure. In the actual control, the three-phase voltage unbalance degree and amplitude are usually used for judgment. In the electromechanical transient model, since the instantaneous value of the three-phase AC voltage cannot be obtained, it needs to be equivalent and judged according to the RMS of the AC voltage. The control block diagram is shown in Figure 7.12.

In Figure 7.12:

U_{ac}— AC bus positive sequence voltage;
U_{ac0}— AC bus positive sequence voltage at initial steady-state;
G_{cf}— Voltage gain for commutation failure;
Kcf— Voltage threshold for commutation failure;
T_{dncf}— Filter time constant for output angle drop.

7.2.2.7 Extinction angle control module

The basic goal of extinction angle control is to control the certain extinction angle of the inverter side to prevent the occurrence of commutation failure. At present, there are two kinds of control methods: predictive control and measured control. The measured extinction angle control is to measure the extinction angle of the converter valve, forming a closed loop with the target value; the predicted extinction angle control is to calculate the acceptable maximum trigger angle of the inverter side by the physical quantities such as the current DC voltage, current, and AC voltage.

The predicted extinction angle control is shown in Figure 7.13.

In Figure 7.13:

U_{di0}— Ideal no-load DC voltage setting;
d_x— Commutation reactance setting;
α_{max}— Maximum trigger angle;
γ_{ref}— Extinction angle reference;
k— Proportional gain of current difference;
$\Delta\alpha$— Output angle of commutation failure prediction module;

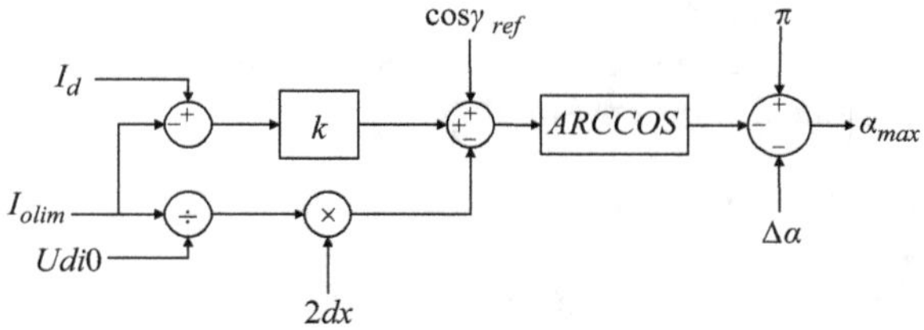

Figure 7.13 Extinction angle control block diagram

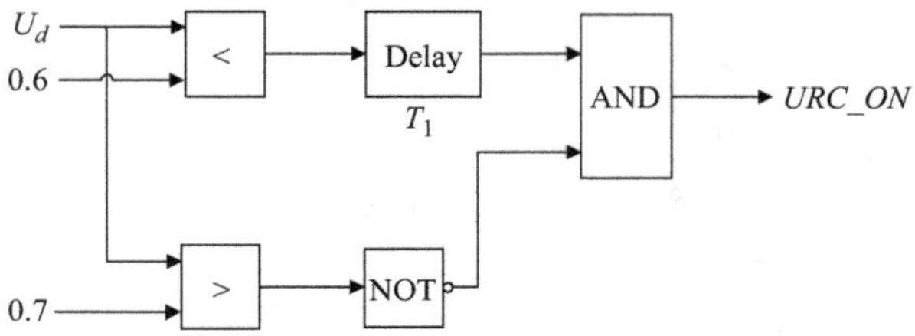

Figure 7.14 γ_0 control block diagram

I_{olim}—— Current command after low-voltage current-limiting control.

$\Delta\alpha$, the output of the commutation failure prediction, directly acts on the extinction angle control, reducing its output by phase margin. In normal operation, $\Delta\alpha = 0$. The dynamics of the slope of the volt-ampere characteristic curve is included in the extinction angle control strategy.

7.2.2.8 γ_0 control module

The logic of this control is shown in Figure 7.14.

The module realizes the accelerated recovery of voltage in the transient process. If the DC voltage is lower than 0.6 p.u. and keeps over the time T1, the link starts and the output symbol *URC_ON* is set to 1, which indicates that the trigger angle command of the inverter side is forced to be the value calculated by the extinction angle control to improve the recovery speed of the DC voltage. If the DC voltage is restored to above 0.7 p.u., the link stops. Then the trigger angle command of the inverter side is controlled according to the conventional dynamic limiting function.

7.2.3 Equivalent simulation method for commutation failure

Commutation failure of HVDC has a greater impact on the grid. In the electromechanical transient simulation, the fundamental wave and phasor are used to represent the variables. The converters are modeled by equivalent circuits on the AC and DC sides, respectively, and the switching process of the valve group cannot be calculated. Therefore, the occurrence or the finish of the commutation failure cannot be accurately predicted. However, DC commutation failure has a relatively large impact on the electromechanical transient process, which cannot be ignored in the electromechanical transient simulation.

7.2.3.1 Proposal of the equivalent trigger angle

In practical applications, a certain criterion is usually used to judge the occurrence or finishing of the commutation failure, such as the extinction angle γ, the AC voltage drop, the current. But these values are basically based on empirical values, and the accuracy is poor. When the γ angle criterion is used to judge the commutation failure, accurate calculation of γ angle is the key. The formula for calculating the γ angle is

$$\cos\,\gamma = \cos\,\beta + \frac{2d_{xi}I_d}{U_{di0i}} \tag{7.9}$$

In the above formula, β angle is the trigger leading angle, and the relationship with the trigger angle α is $\beta = 180° - \alpha$.

For the trigger angle α, the value used in the calculation is the command value calculated by the pole-control system, which can be written as αord. When the three-phase voltage waveform of the commutating bus is intact, smooth, and undistorted, the γ angle calculated by α_{ord} is accurate. However, if the three-phase voltage has a zero offset due to distortion during the short circuit, that is, the commutation voltage of valve group has a zero offset, the DC trigger system cannot adapt to this in a short time due to the trigger method of equal phase interval. If α_{ord} is still used to calculate the γ angle, a larger error will be generated, which makes the commutation failure time in simulation more different from the actual situation. Figure 7.15 shows the comparison between the electromechanical transient simulation and the electromagnetic transient results without considering the commutation voltage zero offset when the short-circuit fault on the inverter side causes the AC voltage to drop to approximately 0.6 p.u.

Figure 7.15(a) is the extinction angle curve. It can be seen that due to the zero offset of the commutation voltage, the end time of the commutation failure of the electromagnetic transient is earlier than that of the electromechanical transient. The electromagnetic transient current curve shown in Figure 7.15(b) also recovered earlier.

In order to consider this situation, the concept of the equivalent trigger angle is proposed, which means that the trigger angle obtained by the electromechanical transient is consistent with the actual recovery characteristics after the short-circuit fault occurs. Using the equivalent trigger angle α_{equ} instead of α_{ord} to calculate the γ angle, the commutation failure time can be judged more accurately, and the response characteristics of the electromechanical transient model are more accurate. The relationship between α_{equ} and α_{ord} is shown in Figure 7.16.

In Figure 7.16, the short-circuit occurs at 0.1 s. Assume that the negative zero-crossing time of the commutation voltage of valve 5 to valve 1 of C phase is t_1 (before the short circuit), and the positive zero-crossing time is t_2 (after the short-

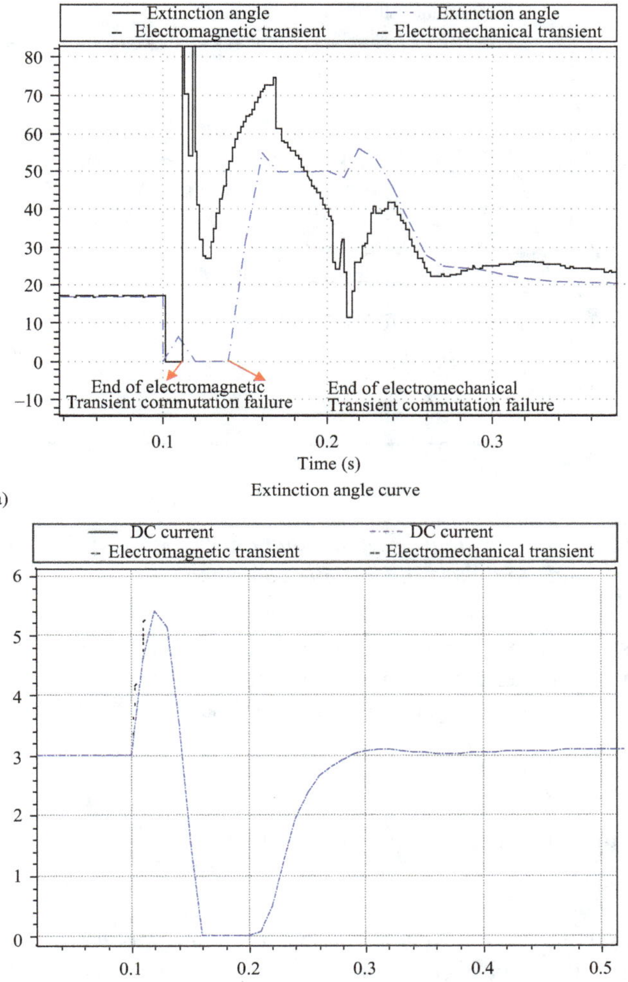

(a) Extinction angle curve

(b) DC current curve

Figure 7.15 Comparison of electromechanical transient and electromagnetic transient simulation without considering the commutation voltage zero offset

circuit). Due to the zero offset, the difference between the two times is greater than half of the cycle. The trigger time from valve 5 to valve 1 is t_f. The trigger mode of the equal phase interval adopted by the DC system starts counting on the basis of t_1 and emits a trigger angle pulse when the accumulated time reaches α_{ord}. However, due to the offset of the positive zero-crossing, valve 5 has a longer arc-extinguishing time after the end of the commutation. The trigger system cannot

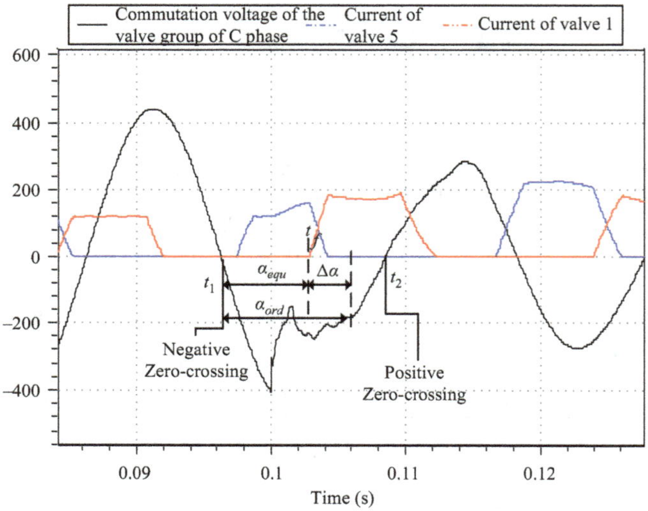

Figure 7.16 Schematic diagram of the relationship between the equivalent trigger angle and the trigger angle command

predict this, so in fact, an early trigger occurs, which can be equivalent to a reduction of the trigger angle. If the angle between the two zero-crossing points is kept at π, the equivalent trigger angle can be described as

$$\alpha_{equ} = \frac{t_f - t_1}{t_2 - t_1}\pi \tag{7.10}$$

Define the difference between the two angles as

$$\Delta\alpha = \alpha_{ord} - \alpha_{equ} \tag{7.11}$$

As long as the $\Delta\alpha$ value can be known, α_{equ} can be obtained. Before the fault, $\Delta\alpha = 0$; after the fault occurs, $\Delta\alpha$ is a quantity that changes with time. As the system returns to the steady state, $\Delta\alpha$ will gradually return to zero.

7.2.3.2 The relationship between equivalent trigger angle and trigger angle command deviation and voltage drop level

Based on the electromagnetic transient DC transmission system, the three-phase short-circuit test on the inverter side is performed. The short-circuit time is set to 0.1 s, and different short-circuit transition resistances are adopted to make the commutation bus voltage drop in different ranges, simulating AC faults of different severity. According to the simulation data, the curve of $\Delta\alpha$ with time under different voltage drop levels is shown in Figure 7.17.

In Figure 7.17, the abscissa is the times n of triggers and the ordinate is $\Delta\alpha$. It can be seen that, in general, $\Delta\alpha$ tends to increase first and then decrease with time.

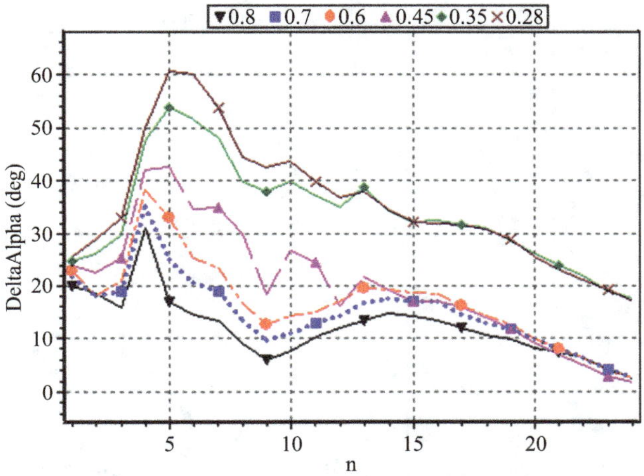

Figure 7.17 Variation trend of equivalent trigger angle and trigger angle command deviation with voltage drop level

$N = 1$ indicates the first trigger pulse after a short-circuit, and the initial value of $\Delta\alpha$ is $\Delta\alpha(1)$. The following conclusions can be drawn:

1. The initial value $\Delta\alpha(1)$ increases as the voltage drop increases, however, the difference is close to zero when n is between 20 and 25.
2. The peak value of $\Delta\alpha$ is related to the voltage drop level. The more serious the fault, the larger the voltage drop, and the higher the peak value of $\Delta\alpha$.
3. The occurrence moment of $\Delta\alpha$ peak also increases with the increase of voltage drop amplitude, but it is equivalent in general, and the difference between peak moments of different voltage levels does not exceed two trigger cycles.
4. The rate of falling from the peak is roughly the same under different voltage drop levels.
5. Fitting method of equivalent trigger angle and trigger angle command deviation based on statistical results.

According to the above rules, the change of $\Delta\alpha$ can be fitted by two straight lines, as shown in Figure 7.18.

When the $\Delta\alpha$ is fitted by the piecewise method, the formula is

$$\Delta\alpha = \begin{cases} \Delta\alpha(1) + \dfrac{\Delta\alpha_{\max} - \Delta\alpha(1)}{t_{\max} - 3.33}(t - 3.33) & t < t_{\max} \\ \Delta\alpha_{\max} + k_1(t - t_{\max}) & t \geq t_{\max} \end{cases} \tag{7.12}$$

The variables that need to be determined include

- $\Delta\alpha(1)$—— Initial value of $\Delta\alpha$;
- $t_{\max}$—— Peak time;

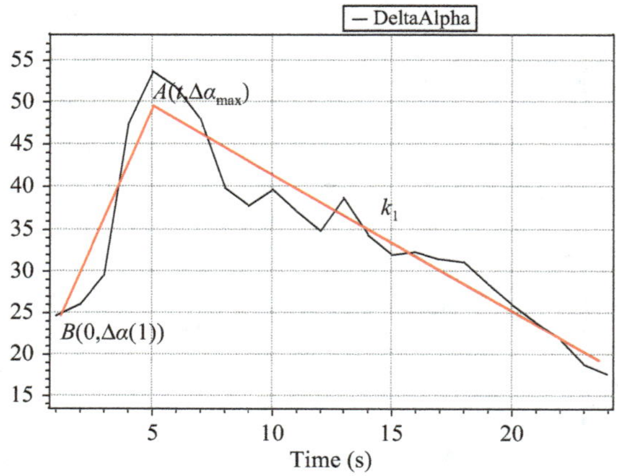

Figure 7.18 Piecewise linear fitting of equivalent trigger angle and trigger angle command deviation

- $\Delta\alpha_{max}$—— The peak of the fitting;
- k_1—— Falling slope.

1. According to the previous analysis, the initial values are not much different, so $\Delta\alpha(1)$ can take a fixed value, where $\Delta\alpha(1) = 23°$.
2. According to Figure 7.16, it can be seen that the peak time t of $\Delta\alpha$ under different voltage drop levels is not much different. The trigger interval is set to $10/3 = 3.33$ ms, and the relationship between the t_{max} and the voltage level U is as follows:

$$t_{max} = \begin{cases} \dfrac{40}{3}\text{ms} & U \geq 0.4 \text{ pu} \\ \dfrac{40}{3}\text{ms} & U < 0.4 \text{ pu} \end{cases} \tag{7.13}$$

3. The relationship between $\Delta\alpha_{max}$ and U is shown in Figure 7.19.

   The following equation can be obtained by approximately fitting the $\Delta\alpha_{max}$ with a straight line.

$$\Delta\alpha_{max} = 82 - 67.3U \tag{7.14}$$

4. As can be seen from Figure 7.17, k_1 can take a fixed value, where $k_1 = -0.5$ (degrees / ms).

In summary, $\Delta\alpha$ can be calculated according to the above parameters, and the equivalent trigger angle α_{equ} can be obtained further. The γ angle is then calculated from α_{equ}, and the simulation results are shown in Figure 7.20.

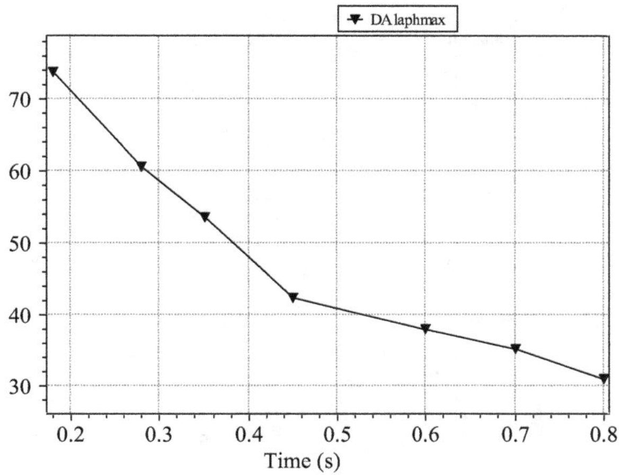

Figure 7.19 Δα peak varies with voltage level

It can be seen that when the equivalent trigger angle is used, the commutation failure time can be judged more accurately, so as to improve the accuracy of simulation.

7.2.4 Parameter identification principle of actual DC engineering model

The electromechanical transient modeling of the HVDC control system is consisted of a group of modules. There are about 18 key parameters from five modules in the control blocks, as shown in Table 7.1.

Some parameters in the model can be found directly in the actual controller, such as the proportional and integral (PI) parameter of the current control block. But, for some parameters, if the design parameters in the actual controller are directly used, it could not reach the best pertinent simulation results for the following reasons:

1. The primary system model of an electromechanical transient is different from the actual system, which is mainly reflected in the quasi-steady model of the DC converter. The DC line uses a series model of resistors and inductors, not the most detailed model. Therefore, the electrical quantity obtained by the simulation, such as DC current and voltage, must have a deviation from the actual system. If the design parameters are directly adopted, the deviation cannot be compensated, not the mathematical optimal value.
2. Although the main control logic of the electromechanical transient control system is consistent with the actual controller, the control details are approximated or ignored. Therefore, the design parameters can be corrected to achieve better simulation results.

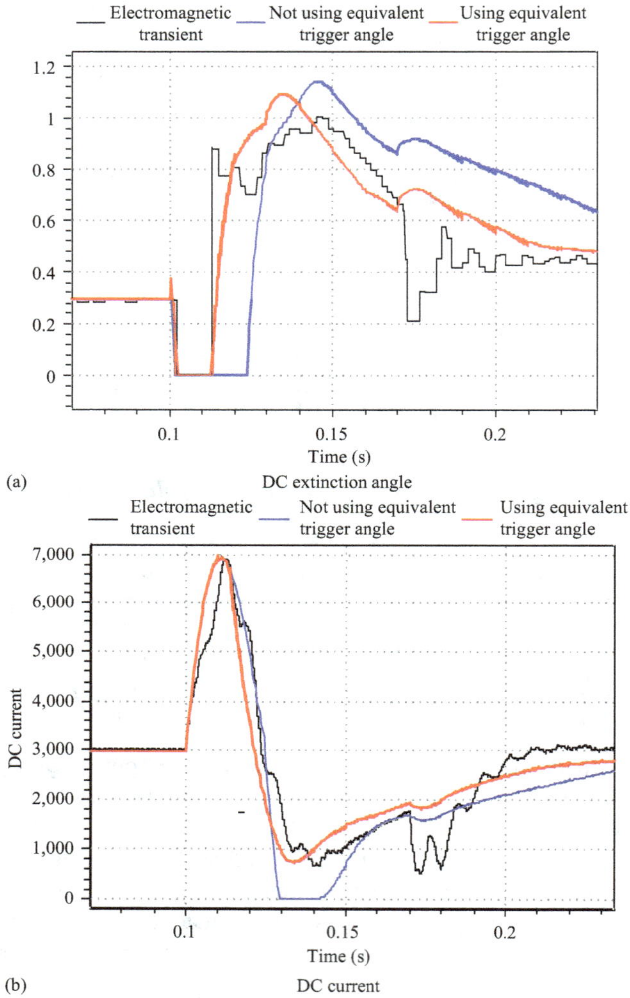

Figure 7.20 The comparison between whether the equivalent trigger angle is used to calculate the commutation failure

Some parameters in the model could be equivalent parameters, such as the relevant parameters of the commutation failure prediction module and the minimum trigger angle control module on the rectifier side. Since these modules are equivalent when modeling, the calculation parameters cannot be found in the actual controller.

Therefore, it is necessary to identify the parameters of the electromechanical transient simplified model based on the actual DC detailed control system [13]. The basic ideas are

Table 7.1 Model parameters list

Control blocks	Parameters
1. Current control	1. *Gain* 2. *Kp_I* 3. *Ti_I*
2. Voltage control	4. *Kp_V* 5. *Ti_V*
3. Minimum α control on the rectifier side	6. *K1_ra* 7. *K2_ra* 8. *Cdl* 9. *Dl* 10. *Decr*
4. Commutation failure prediction	11. *K_cf* 12. *G_cf* 13. *Tdn_cf*
5. Low-voltage current-limiting control	14. *Udlow* 15. *Udhigh* 16. *Tup* 17. *Tdn* 18. *Iomin*

Table 7.2 The corresponding relationship between the control link and the test project

Control block	**Test project**
1. Current control	Current step
2. Voltage control	Voltage step
3. Minimum trigger control on the rectifier side	Three-phase short circuit of commutation bus on the rectifier side
4. Commutation failure prediction 5. Low-voltage current-limiting control	Three-phase short circuit of commutation bus on the inverter side

- Establish an electromagnetic transient test system. The DC model uses the manufacturer's detailed model.
- Carry out corresponding simulations. Fitting the parameters of electromechanical transient model based on wave recording curves.

Due to a large number of parameters, it is necessary to follow the principle of sub-section identification. It is to decouple the control system and design the corresponding relationship between the control link and the test project. All parameters of a control link are obtained through an experiment. After obtaining the parameters through the test, the parameters of the model are compared with the curves of the DC engineering integrated test and the system commissioning to check the model parameters further. The corresponding relationship between the test project and the control link is shown in Table 7.2.

References

[1] Dommel H.W. *EMTP Theory Book.* Vancouver: Microtran Power System Analysis Corporation; 1996.

[2] Watson N., and Arrillage J. *Power Systems Electromagnetic Transient Simulation*. London: The Institute of Engineering Technology; 2007.

[3] Kundur P. *Power System Stability and Control.* New York: McGraw-Hill Inc.; 1994.

[4] Manitoba HVDC Research Centre (Canada). *PSCAD User Guide*. Manitoba: Manitoba HVDC Research Centre; 2003.

[5] Mahseredjian J., Lefebvre S., and Mukhedkar D. 'Power converter simulation module connected to the EMTP'. *Journal of IEEE Transactions on Power Systems*. 1991;6(2): 501–510.

[6] Goldsworthy D., and Vithayarhil J. 'EMTP model of an HVDC System'. *Proceedings of IEEE Montech 86 Conference on HVDC Power Transmission*; Montreal, Canada, Sep 1986. IEEE; pp. 39–46.

[7] Zhen R., Kaijian O., Yong J., and Jun Y. 'Digital simulation of HVDC transmission system based on PSCAD/EMTDC'. *Journal of Electric Power Automation Equipment*. 2002;22(9):11–12.

[8] Zhiling H., and Jie T. 'Simulation with EMTDC based on the detailed HVDC control system model'. *Journal of Automation of Electric Power Systems*. 2008;32(2):45–48.

[9] Karlsson J. 'Simplified control model for HVDC classic'. Stockholm: Royal Institute of Technology; 2006.

[10] Ni Y.X. 'A simplified two-terminal HVDC model and its use in direct transient stability assessment'. *Journal of IEEE Transactions on Power Systems*. 1987;2(4):1006–1012.

[11] Xinli S., Xiaochen W., Wenzhuo L., Aidong X., Guangquan B., and Xiaoming J. 'New quasi-steady-state HVDC models for PSD-BPA power system transient stability simulation program'. *Journal of Power System Technology*. 2010;34(1):62–67.

[12] Lei W., Hui D., and Wenzhuo L. 'Simulation model of control system for HVDC power transmission based on actual project'. *Journal of Power System Technology*. 2013;37(3): 629–634.

[13] Lei W., Yong T., Wenchuan W., and Tiezhu W. 'Equivalent modeling and real parameter measurement methods of control systems of UHVDC transmission systems'. *Journal of Power System Technology*. 2017;41(3):708–714.

Chapter 8

Electromechanical transient simulation of VSC HVDC

Junxian Hou[1], *Lei Wan*[1] *and Jian Zhang*[1]

8.1 General introduction of electromechanical transient simulation for VSC HVDC

Voltage source converter based high-voltage direct current (VSC HVDC) has been developed rapidly all over the world [1–3], and several VSC HVDC projects have been commissioned in China, in which the most complex project is 500 kV and four terminals. VSC HVDC is usually privileged to transmit renewable energy power [4,5] or supply power to the load center and can be effectively used for power system stability control.

Be the same with line commutated converter based high-voltage direct current (LCC HVDC) electromechanical model in Chapter 7, the VSC HVDC also requires the building of the electromechanical simulation model for a large scale of power system with VSC HVDC. Because the main purpose is to simulate the interaction between VSC HVDC and the power system, the external characteristics of VSC HVDC should be kept as accurate as possible while the internal characteristics can be simplified. A quasi-steady-state model will also be employed for the electromechanical simulation of VSC HVDC, which includes the main converter circuit, DC line circuit and converter control strategy.

8.2 Electromechanical transient simulation of VSC HVDC

The electromechanical transient model of VSC HVDC mainly includes the VSC circuit model, the DC network model, the control system model and simulation methods of DC faults.

8.2.1 VSC circuit electromechanical transient model

Similar to the conventional LCC HVDC, the electromechanical transient model of VSC is also an external equivalent model. As the physical variables are

[1]Electric Power Research Institute, Beijing, China

demonstrated by the three-phase fundamental phasors (positive, negative and zero sequences), the three-phase voltages and currents of AC interface nodes are not available; therefore, the topologies of the internal valve groups and the switching progress and trigger control will be ignored during the modeling.

8.2.1.1 The AC-side model of VSC

On the AC side, the VSC is considered as an AC voltage source whose magnitude and phase angle are controllable. This source is connected to the AC system through a series of the transformer–inductor circuit. The reactor called 'commutation reactor' could supply access impedance for the injecting voltage of VSC and be used as a filter at the same time. For lower levels of VSC, such as two-level converter or three-level converter, AC filters should be installed on the secondary side of the connection transformer. For multilevel VSC (e.g. modular multilevel converter (MMC)), filters are not necessary in general. Therefore, the AC-side circuit has two typical forms: with filters and without filters.

- No parallel AC filters

The circuit configuration is shown in Figure 8.1.

The total resistance and inductance combined with the connection transformer and phase inductor are R and L, respectively. $\dot{U}_s$ is the node voltage phasor with its angle to be zero. $\dot{U}_c$ is the injecting voltage phasor with its angle lagging $\dot{U}_s$ to be δ. $\dot{I}_s$ is branch current phasor. Assume the power flow direction from the AC side to the DC side is positive, and the active power and reactive power of the interface node are P_s and Q_s, the active and reactive power of the node at the converter side are P_c and Q_c.

Writing down the branch equation between $\dot{U}_s$ and $\dot{U}_c$:

$$\dot{U}_c = \dot{U}_s - (R + j\omega L)\dot{I}_s \tag{8.1}$$

Map each phasor to the synchronous dq0 coordinate system. Equation (8.1) can be rewritten as:

$$u_{cd} + ju_{cq} = u_{sd} + ju_{sq} - (R + j\omega L)(i_{sd} + ji_{sq}) \tag{8.2}$$

Assembled:

$$\begin{cases} u_{cd} = u_{sd} - Ri_{sd} + \omega Li_{sq} \\ u_{cq} = u_{sq} - Ri_{sq} - \omega Li_{sd} \end{cases} \tag{8.3}$$

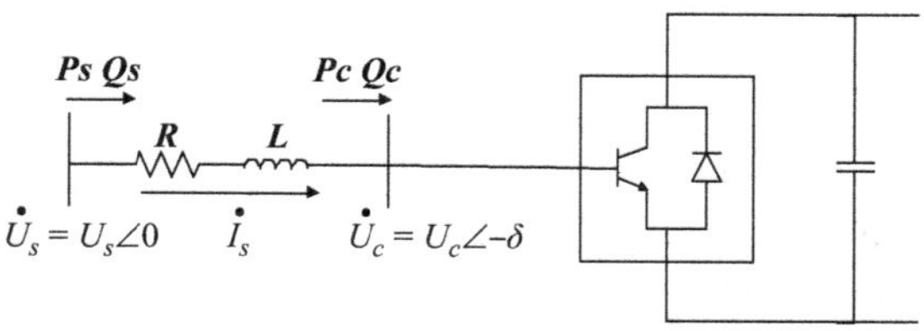

Figure 8.1 Circuit configuration of VSC without AC filters

Equation (8.16) presented the characteristics of the AC branch in a steady state.

- With parallel AC filters

The circuit is shown in Figure 8.2.

Assume that the secondary node voltage of the connection transformer is $\dot{U}_{s1}$, the branch current of the transformer is $\dot{I}_t$, the branch current of phase reactance is $\dot{I}_s$, and the equivalent fundamental susceptance of filters is B_c. Ignore the excitation branch, and use a resistor and an inductor in series instead of a transformer.

The branch equation set is given:

$$\begin{cases} \dot{U}_s - (R_t + jX_t)\dot{I}_t = \dot{U}_{s1} \\ \dot{U}_{s1} - (R_c + jX_c)\dot{I}_s = \dot{U}_c \\ \dot{I}_s = \dot{I}_t - \dot{U}_{s1} jB_c \end{cases} \tag{8.4}$$

where $X_t = \omega L_t$, $X_c = \omega L_c$.

Eliminate $\dot{I}_t$ and $\dot{U}_s$ in (8.4), and (8.5) can be obtained:

$$\begin{aligned} &\dot{U}_s - [(R_t + R_c - R_t X_c B_c - R_c X_t B_c) + j(X_t + X_c - X_t X_c B_c + R_t R_c B_c)]\dot{I}_s \\ &= (1 - X_t B_c + jR_t B_c)\dot{U}_c \end{aligned} \tag{8.5}$$

Transfer, (8.5) into dq-axis:

$$\begin{cases} U_{sd} - (R_t + R_c - R_t X_c B_c - R_c X_t B_c)I_{sd} + (X_t + X_c - X_t X_c B_c + R_t R_c B_c)I_{sq} \\ = (1 - X_t B_c)U_{cd} - R_t B_c U_{cq} \\ U_{sq} - (X_t + X_c - X_t X_c B_c + R_t R_c B_c)I_{sd} - (R_t + R_c - R_t X_c B_c - R_c X_t B_c)I_{sq} \\ = R_t B_c U_{cd} + (1 - X_t B_c)U_{cq} \end{cases} \tag{8.6}$$

Therefore, after AC filters are considered, the AC branch equations are more complicated than without filters as there are more coupling elements under dq-axis. As the resistance is small, the product of resistance, inductor and susceptance is

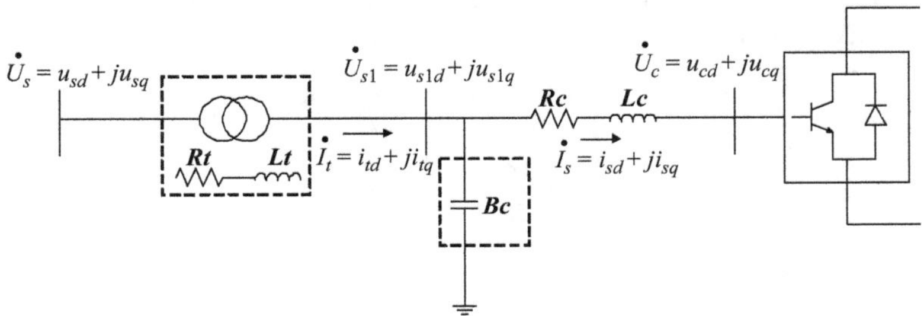

Figure 8.2 Circuit configuration of VSC with AC filters

negligible, and (8.6) can be simplified as follows:

$$\begin{cases} u_{sd} - (R_t + R_c)i_{sd} + (X_t + X_c)i_{sq} = (1 - X_tB_c)u_{cd} \\ u_{sq} - (R_t + R_c)i_{sq} - (X_t + X_c)i_{sd} = (1 - X_tB_c)u_{cq} \end{cases} \tag{8.7}$$

Comparing (8.6) with (8.3), we can see that the u_c term has a multiplier factor 1-X_tB_c in the steady-state equation when there is an AC filter in the circuit, which reflects the effect of the filter on the characteristics of the branch. When the filtering capacity is small (that is, B_c is small), this factor can be considered to be close to 1, which is completely consistent with the steady-state characteristics of the non-filter case.

8.2.1.2 The DC-side model of VSC

The AC phase current is decomposed into the upper bridge arm current and the lower bridge arm current under the action of the valve group switching modulation. After the three-phase synthesis, it is injected into the DC side. Therefore, VSC in the DC side is equivalent to a controlled current source. According to KCL, the magnitude and direction of currents flowing into the positive and negative terminals of the converter must be equal and opposite. Therefore, the converter is represented by an equivalent current source and its two ends are directly connected to the positive and negative bus, as shown in Figure 8.3.

The structure in Figure 8.3 indicates that the DC-side concentrated capacitor is connected to the positive and negative outlets, such as VSC in the two- or three-level structure. For the voltage source converter of the MMC structure, DC capacitors are connected in parallel in each submodule, that is, the capacitor is distributed. When modeling as the valve group structure is ignored, an equivalent concentrated capacitor in the DC side can be used to simulate the distributed capacitor. Therefore, the DC-side model of MMC can still be shown in Figure 8.3, and there is a certain relationship function between the size of the equivalent capacitor and each module capacitor. The two cases are discussed below.

- Concentrated capacitor

  Let the current of the equivalent current source injects into the positive pole is i_d, the capacitor current is i_c (the positive direction is from positive pole to negative pole), the current behind the capacitor (namely injecting into DC

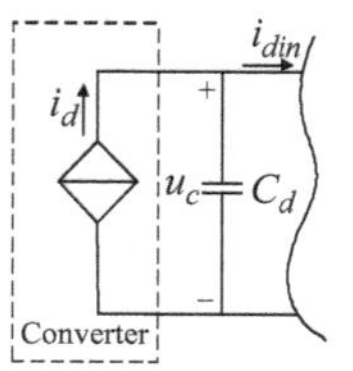

Figure 8.3 Equivalent model in DC side of VSC

network) is i_{din}, the capacitor voltage is u_c, and the capacitor capacity is C_d. i_{din} can be given:

$$i_{din} = i_d - C_d \frac{du_c}{dt} \tag{8.8}$$

This equation represents the DC-side model of the concentrated capacitor accessed converter.

- Distributed capacitor (MMC)

  The topology of MMC [6,7] is shown in Figure 8.4. Ignoring redundant submodules, assume that the number of one bridge arm is n, the capacitor of each submodule is C_0. u_1 is the upper bridge arm voltage, and u_2 is the lower bridge arm voltage.

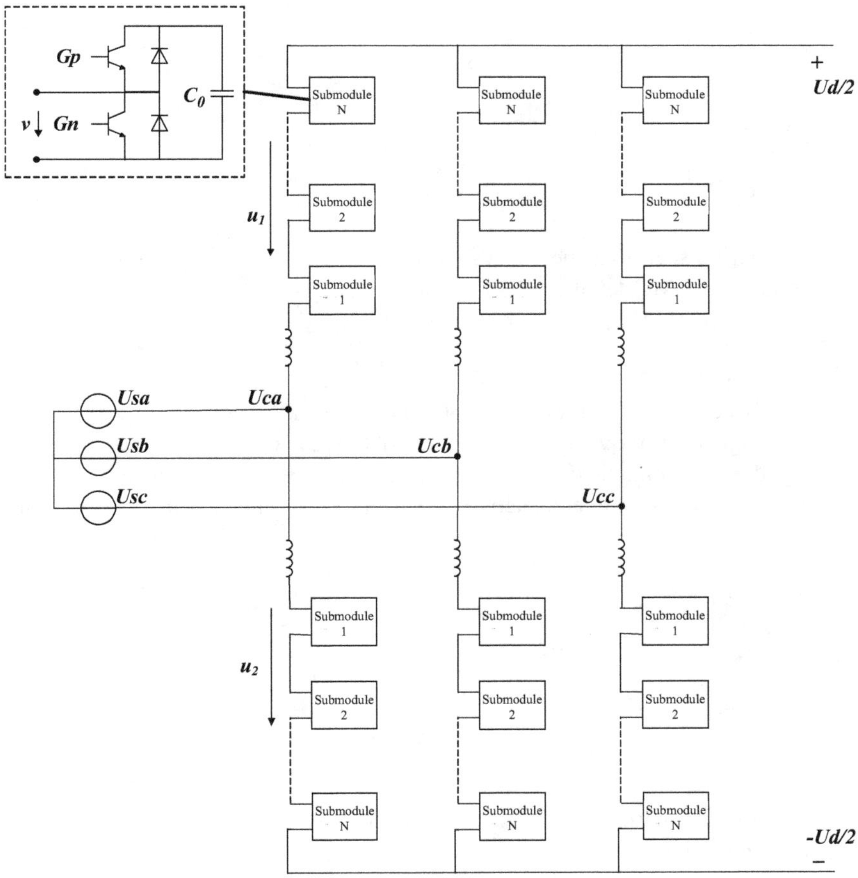

Figure 8.4 Topology of MMC

So, the DC voltage can be shown as

$$u_d = u_1 + u_2 \tag{8.9}$$

If MMC uses the fundamental modulation, then the number of insertion submodules of one phase at any time is n. And if the periodic fluctuation of submodule voltage is ignored and only the DC component (namely, the average value) is counted, then the capacitor voltage of each submodule is u_d/n. Assume that the MMC operates symmetrically and all submodules have equal capacitor voltages. Let the equivalent concentrated capacitor size be C_d', then this requirement must be fulfilled no matter before or after equivalent: when the change of the DC voltage is the same, the total energy change stored in distributed capacitors is the same as the energy change of the concentrated capacitor. According to this principle, when the DC voltage changes by Δu:

$$\frac{1}{2}C_0\left(\frac{u_d}{n}+\frac{\Delta u}{n}\right)^2 \times 6n = \frac{1}{2}C_d'(u_d+\Delta u)^2 \tag{8.10}$$

Equation (8.11) can be obtained:

$$C_d' = \frac{6}{n}C_0 \tag{8.11}$$

It can be seen that the equivalent capacitor is inversely proportional to the number of MMC levels. By substituting (8.11) into (8.8), the equivalent DC-side model of MMC can be obtained.

8.2.1.3 Power balance equation

In the equivalent DC-side model of VSC, the value of the controlled current source embodies injection active power from the AC side to the DC network. Ideally, ignoring the switching loss of the converter valve group, by the law of energy conservation, the instantaneous active power of the AC side and the DC side should be equal:

$$P_{ac} = P_d \tag{8.12}$$

where

$$P_{ac} = u_{cd}i_{sd} + u_{cq}i_{sq} \tag{8.13}$$

$$P_d = u_d i_d \tag{8.14}$$

The power equation (8.12) is the connection of AC-side model and DC-side model of converter.

Among the above models, the values in AC side are represented using per-unit values, and these in DC side are using actual values.

8.2.2 DC-side network model

In voltage source converter based multi-terminal direct current (VSC MTDC), the DC side consists of DC transmission lines (cables or overhead lines) to form a network with a certain topology. Unlike the AC grid, the physical quantities such as voltages and currents of the DC-side network are DC, which changes with the AC power during the transient process. Therefore, these physical quantities are represented by instantaneous values, which are in accordance with electromagnetic transient modeling. The general idea is to discretize the differential equation of the line model to form a circuit in which the current source and the admittance are connected in parallel. The current source reflects the calculation history in the line model, and the admittance reflects the line parameters. Set the differential equations of each circuit line to form the total node voltage equation. The admittance array is composed of the relevant line admittances, and the terms at the right end are composed of the node injection currents and the history terms of the lines connected to the node. The node voltage of the DC network can be obtained by solving the equation.

The line model is replaced by a single PI model, as shown in Figure 8.5.

Where R_l is the line equivalent resistance, L_l is the line equivalent inductance and C_l is the line equivalent capacitance. If considered the distribution characteristics, the resistance, inductance and capacitance of the line are r_l, l_l and c_l, respectively. Assume the total length of the line is l, then

$$\begin{cases} R_l = r_l l \\ L_l = l_l l \\ C_l = c_l l \end{cases} \tag{8.15}$$

The differential equation of the series branch in line model is as follows:

$$R_l i + L_l \frac{di}{dt} = u_i - u_j \tag{8.16}$$

After differentiating the equation (the specific form after the differentiation is related to the simulation method used, which is not exactly the same, but the process and the overall structure are similar):

$$i(t) = \frac{\Delta t[u_i(t) - u_j(t)] + L_l i(t - \Delta t)}{R_l \Delta t + L_l} \tag{8.17}$$

Its circuit form is shown in Figure 8.6.

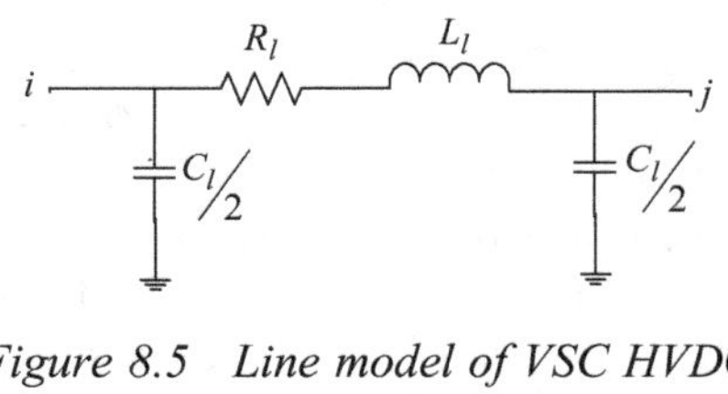

Figure 8.5 Line model of VSC HVDC

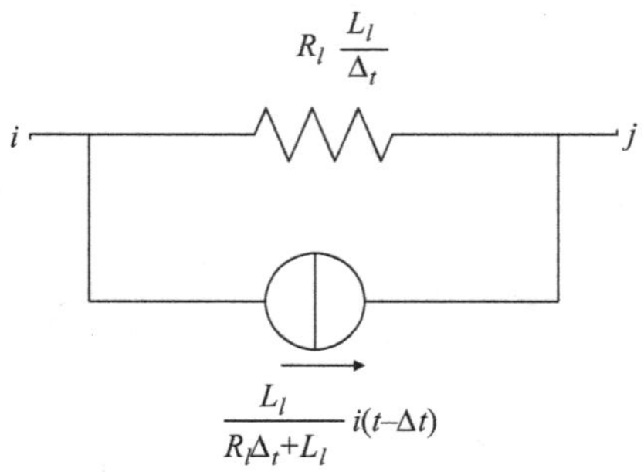

Figure 8.6 Difference circuit model of series impedance branch in DC line

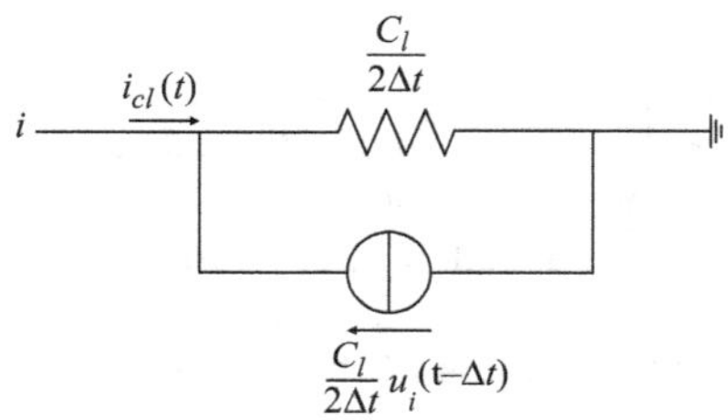

Figure 8.7 Difference circuit model of parallel capacitor branch in DC line

The differential equation of the parallel capacitor is as follows:

$$i_{cl} = \frac{C_l}{2}\frac{du_i}{dt} \tag{8.18}$$

After differentiating the equation:

$$i_{cl}(t) = \frac{C_l}{2\Delta t}u_i(t) - \frac{C_l}{2\Delta t}u_i(t-\Delta t) \tag{8.19}$$

Its circuit form is shown in Figure 8.7.

The DC-side capacitors of each converter are merged into the DC network, and the differential and difference equations are

$$i_{cd} = C_d\frac{d(u_i - u_j)}{dt} \tag{8.20}$$

$$i_{cd}(t) = \frac{C_d}{\Delta t}[u_i(t) - u_j(t)] - \frac{C_d}{\Delta t}[u_i(t-\Delta t) - u_j(t-\Delta t)] \tag{8.21}$$

Discuss any form of DC network with intermediate nodes. Let the DC network contains n converters and m intermediate nodes. Consider the positive and negative poles, and the total number of nodes is $2(n+m)$. The positive nodes of the n converters are sequentially numbered as 0, 1, ..., n-1; the number of the intermediate positive node is n, $n+1$, $n+m-1$; and the corresponding negative nodes of the converters are sequentially numbered as $n+m$, $n+m+1$, ..., $2(n+m)-1$.

Combine (8.17), (8.19) and (8.21) to obtain the node voltage equation in matrix form:

$$Gu(t) = i(t) + hist(t - \Delta t) \tag{8.22}$$

where G is the total admittance array of the network, $i(t)$ is the node injection current and $hist(t\text{-}\Delta t)$ is the node current history term.

8.2.2.1 Admittance matrix

According to the above numbering rules, the structure of the array G is

$$G = \begin{bmatrix} G_{RL} + G_{Cl} & -G_{Cd} \\ -G_{Cd} & G_{RL} + G_{Cl} \end{bmatrix} \tag{8.23}$$

In the above formula, G_{RL} is the total admittance matrix of the impedance branch, G_{Cl} is the total admittance matrix of the capacitance branch and G_{Cd} is the total admittance matrix of the capacitance of the converter in DC side.

The non-diagonal elements of G_{RL} is

$$G_{RL}[i][j] = \begin{cases} -\dfrac{1}{R_{ij} + 2L_{ij}/\Delta t} & i,j \text{ have a line connection} \\ 0 & i,j \text{ have no line connection} \end{cases} \tag{8.24}$$

The diagonal elements of G_{RL} are

$$G_{RL}[i][i] = -\sum_{\substack{j=0 \\ j \neq i}}^{n+m-1} G_{RL}[i][j] \tag{8.25}$$

G_{Cl} is a diagonal matrix:

$$G_{Cl} = \begin{bmatrix} \sum_{j=0}^{n+m-1} C_{l0j}/\Delta t & & & \\ & \sum_{j=0}^{n+m-1} C_{l1j}/\Delta t & & \\ & & \ddots & \\ & & & \sum_{j=0}^{n+m-1} C_{l(n+m-1)j}/\Delta t \end{bmatrix} \tag{8.26}$$

G_{Cd} is a partitioned matrix:

$$G_{Cd}=\left[\begin{array}{cccc|c} 2C_{d0(n+m)}/\Delta t & & & & \\ & 2C_{d1(n+m+1)}/\Delta t & & & 0 \\ & & \ddots & & \\ & & & 2C_{d(n-1)(2n+m-1)}/\Delta t & \\ \hline & & 0 & & 0 \end{array}\right] \tag{8.27}$$

In formula (8.27), the partitioned part of 0 represents the mutual admittance between the intermediate nodes and other nodes of the other pole. Since the mutual admittance of intermediate nodes is completely reflected in G_{RL}, the other terms are all zero.

8.2.2.2 Injection current

For converter nodes, the injection current is the power exchange between AC and DC, which is obtained according to (8.14). For intermediate nodes, the injection current is always equal to zero.

8.2.2.3 History terms of injection current

The history terms of injection current are divided into the history terms of the series impedance part and the parallel capacitance part of the transmission line, and the history terms of the capacitance in the DC side. For the converter node i of the positive pole, the total history term can be obtained by (8.28):

$$\begin{aligned} hist_i(t-\Delta t) = & -\sum_{\substack{j=0 \\ ij\ Connect}}^{n+m-1} \frac{u_{ij}(t-\Delta t)+(2L_{ij}/\Delta t-R_{ij})i_{ij}(t-\Delta t)}{R_{ij}+2L_{ij}/\Delta t} \\ & +\sum_{\substack{j=0 \\ ij\ Connect}}^{n+m-1} \frac{C_{lij}}{\Delta t}u_i(t-\Delta t)+i_{clij}(t-\Delta t) \\ & +\frac{2C_{di(i+n+m)}}{\Delta t}[u_{i(i+n+m)}(t-\Delta t)]+i_{cdi(i+n+m)}(t-\Delta t) \end{aligned} \tag{8.28}$$

$$\begin{aligned} hist_i(t-\Delta t) = & -\sum_{\substack{j=0 \\ ij\ Connect}}^{n+m-1} \frac{u_{ij}(t-\Delta t)+(2L_{ij}/\Delta t-R_{ij})i_{ij}(t-\Delta t)}{R_{ij}+2L_{ij}/\Delta t} \\ & +\sum_{\substack{j=0 \\ ij\ Connect}}^{n+m-1} \frac{C_{lij}}{\Delta t}u_i(t-\Delta t)+i_{clij}(t-\Delta t) \end{aligned} \tag{8.29}$$

For the intermediate node i of the positive pole, the total history item does not include the capacitance in the DC side, as shown in (8.29).

For negative nodes, the history items have the same form as the positive nodes, and will not be described here.

After obtaining the above-mentioned admittance matrices, injection current and history terms, (8.22) can be solved to obtain the current DC node voltage. The calculation step Δt can generally be taken as a fraction of the calculation step of the electromechanical transient program. The smaller the step size, the more accurate the calculation result, but the more time it takes.

8.2.3 Control strategies of VSC HVDC

The control system [8–10] of the converter in the electromechanical transient model refers to the local control of the converter, excluding the modulation of valve groups, and does not include coordinated control between stations. The overall structure of the control system model is shown in Figure 8.8.

The control strategy is divided into two levels: system level and converter level. The latter includes the outer control loop and the inner control loop. Modern large-capacity VSCs widely adopt direct current control based on active and reactive power (PQ) decoupling, so the total control structure can be divided into two parts: active and reactive. In the active part, the system level is frequency control of the AC system, and the outer control loop is active power control, DC voltage control or DC voltage margin control. In the reactive part, the system level is AC voltage control, and the outer control loop is reactive power control or AC voltage limiting control.

As can be seen from Figure 8.8, the frequency control will generate an active power command value to the active power control. Set a control signal

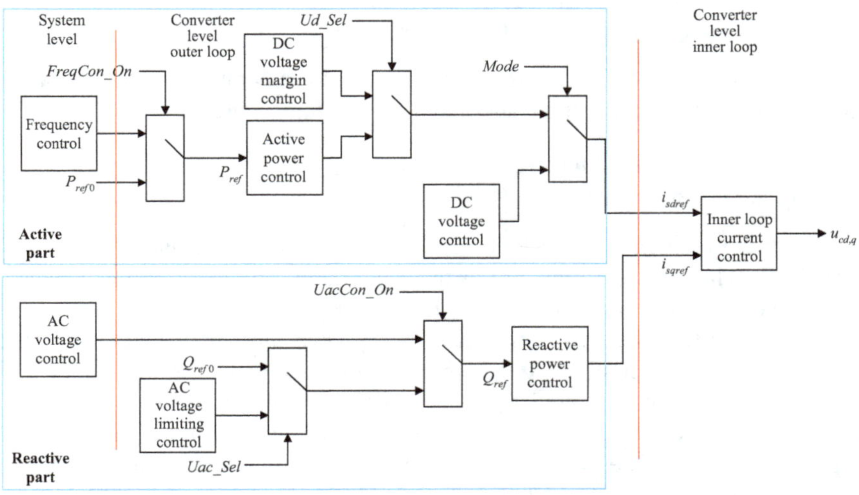

Figure 8.8 Overall structure of the control system model of the converter

FreqCon_On to determine whether the frequency control is activated. When starting, the frequency control output is valid. Otherwise, the active power command is a specified value. The active part of the outer control loop will select one of the d-axis current commands of the active power control and the DC voltage control through the mode selection signal mode. When the active power control mode is selected, under normal operating conditions, the output is the d-axis current command of active power control. If the DC voltage is too high or too low (beyond the set threshold range), the signal Ud_Sel is set. In the meantime, DC voltage margin control will work, and its output will replace the output of active power control.

The reactive part is slightly different from the active part. AC voltage control will generate a reactive power command value to reactive power control. The control signal UacCon_On determines its startup state. When starting, the reactive power command output by the AC voltage control is the command of reactive power control. When it is not started, the reactive power control command of the outer control loop takes a specified value under normal operating conditions. If the AC voltage is too high or too low (beyond the set threshold range), the signal Uac_Sel is set. The AC voltage limiting control will be active, and the output will be used as the command value for reactive power control. The reactive power control will output the q-axis current command.

The inner loop is the current control. It receives the dq-axis current commands obtained by the outer control loop to calculate the injection voltage of the converter.

8.2.4 Converter fault treatment

Three typical converter-related faults are considered in fault modeling:

1. Converter blocking;
2. Circuit breaker in AC-side disconnecting;
3. Switch (circuit breaker) in DC-side disconnecting.

All three faults will cause the converter to stop operation. In the actual protection system design, the three actions have a certain sequence of timing.

8.2.4.1 Converter blocking

- Two-level converter

After the converter is blocked, insulated gate bipolar translators (IGBTs) are turned off, which can be removed in the model and only anti-parallel diodes are considered. The converter structure is shown in Figure 8.9.

Assuming that the DC-side voltage is U_d when blocking, the converter still exhibits the characteristics of the voltage source due to the presence of the capacitor voltage, but it is no longer controllable at this time. The amplitude and phase angle of the voltage injected into the AC side will be related to U_d. Figure 8.10 shows the waveform relationship between the system voltage of phase a u_{sa} (black curve) and the voltage of the injection point u_{ca} (blue curve). There are two cases.

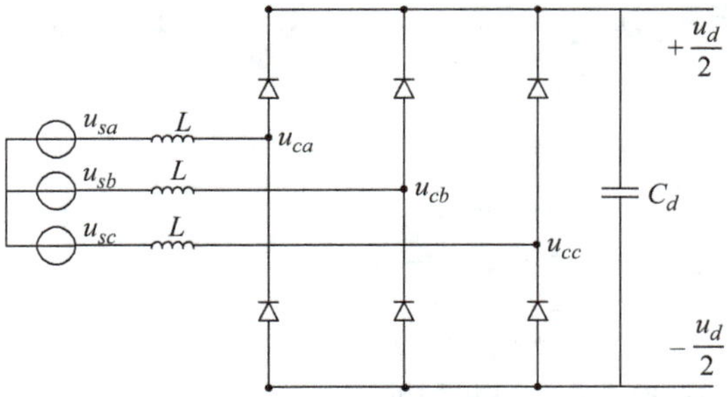

Figure 8.9 Circuit topology after blocking of two-level converter

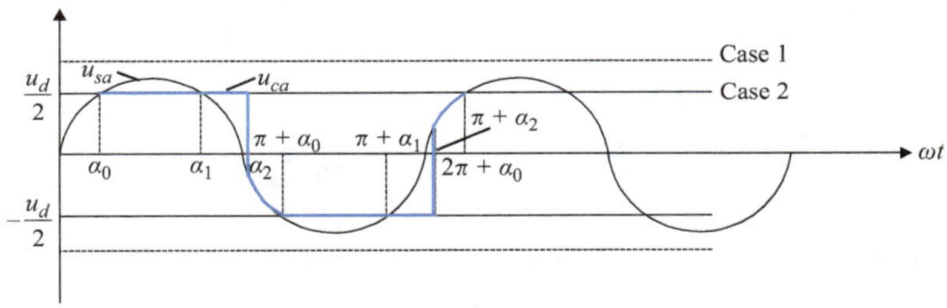

Figure 8.10 System voltage and injection voltage waveform after blocking of two-level converter

Case 1, the DC voltage exceeds the peak-to-peak voltage of the phase voltage, as shown by the dotted line in Figure 8.10. At this time, all the diodes of the upper and lower bridge arms are not turned on at any time. The injection current is zero and there is no power exchange between the converter and the system.

Case 2, the DC voltage is less than the peak-to-peak voltage of the phase voltage. At this time, the converter can inject voltage into the system. This process can be divided into the following stages: When u_{sa} is greater than $U_d/2$, the diode of the upper arm is forward-conducting, and the injection point voltage is $U_d/2$, corresponding to the angular interval $[\alpha_0, \alpha_1]$. When u_{sa} is less than $U_d/2$, since the current of the inductor branch cannot be abruptly changed, it will gradually decrease to zero, corresponding to the interval $[\alpha_1, \alpha_2]$, and the injection voltage is still $U_d/2$. Then, the diode of the upper bridge arm is cut-off. The branch current is kept at zero, and the injection point voltage is equal to the system voltage u_{sa}, corresponding to the interval $[\alpha_2, \pi + \alpha_0]$. When u_{sa} is less than $-U_d/2$, the diode of the lower bridge arm is forward-conducting, and the injection point voltage is $-U_d/2$, corresponding to the interval $[\pi + \alpha_0, \pi + \alpha_1]$. Similar to the upper bridge arm, when u_{sa} is greater than $-U_d/2$, the current of the lower bridge arm will gradually decrease to zero, corresponding to the interval $[\pi + \alpha_1, \pi + \alpha_2]$. After that, the zero current state is maintained, and the injection voltage is u_{sa}, corresponding to the interval $[\pi + \alpha_2, 2\pi + \alpha_0]$. The injection voltage segmentation expression can be written as

$$\begin{cases} u_{ca} = \dfrac{u_d}{2} & \alpha \in [\alpha_0, \alpha_2] \\ u_{ca} = u_{sa} & \alpha \in [\alpha_2, \pi + \alpha_0] \\ u_{ca} = -\dfrac{u_d}{2} & \alpha \in [\pi + \alpha_0, \pi + \alpha_2] \\ u_{ca} = u_{sa} & \alpha \in [\pi + \alpha_2, 2\pi + \alpha_0] \end{cases} \tag{8.30}$$

where α_0 is related to U_d. Assume that $u_{sa} = U_s \sin \omega t$, and we can get the following equations:

$$u_{sa}(\alpha_0) = U_s \sin \alpha_0 = \frac{u_d}{2} \tag{8.31}$$

$$\alpha_0 = \sin^{-1}\left(\frac{u_d}{2U_s}\right) \tag{8.32}$$

α_0 and α_1 have the following relationship:

$$\alpha_1 = \pi - \alpha_0 \tag{8.33}$$

Ignore the branch resistance and get the following equation:

$$i_{sa}(\alpha_1) = \frac{1}{L}\int_{\alpha_0}^{\alpha_1}(u_{sa} - u_{ca})dt = -\frac{1}{L}\int_{\alpha_1}^{\alpha_2}(u_{sa} - u_{ca})dt \tag{8.34}$$

α_2 can be obtained as follows:

$$U_s \cos \alpha_2 + \frac{u_d}{2}\alpha_2 = U_s \cos \alpha_0 + \frac{u_d}{2}\alpha_0 \tag{8.35}$$

The injection voltage u_{ca} is a periodic signal, which is expanded into a Fourier series, and its fundamental frequency component is

$$\begin{cases} a_1 = \dfrac{2}{T}\displaystyle\int_{\alpha_0}^{2\pi+\alpha_0} u_{ca}(t)\cos \omega t dt = \dfrac{u_d}{\pi}(\sin \alpha_2 - \sin \alpha_0) - \dfrac{U_s}{2\pi}(\cos 2\alpha_0 - \cos 2\alpha_2) \\ b_1 = \dfrac{2}{T}\displaystyle\int_{\alpha_0}^{2\pi+\alpha_0} u_{ca}(t)\sin \omega t dt = \dfrac{u_d}{\pi}(\cos \alpha_0 - \cos \alpha_2) + \dfrac{U_s}{\pi}\left[\pi + \alpha_0 - \alpha_2 - \dfrac{1}{2}(\sin 2\alpha_0 - \sin 2\alpha_2)\right] \end{cases} \tag{8.36}$$

Its amplitude is as follows:

$$U_c = \sqrt{a_1^2 + b_1^2} \tag{8.37}$$

The phase angle of u_{ca} lagging u_{sa} is

$$\delta = \tan^{-1}\left(-\frac{a_1}{b_1}\right) \tag{8.38}$$

Taking the direction of the system voltage u_{sa} as the d-axis direction, the dq-axis components of the injection voltage can be expressed as

$$\begin{cases} u_{cd} = U_c \cos\,\delta \\ u_{cq} = -U_c \sin\,\delta \end{cases} \tag{8.39}$$

- Modular multilevel converter (MMC)

Take a three-level MMC converter as an example to illustrate the external characteristics of the converter after the valves in it exit, as shown in Figure 8.11.

Assuming that the DC voltage is u_d when blocking, the capacitor voltages of the four submodules of the upper and lower bridge arms are $u_d/2$ according to the MMC characteristics. Similar to a two-level converter, there are two cases.

Case 1, if the DC voltage exceeds the peak-to-peak value of the system phase voltage, diodes cannot conduct at any time, and there is no power exchange between the converter and the system, which is consistent with the two-level converter.

Case 2, the DC voltage is less than the peak-to-peak value of the system phase voltage. When u_{sa} is greater than $U_d/2$, the difference between the system voltage and the negative voltage exceeds the sum of the capacitor voltages of the lower bridge arm, then the two diodes D1 in the submodules of the lower bridge arm are

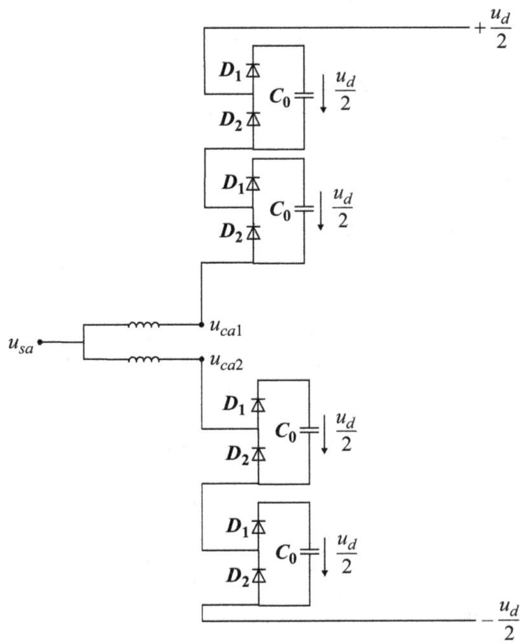

Figure 8.11 Circuit topology after MMC blocking

forward-conducting, and the two diodes D2 of the upper bridge arm conduct. The injection point voltage of the converter is $U_d/2$, and the branch current flows from the system to the converter, corresponding to the angular interval $[\alpha_0, \alpha_1]$. Similar to the two-level converter, when u_{sa} is less than $U_d/2$, the branch current is gradually reduced to zero, and the injection voltage is still $U_d/2$, corresponding to the interval $[\alpha_1, \alpha_2]$. Then all diodes are cut-off and the injection point voltage is equal to the system voltage u_{sa}, corresponding to the interval $[\alpha_2, \pi + \alpha_0]$. When u_{sa} is less than $-U_d/2$, the two diodes D1 of the upper bridge arm conduct, and the two diodes D2 of the lower bridge arm conduct. The injection voltage of the converter is $-U_d/2$, and the branch current flows from the converter to the system, corresponding to the interval $[\pi + \alpha_0, \pi + \alpha_1]$. Then the current is reduced to zero, corresponding to the interval $[\pi + \alpha_1, \pi + \alpha_2]$.

It can be seen that the injection voltage of the MMC changes exactly the same as the two-level converter in one period. A unified equation can be used, and the derivation process is also the same as the above two-level converter, and will not be described here.

8.2.4.2 Circuit breaker in AC-side disconnecting

The main wiring of the converter station is shown in Figure 8.12. The AC circuit breaker (ACCB) is installed on the primary side of the connection transformer. The disconnection of the AC circuit breaker cuts off the electrical connection between AC and DC, that is, there is no more power exchange between AC and DC.

The model is set as follows:

(a) AC injection power $P_s = 0$, $Q_s = 0$;
(b) AC branch current $i_{sd} = 0$, $i_{sq} = 0$;
(c) The state of the primary- and secondary-side voltage of the transformer is kept unchanged;
(d) Current injected into the DC network in DC side $i_d = 0$.

8.2.4.3 Switch in DC-side disconnecting

The switch in the DC side is installed after the DC-side equivalent capacitor (DCSW) as in Figure 8.12. When the switch is disconnected, the converter station is separated from the DC network and becomes an isolated station. As the four-terminal system is shown in Figure 8.13, since the switch in DC side of the

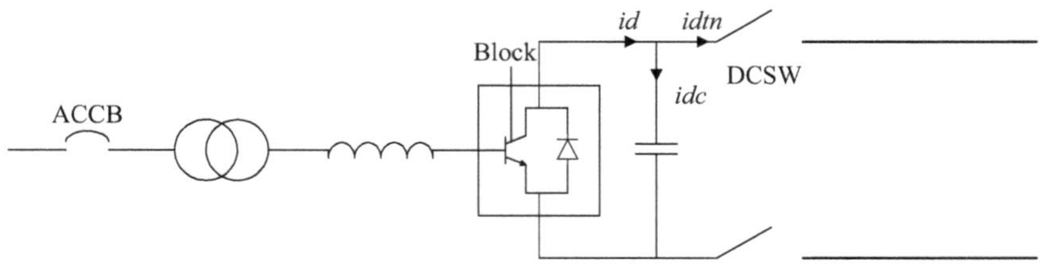

Figure 8.12 Positions of the main circuit breakers and switches of the converter station

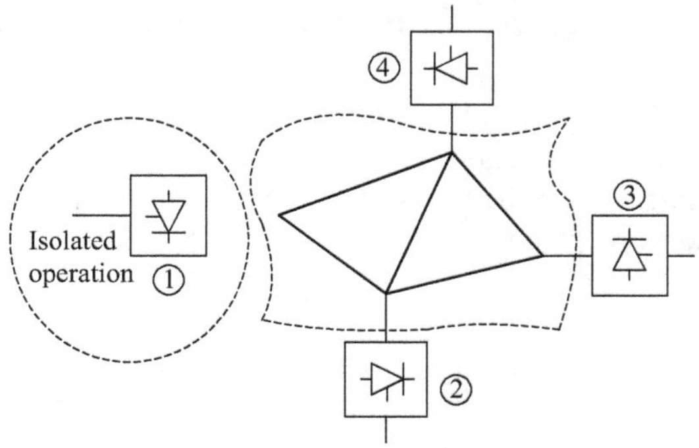

Figure 8.13 Operating state of the four-terminal system after the switch in DC side is disconnected

converter station 1 is disconnected, the original system is divided into two parts, which should be solved separately.

For the isolated converter station, when the DC switch is disconnected, the AC-side current i_{sd} and i_{sq} will remain unchanged due to the continuity of the AC branch current, so the current i_d injected into the DC side also remains unchanged. Since the current in DC side (the current after the capacitor) i_{dtn} abruptly changes to zero, all of the current flows into the DC-side capacitor, that is, $i_d = i_{dc}$. The converter will operate according to its own control law. At this time, the converter is similar to a parallel compensation device.

For the rest of the system, in the DC network, the exiting converter node remains and its injection current is constant at zero. And, the equivalent capacitance in DC side of the exiting converter will be removed from the admittance matrix of the DC network, that is, the admittance matrix needs to be modified.

If the converter station i exits, the modified admittance matrix G_{cd} is

$$G_{Cd} = \left[\begin{array}{ccccc|c} 2C_{d0(n+m)}/\Delta t & & & & & \\ & 2C_{d1(n+m+1)}/\Delta t & & & & 0 \\ & & \ddots & & & \\ & & \mathrm{i} \rightarrow 0 & & & \\ & & & \ddots & & \\ & & & & 2C_{d(n-1)(2n+m-1)}/\Delta t & \\ \hline & & & 0 & & 0 \end{array}\right] \tag{8.40}$$

Then regenerate the total admittance matrix. The rest of the converters remain in normal operation.

8.3 Comparison and verification of simulation result

8.3.1 Test system and its simulation

The test system is shown in Figure 8.14.

The main parameters in the test system are shown in Table 8.1.

Test system operating mode is shown in Table 8.2.

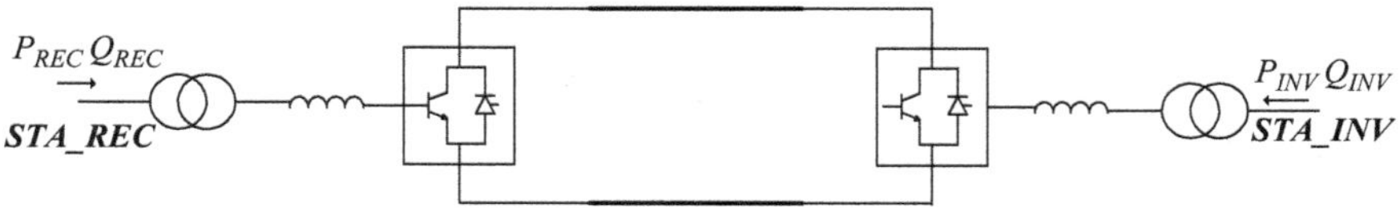

Figure 8.14 Two-terminal VSC HVDC system

Table 8.1 The main parameters of the test HVDC system

Converter station	Note 1							
	K	***Xt***	***Xc***	***Bc***	***n***	***Cd***	***SB***	***UdB***
STA_REC	230/100	0.15	0.15	0	30	2000	200	100
STA_INV	230/100	0.15	0.15	0	30	2000	200	100
Power line	Parameter R(Ω): equivalent resistance; L(H):equivalent inductance; C/2(μF):equivalent capacity							
STA_REC-STA_INV	R = 6.67 Ω, L = 0.2 H, C/2 = 50 μF							

Note 1:

K: coupling transformer ratio;
Xt (pu): transformer leakage reactance;
Xc (pu): reactance;
Bc (pu): filter susceptance;
n: the number of MMC single bridge-arm module;
Cd (μF): capacity of sub-module;
SB (MVA): rated power of converter and
UdB (kV): rated DC voltage.

Table 8.2 Test system operating mode of the test HVDC system

Scheme	Note 2	
	P_{REC} Q_{REC}	P_{INV} Q_{INV}
Operating mode	100, –20	–100, –60

The reference direction of active power, P(MW), and reactive power, Q (MVar), is from AC to DC.

8.3.2 Power step in rectifier side

The simulation comparison of electromechanical transient model and electromagnetic transient model is shown in Figure 8.15.

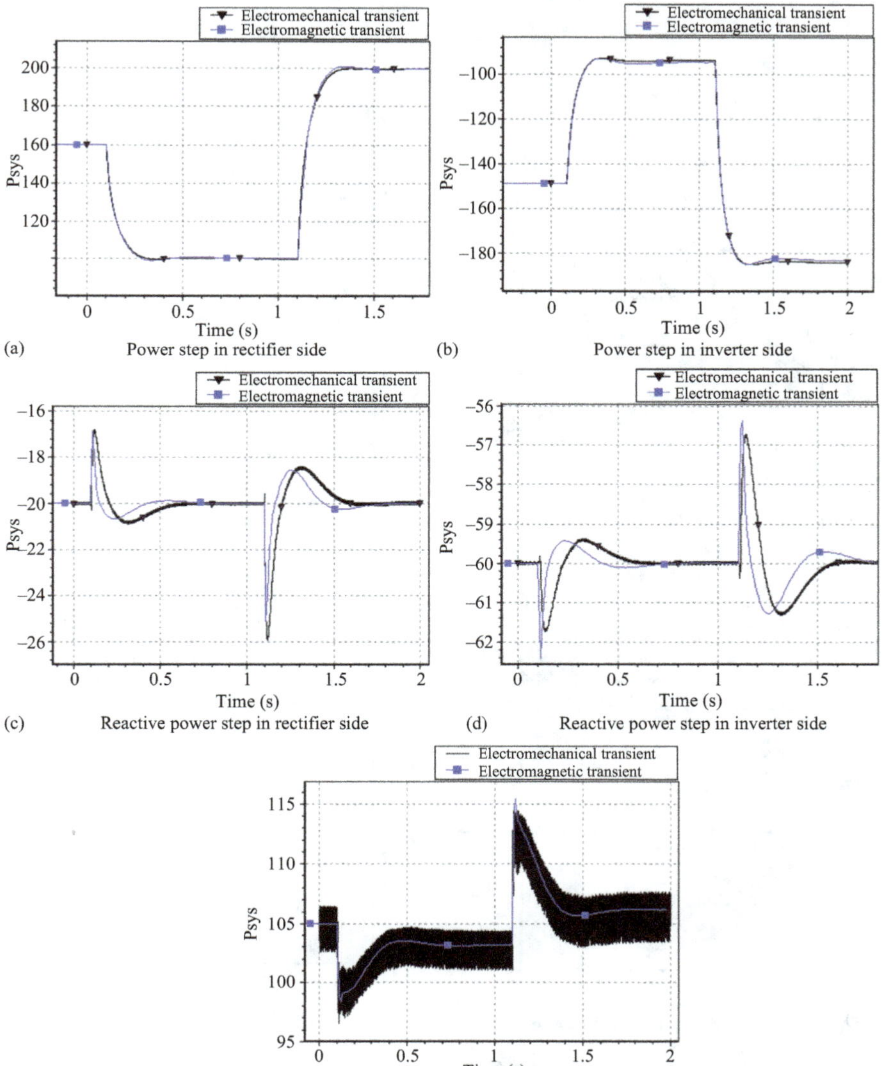

(a) Power step in rectifier side (b) Power step in inverter side (c) Reactive power step in rectifier side (d) Reactive power step in inverter side (e) DC voltage in rectifier side

Figure 8.15 Simulation results of power step in the test system

8.3.3 Short-circuit fault in AC system

The AC short-circuit fault simulated with the comparison of the electromagnetic transient model is shown in Figure 8.16.

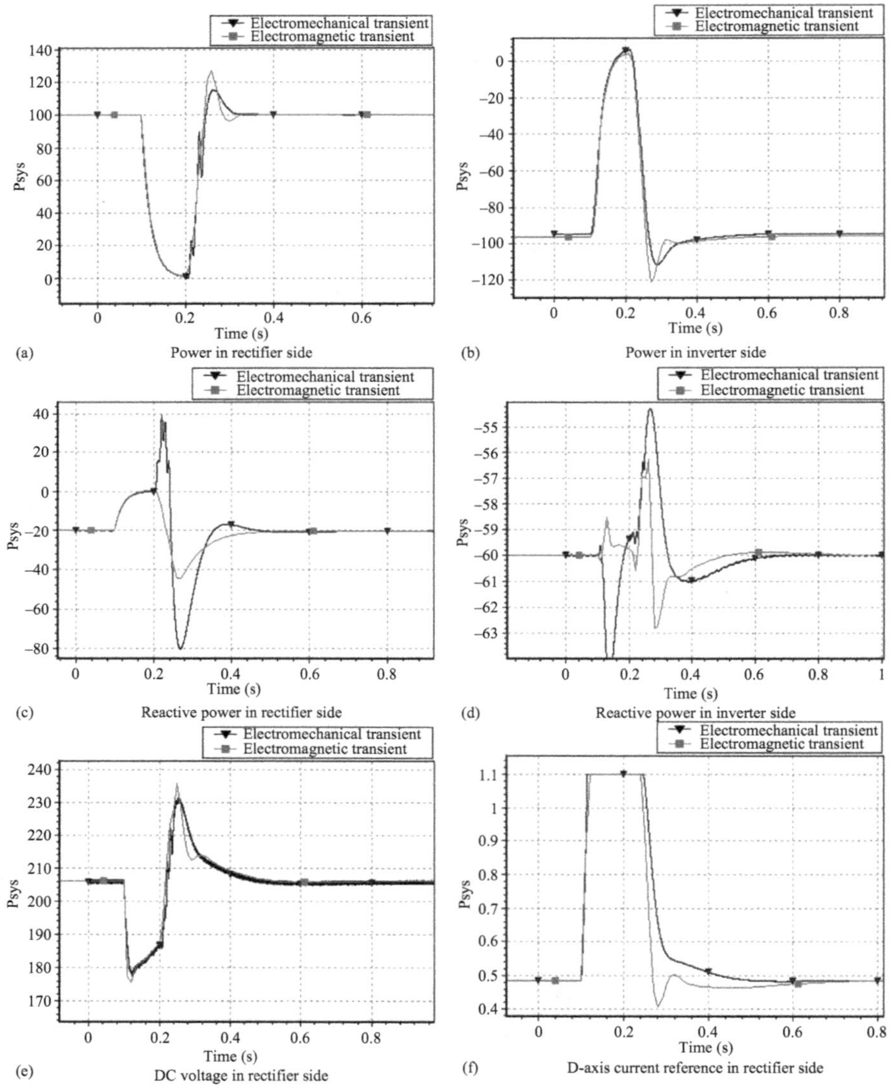

Figure 8.16 Simulation of AC system short-circuit fault in rectifier side

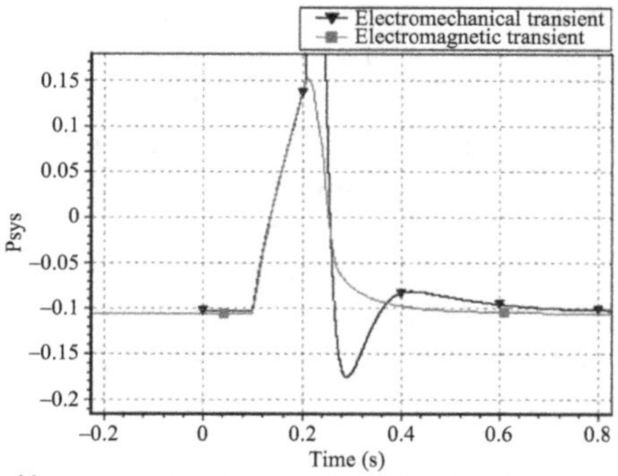

(g) Q-axis current reference in rectifier side

Figure 8.16 (Continued)

The simulation results and comparisons illustrate the satisfied match of the electromechanical model with the electromagnetic model for the test HVDC.

References

[1] Schettler F., Huang H., and Christl N. 'HVDC Transmission Systems Using Voltage Sourced Converters – Design and Applications'. Proceedings of Power Engineering Society Summer Meeting IEEE; Seattle, WA, USA, July 2000. IEEE; 2002

[2] Asplund G., Eriksson K., and Stevenson K. 'DC Transmission Based on voltage Source Converters'. Presented at CIGRE SC14 Colloquium; South Africa, 1997

[3] Flourentzou N., Agelidis V.G., and Demetriades G.D. 'VSC-Based HVDC Power Transmission Systems: An Overview'. *Journal of IEEE Transactions on Power Electronics*. 2009;24(3):592–602

[4] Xu L., Yao L., and Sasse C. 'Grid Integration of Large DFIG-Based Wind Farms Using VSC Transmission'. *Journal of IEEE Transactions on Power Systems*. 2007;22(3):976–984

[5] Reidy A., and Watson R. 'Comparison of VSC Based HVDC and HVAC Interconnections to a Large Offshore WindFarm'. Proceedings of 2005 IEEE Power Engineering Society General Meeting; San Francisco, CA, USA, June 2005. IEEE; 2005. pp. 1–8

[6] Lesnicar A., and Marquardt R. 'An Innovative Modular Multilevel Converter Topology Suitable for a Wide Power Range'. Proceedings of IEEE Power Tech Conference; Bologna, Italy, June 2003. IEEE; 2004

[7] Peralta J., Saad H., Dennetière S., Mahseredjian J., and Nguefeu S. 'Detailed and Averaged Models fora 401-Level MMC–HVDC System'. *Journal of IEEE Transactions on Power Delivery*. 2012;27(3):1501–1508

[8] Saad H., Peralta J., Dennetière S., *et al.* 'Dynamic Averaged and Simplified Models for MMC-Based HVDC Transmission Systems'. *Journal of IEEE Transactions on Power Delivery*. 2013;28(3):1723–1730
[9] Cole S., Beerten J., and Belmans R. 'Generalized Dynamic VSC MTDC Model for Power System Stability Studies'. *Journal of IEEE Transactions on Power Systems*. 2010;25(3): 1655–1662
[10] Nikkhajoei H., and Iravani R. 'Dynamic Model and Control of AC–DC–AC Voltage-Sourced Converter System for Distributed Resources'. *Journal of IEEE Transactions on Power Delivery*. 2007;22(2):1169–1178

Chapter 9

Dynamic phasor modeling of HVDC systems

U.D. Annakkage[1], C. Karawita[2], S. Arunprasanth[3] and H. Konara[3]

9.1 Introduction

Power system analysis and simulation technologies can be broadly categorized into three types [1]: (a) steady-state analysis, (b) quasi-steady-state analysis and (c) electromagnetic transient (EMT) analysis.

In the steady-state analysis, the behavior of the power system is modeled using algebraic equations related to steady-state voltages and currents. It involves representing the sinusoidal currents and voltages by a constant magnitude and a phase angle. Hence, the power system variables are expressed as complex numbers (phasors). Time-varying sinusoidal signals are constant at the steady state as a result of removing the rotational term from the signals. This can be considered as a demodulation of the signal from the carrier signal of 60 Hz [2]. Power flow analysis and fault analysis fall into this category.

The same phasors are used to model the system in quasi-steady-state analysis. The stationary assumption is extended for a quasi-stationary condition, where it allows the system voltages and currents to vary slowly. The transient stability (TS), or, more preciously, the transient rotor angle stability simulation, falls into this category. This approach neglects the EMTs in the network. Hence, the bandwidth of the dynamics that these type of studies can capture falls into a narrow range (up to 5 Hz). The simulation time-step of the quasi-steady-state approach is typically around half a cycle of the nominal frequency [3,4]. In quasi-stationary assumption, the EMTs of the systems will be filtered out and, hence, it is assumed that the power transfer occurs only at the system frequency [5]. Since the electrical variables of the system such as phasors of currents and voltages are considered as algebraic variables (not as state variables), it can only model the system in a quasi-steady-state manner. The interactions between generators and other dynamic equipment are studied only around the system frequency.

[1]University of Manitoba, Winnipeg, MB, Canada
[2]Transgrid Solutions Inc., Winnipeg, MB, Canada
[3]RTDS Technologies Inc., Winnipeg, MB, Canada

The EMT studies model the power system more accurately and capture the dynamics at higher frequencies. It works with instantaneous time-domain values and it continuously follows the trajectories of the states of the system. These types of simulation programs need a very small time-step to follow the high-frequency components of the waveforms. In order to produce low-frequency oscillations in a simulation, the period of simulation has to increase appropriately (typically, 5 s). Therefore, the EMT with small integration time steps is slow to study such phenomena [5]. Typically, EMT studies are not used in studying electromechanical oscillations for many reasons such as (a) two programs have two different objectives where one is designed for fast EMTs and the other one for slow dynamics, (b) simulation time required for stability studies can be significantly large and an EMT simulation needs a significant amount of computational power for it to simulate it with a small time step like 50 μs and (c) TS studies are typically done for large power systems and EMT programs are computationally demanding for this purpose.

One solution suggested to circumvent this problem is to interface electromechanical and electromagnetic simulations as a hybrid simulator. The focus of this type of a study goes to a small part of the network, where detailed modeling is required. The rest of the system is analyzed using a TS-type model. However, integrating two different programs where EMT covers a wide bandwidth and TS covers a narrow bandwidth is challenging and needs sophisticated frequency-mapping algorithms [6].

In general, it can be stated that there is a need of a simulation tool that is capable of studying the dynamics in the quasi-stationary range while not compromising the speed of the simulation. The concept of dynamic phasors was introduced by Venkatasubramanian [2,7] in the mid-90s to overcome the limitations imposed by quasi-stationary assumption made in TS studies. In dynamic phasors, currents and voltages are multiplied by the term $e^{j\omega_0 t}$ to remove the periodical oscillatory movement around the base frequency ω_0 [5]. This demodulation results in a constant magnitude and an angle similar to a TS model. However, during a transient the currents and voltages are allowed to change continuously over time [2].

The concept of dynamic phasors can be explained by using the theory of shifted frequency analysis discussed in [8,9]. For example, if we are interested in transients up to 70 Hz, by shifting the frequency by the nominal frequency of 60 Hz, 70 Hz becomes 10 Hz and the simulation time step can be chosen to cover 10 Hz (not 70 Hz). Hence, if the frequency spectrum of the system is shifted by its fundamental frequency, the system becomes a narrowband. This will allow the dynamic phasors to use larger integration time step than the EMT simulations resulting in faster simulations. The frequency spectrum of a bandpass signal is shown in Figure 9.1 and, after shifting the frequency from the carrier frequency (f_c), the frequency spectrum becomes narrow banded as shown in Figure 9.2.

Compared to TS-type models, the dynamic phasors represent the currents and voltages in the system as state variables [2]. Hence, the bandwidth of the dynamic phasor-type simulations is higher compared to TS. Therefore, it can deliver faster simulations than the EMT simulations. The dynamic phasor approach has attracted attention as a method that can be used in parallel with electromechanical

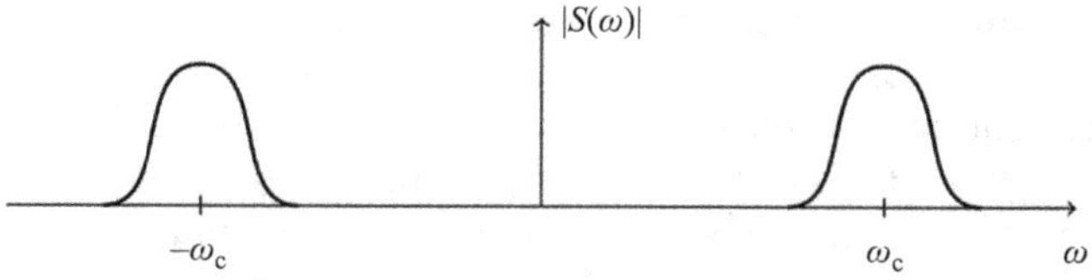

Figure 9.1 Frequency spectrum of a bandpass signal

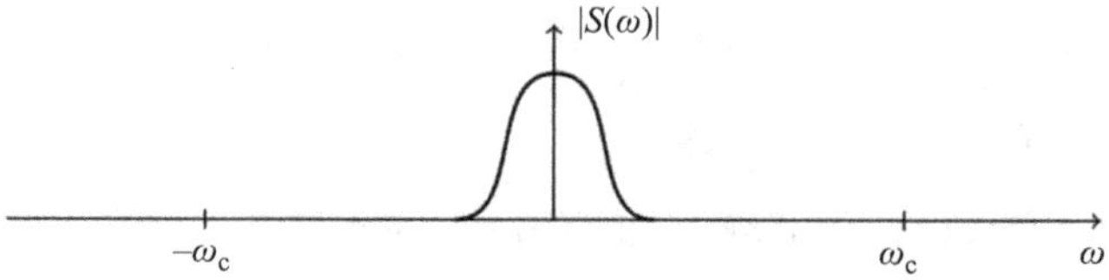

Figure 9.2 Frequency spectrum of dynamic phasors

simulations to accommodate high-frequency switching components [10–13]. Reference [10] shows that dynamic phasors significantly improve the accuracy of electromechanical simulations with power electronic devices. The other advantage of dynamic phasors is their ability to change the level of complexity of the model. The level of complexity can be altered by selecting the number of harmonics [14] and also the number of sequence components in an unbalanced situation [15,16]. Some of the work done on modeling higher order harmonics using dynamic phasors are presented in [14,17–19].

9.1.1 Concept of dynamic phasors

The concept and development of dynamic phasor-based models are presented in this section and they are based on the paper written by Venkatasubramanian [7] in 1994. Time-varying phasors have been used in communication theory over a long time and the modulation of signals is used for analyzing narrowband signals with signal bandwidth that is much less compared to the carrier frequency. This kind of phasor representation is very useful since the dynamics of a sinusoidal signal can be represented using the envelope of the modulated signal and the carrier signal can be filtered out. The same narrowband assumption can be made on the phasor signals of power system. However, bandwidth of the signal has to be much less than the carrier signal frequency, which is equal to 60 Hz. With this assumption, the model simplifies to the same model used in a quasi-stationary analysis and, therefore, there won't be much of a new information that can be gained from such a study. The methods like Hilbert transformation and the analytic signal approach used in communication theory for establishing the concept of time varying phasors are not limited for narrowband signals. These techniques can therefore apply to signals that have frequency content up to the carrier frequency. This will allow capturing most of the dynamics in power system transients. Phasor signals where the bandwidths are smaller than carrier frequency (60 Hz) are defined as “low-pass” phasor signals [7]. It has been shown in [20] that band-limited signal assumption is the fundamental requirement for modulating signals. It can be shown that the band-limited assumption, along with the low-pass phasor signals, preserves the non-linear power balance equations. Let us consider the modulated signal $u(t)$, where the carrier frequency equals to ω_c. The magnitude of the $u(t)$ is denoted as $|u(t)|$ and the angle is denoted by $\delta(t)$

$$u(t) = |u(t)| \cos(\omega_c t + \delta(t)) \tag{9.1}$$

The phasor representation of the signal can be written as,

$$\widehat{u}(t) = |u(t)| e^{j\delta(t)} = |u(t)| \angle \delta(t) \tag{9.2}$$

Let us say that the signal $u(t)$ is a bandpass signal. Then, it can be expressed as (9.3) [5].

$$u(t) = u_I(t)\cos(\omega_c t) - u_Q(t)\sin(\omega_c t) \tag{9.3}$$

This kind of representation suggests that the dynamic phasors (time-varying phasor) are equal to $\widetilde{u}(t) = u_I(t) + ju_Q(t)$. Let us calculate the pre-envelop or the analytic signal $u_+(t)$ using the Hilbert transform of $u(t)$.

$$u_+(t) = u(t) + jH[u(t)] \tag{9.4}$$

$$u_+(t) = \left(u_I(t)\cos(\omega_c t) - u_Q(t)\sin(\omega_c t)\right) + j\left(u_I(t)\sin(\omega_c t) - u_Q(t)\cos(\omega_c t)\right) \tag{9.5}$$

$$u_+(t) = \widetilde{u}(t)e^{j\omega_c t} = \left(u_I(t) + ju_Q(t)\right)e^{j\omega_c t} \tag{9.6}$$

The dynamic phasor of the signal $u(t)$ is the complex envelope $u_+(t)$ of the signal $u(t)$ shifted from the carrier frequency of ω_c. Hence, the dynamic phasors are obtained relaxing the oscillatory movement around the carrier frequency of a complex signal. It is evident that (9.6) can only be defined for a bandpass signal (9.3). Hence, the dynamic phasor of the signal $u(t)$ equals to $\widetilde{u}(t)$ and, from this point onwards, it will be referred to as $U(t)$.

$$U(t) = u_I(t) + ju_Q(t) = u_+(t)e^{-j\omega_c t} \tag{9.7}$$

The components $u_I(t)$ and $u_Q(t)$ of $U(t)$ are denoted as the in-phase component and quadrature component, respectively. Signals $u_I(t)$ and $u_Q(t)$ are the low-pass signal content of the signal $u(t)$ as shown in (9.3). Hence, the in-phase and quadrature components can be re-constructed by passing the product of $u(t)cos(\omega_c t)$ and $u(t)\sin(\omega_c t)$, respectively, through a low-pass filter which has a cut-off frequency of $\omega = \omega_c$. To construct the phasor signal in a linear way, the phasors have to be low-pass signals. Else, the construction of $u_I(t)$ and $u_Q(t)$ is possible only for a linear set of network equations. In a nodal analysis-type approach for network solving, the objective of the simulation is to observe the voltages and currents in the system for a sinusoidal excitation of a source. Hence, the problem becomes simple, where a set of linear differential equations are solved for a time-varying sinusoidal input. There is no restriction on the bandwidth on the validity of the phasor representation for such a purpose. However, the equations are written in the complex phasor domain and the dimension of the system becomes the twice the size of the original system.

In the traditional TS-type studies, the derivatives are substituted by the constant term $j\omega_c$. This rule is modified in dynamic phasors to model the dynamics [7]. Let the dynamic phasor of signal $u(t)$ be equal to $U(t)$:

$$u(t) = \text{Re}(U(t)e^{j\omega_c t}) \tag{9.8}$$

$$\frac{d}{dt}u(t) = \text{Re}\left(\frac{d}{dt}U(t)e^{j\omega_c t}\right) \tag{9.9}$$

$$\frac{d}{dt}u(t) = \text{Re}\left(\left(\frac{d}{dt}U(t) + j\omega_c U(t)\right)e^{j\omega_c t}\right) \tag{9.10}$$

Hence, in dynamic phasors, the derivative is equal to the addition of the derivative of the phasor signal and the $j\omega_c$ multiplied by the phasor signal. Using this property, differential equations for a capacitor and an inductor can be written as (9.11) and (9.12), respectively.

$$I_c(t) = C\frac{d}{dt}V_c(t) + j\omega_c CV_c(t) \tag{9.11}$$

$$V_L(t) = L\frac{d}{dt}I_L(t) + j\omega_c LI_L(t) \tag{9.12}$$

where signals $I_c(t)$, $V_c(t)$, $V_L(t)$ and $I_L(t)$ are dynamic phasor voltages and currents. It should be noted that the above equations are derived for the fundamental frequency component. As we can see, in dynamic phasors we have voltages and currents in the network as state variables, which allows capturing network dynamics. In TS studies, currents and voltages are allowed to change in discrete steps and, for the above reason in dynamic phasors, they change continuously through time. The use of dynamic phasor representation allows capturing network dynamics accurately and thus making the small-signal (SS) model valid for sub-synchronous oscillations.

9.2 Dynamic phasor representation of an AC network

This section presents how dynamics phasor modeling presented in Section 9.1.1 is used to model transmission lines. The representation of RL series branch and RC parallel branch is presented first as the building blocks of the transmission line model. Then, the T and π representations of the transmission line are presented.

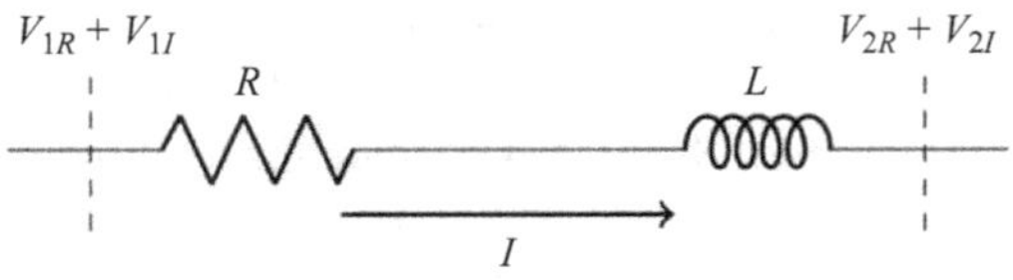

Figure 9.3 Series RL branch

9.2.1 Representation of a series RL branch

Let us consider a series connected RL branch shown in Figure 9.3, where the voltages of the opposite sides are equal to $V_1 = V_{1R} + V_{1I}$ and $V_2 = V_{2R} + V_{2I}$.

$$\widetilde{V}_{12} = L\frac{d}{dt}\widetilde{I} + (R + j\omega_0 L)\widetilde{I} \tag{9.13}$$

$$\frac{d}{dt}\widetilde{I} = -\left(\frac{R}{L} + j\omega_0\right)\widetilde{I} + \frac{1}{V}\widetilde{V}_{12} \tag{9.14}$$

Separating the real and the imaginary parts of (9.14) results in

$$\frac{d}{dt}\begin{pmatrix} I_R \\ I_I \end{pmatrix} = \begin{pmatrix} \dfrac{-R}{L} & \omega_0 \\ -\omega_0 & \dfrac{-R}{L} \end{pmatrix}\begin{pmatrix} I_R \\ I_I \end{pmatrix} + \begin{pmatrix} \dfrac{1}{L} & 0 & -\dfrac{1}{L} & 0 \\ 0 & \dfrac{1}{L} & 0 & -\dfrac{1}{L} \end{pmatrix}\begin{pmatrix} V_{1R} \\ V_{1I} \\ V_{2R} \\ V_{2I} \end{pmatrix} \tag{9.15}$$

where

$$\dot{x} = Ax + Bu \tag{9.16}$$

The eigenvalues of the system are complex conjugates which are at $-\tau \pm j\omega_0$, where τ is the time constant of the system which equals to $\tau = \frac{R}{L}$. The system oscillates at the frequency ω_0 which is the carrier frequency. Hence, the eigenvalue depicts the interactions between the original system and the carrier.

9.2.2 Representation of a shunt RC branch

Let us analyze the shunt RC branch shown in Figure 9.4.

$$\widetilde{I} = C\frac{d}{dt}\widetilde{V} + (G + j\omega_0 C)\widetilde{V} \tag{9.17}$$

$$\frac{d}{dt}\widetilde{V} = -\left(\frac{G}{C} + j\omega_0\right)\widetilde{V} + \frac{1}{C}\widetilde{I} \tag{9.18}$$

$$\frac{d}{dt}\begin{pmatrix} V_R \\ V_I \end{pmatrix} = \begin{pmatrix} -G/C & \omega_0 \\ -\omega_0 & -G/C \end{pmatrix}\begin{pmatrix} V_R \\ V_I \end{pmatrix} + \begin{pmatrix} 1/C & 0 \\ 0 & 1/C \end{pmatrix}\begin{pmatrix} I_R \\ I_I \end{pmatrix} \tag{9.19}$$

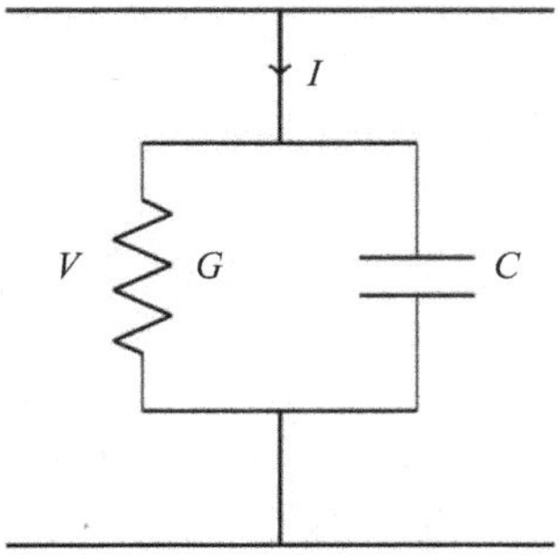

Figure 9.4 Shunt RC branch

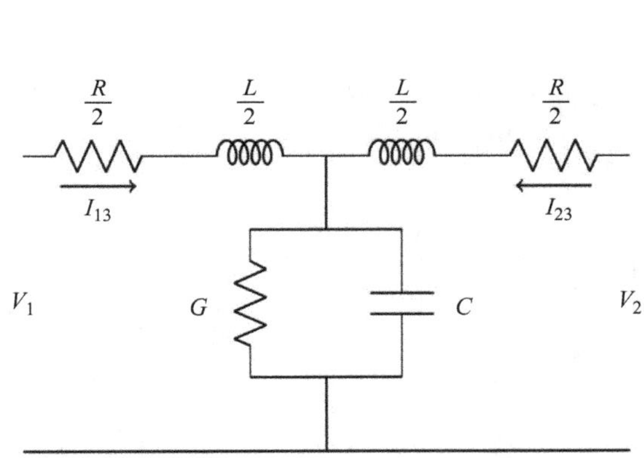

Figure 9.5 A T-section model

The eigenvalues can be calculated for the network and they lie on $-\tau \pm j\omega_0$, where the time constant of the system is equal to $\tau = \frac{G}{C}$. The system oscillates at the frequency of carrier frequency (ω_0).

9.2.3 Representation of a T-section model

A T-section model of a transmission line is shown in Figure 9.5. The parameters R, L, G and C are series resistance, series inductance, shunt conductance and shunt capacitance, respectively. Dynamic phasor equations for series RL branch are as follows:

$$\frac{d}{dt}\begin{pmatrix} I_{13R} \\ I_{13I} \end{pmatrix} = \begin{pmatrix} -R/L & \omega_0 \\ -\omega_0 & -R/L \end{pmatrix}\begin{pmatrix} I_{13R} \\ I_{13I} \end{pmatrix} + \begin{pmatrix} 2/L & 0 \\ 0 & 2/L \end{pmatrix}\begin{pmatrix} V_R \\ V_I \end{pmatrix} + \begin{pmatrix} -2/L & 0 \\ 0 & -2/L \end{pmatrix}\begin{pmatrix} V_{3R} \\ V_{3I} \end{pmatrix} \tag{9.20}$$

Dynamic phasor equations for second series RL branch,

$$\frac{d}{dt}\begin{pmatrix} I_{23R} \\ I_{23I} \end{pmatrix} = \begin{pmatrix} -R/L & \omega_0 \\ -\omega_0 & -R/L \end{pmatrix}\begin{pmatrix} I_{23R} \\ I_{23I} \end{pmatrix} + \begin{pmatrix} 2/L & 0 \\ 0 & 2/L \end{pmatrix}\begin{pmatrix} V_{2R} \\ V_{2I} \end{pmatrix} + \begin{pmatrix} -2/L & 0 \\ 0 & -2/L \end{pmatrix}\begin{pmatrix} V_{3R} \\ V_{3I} \end{pmatrix} \tag{9.21}$$

Dynamic phasor equations for shunt RC branch,

$$\frac{d}{dt}\begin{pmatrix} V_{3R} \\ V_{3I} \end{pmatrix} = \begin{pmatrix} -G/C & \omega_0 \\ -\omega_0 & -G/C \end{pmatrix}\begin{pmatrix} V_{3R} \\ V_{3I} \end{pmatrix} + \begin{pmatrix} 1/C & 0 \\ 0 & 1/C \end{pmatrix}\begin{pmatrix} I_{13R} \\ I_{13I} \end{pmatrix} + \begin{pmatrix} 1/C & 0 \\ 0 & 1/C \end{pmatrix}\begin{pmatrix} I_{23R} \\ I_{23I} \end{pmatrix} \tag{9.22}$$

Combining (9.20), (9.21) and (9.22) into one matrix results in overall state-space system of the network, which is given by (9.23).

$$\frac{d}{dt}\begin{pmatrix} I_{13R} \\ I_{13I} \\ I_{23R} \\ I_{23I} \\ V_{3R} \\ V_{3I} \end{pmatrix} = \begin{pmatrix} -R/L & \omega_0 & 0 & 0 & -2/L & 0 \\ -\omega_0 & -R/L & 0 & 0 & 0 & -2/L \\ 0 & 0 & -R/L & \omega_0 & -2/L & 0 \\ 0 & 0 & -\omega_0 & -R/L & 0 & -2/L \\ 1/C & 0 & 1/C & 0 & -G/C & \omega_0 \\ 0 & 1/C & 0 & 1/C & -\omega_0 & -G/C \end{pmatrix}\begin{pmatrix} I_{13R} \\ I_{13I} \\ I_{23R} \\ I_{23I} \\ V_{3R} \\ V_{3I} \end{pmatrix} + 2\begin{pmatrix} 1/L & 0 & 1/L & 0 \\ 0 & 1/L & 0 & 1/L \end{pmatrix}\begin{pmatrix} V_{1R} \\ V_{1I} \\ V_{2R} \\ V_{2I} \end{pmatrix} \tag{9.23}$$

9.2.4 Representation of a π-section model

A π-section model of a transmission line is shown in Figure 9.6.

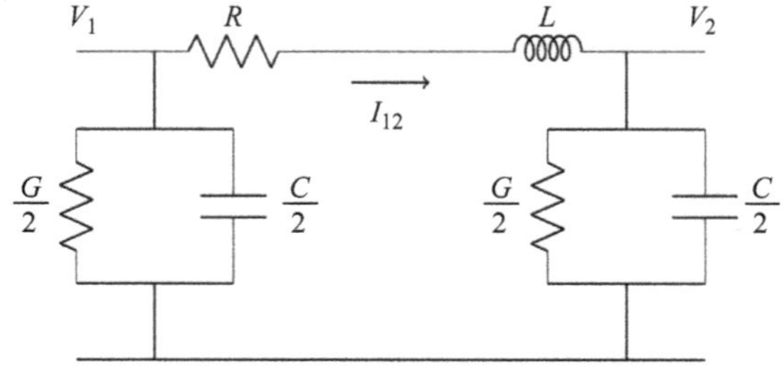

Figure 9.6 A π–section model

Dynamic phasor equations for series RL branch are as follows:

$$\frac{d}{dt}\begin{pmatrix} I_{12R} \\ I_{12I} \end{pmatrix} = \begin{pmatrix} -R/L & \omega_0 \\ -\omega_0 & -R/L \end{pmatrix}\begin{pmatrix} I_{12R} \\ I_{12I} \end{pmatrix} + \begin{pmatrix} 1/L & 0 \\ 0 & 1/L \end{pmatrix}\begin{pmatrix} V_{1R} \\ V_{1I} \end{pmatrix} + \begin{pmatrix} -1/L & 0 \\ 0 & -1/L \end{pmatrix}\begin{pmatrix} V_{2R} \\ V_{2I} \end{pmatrix} \tag{9.24}$$

Dynamic phasor equations for first shunt RC branch are as follows:

$$\frac{d}{dt}\begin{pmatrix} V_{1R} \\ V_{1I} \end{pmatrix} = \begin{pmatrix} -G/C & \omega_0 \\ -\omega_0 & -G/C \end{pmatrix}\begin{pmatrix} V_{1R} \\ V_{1I} \end{pmatrix} + \begin{pmatrix} -2/C & 0 \\ 0 & -2/C \end{pmatrix}\begin{pmatrix} I_{12R} \\ I_{12I} \end{pmatrix} + \begin{pmatrix} 2/C & 0 \\ 0 & 2/C \end{pmatrix}\begin{pmatrix} I_{1R} \\ I_{1I} \end{pmatrix} \tag{9.25}$$

Dynamic phasor equations for second shunt RC branch are as follows:

$$\frac{d}{dt}\begin{pmatrix} V_{2R} \\ V_{2I} \end{pmatrix} = \begin{pmatrix} -G/C & \omega_0 \\ -\omega_0 & -G/C \end{pmatrix}\begin{pmatrix} V_{2R} \\ V_{2I} \end{pmatrix} + \begin{pmatrix} 2/C & 0 \\ 0 & 2/C \end{pmatrix}\begin{pmatrix} I_{12R} \\ I_{12I} \end{pmatrix} + \begin{pmatrix} 2/C & 0 \\ 0 & 2/C \end{pmatrix}\begin{pmatrix} I_{2R} \\ I_{2I} \end{pmatrix} \tag{9.26}$$

9.3 Linearized LCC-HVDC system

The HVDC converter model (LCC) presented in Chapter 3 is summarized below.

9.3.1 Converter model

The 6-pulse Graetz bridge is the basic building block of the LCC-HVDC converters. Each valve in the bridge has a normal conduction period of 120° and two commutation periods ($2 \times \mu$) in each cycle.

The DC-side voltage $(V_{dc1} - V_{dc2})$ is given by,

$$V_{dc} = \frac{3\sqrt{2}B}{\pi T} V_l \cos\alpha - \frac{3X_cB}{\pi I_{dc}} \tag{9.27}$$

The real and the imaginary parts of the fundamental frequency component of the converter AC current are given by,

$$I_R = -\frac{\sqrt{3}BX_cI_{dc}^2V_R}{\pi V_l^2} + \frac{\sqrt{3}BM_1V_I}{2\pi X_cT^2} - \frac{\sqrt{6}BI_{dc}}{\pi T}\cos(\delta - \alpha - \mu) \tag{9.28}$$

$$I_I = -\frac{\sqrt{3}BM_1V_R}{2\pi X_c T^2} - \frac{\sqrt{3}BX_c I_{dc}^2 V_I}{\pi V_l^2} - \frac{\sqrt{6}BI_{dc}}{\pi T}\sin(\delta - \alpha - \mu) \tag{9.29}$$

In the above equations,

$$M_1 = (1 - cos(\mu))sin(2\alpha + \mu) + sin(\mu) - \mu$$

B - Number of 6-pulse bridges
T - Converter transformer turns ratio (AC/DC)
α - Converter firing angle
μ - Commutation angle
X_c - Transformer reactance referred to DC side
V_l - L-L voltage magnitude of the AC bus
δ - Angle of the AC bus voltage with respect to the common reference
V_R - Real component of the AC bus voltage
V_I - Imaginary component of the AC bus voltage

The higher-order harmonics of the DC-side voltage (e.g., 12th harmonic for 12-pulse converter) and the AC-side currents (e.g., 11th and 13th harmonics for 12-pulse converter) are neglected.

In the above steady-state relationships, the derivatives of the DC current are ignored. In the SS stability model, the derivative part is considered separately with the DC transmission system. This is explained in Section 9.3.2.

The DC-side voltage and real and imaginary components of the AC current (9.27), (9.28) and (9.29) depend on the commutation angle. If it is assumed that the change in the DC current during the commutation period is negligible, the commutation angle can be obtained as a function of DC-side current, AC-side voltage and firing angle as given in (9.30).

$$\mu = \cos^{-1}\left(\cos\alpha - \frac{\sqrt{2}X_c T I_{dc}}{V_l}\right) - \alpha \tag{9.30}$$

9.3.2 Linearized converter model

The fundamental frequency relationships (9.27), (9.28) and (9.29) derived from the converter-switching waveforms are used to obtain the linearized model. The commutation angle appeared in these relationships is replaced by (9.30). The converter-linearized relationships can be defined using four inputs (real component and imaginary component of AC-side voltage, DC-side current and firing angle) and three outputs (real component and imaginary component of AC-side current and DC-side voltage). The changes of the outputs for small changes in the inputs are given by (9.31).

$$\begin{bmatrix} \Delta I_R \\ \Delta I_I \\ \Delta V_{dc} \end{bmatrix} = \begin{bmatrix} K_a & K_b & K_c & K_d \\ K_e & K_f & K_g & K_h \\ K_i & K_j & K_k & K_l \end{bmatrix} \begin{bmatrix} \Delta V_R \\ \Delta V_I \\ \Delta I_{dc} \\ \Delta\alpha \end{bmatrix} \tag{9.31}$$

The parameters of the linearized model (9.31) can be obtained using the following equations at a particular steady-state condition.

$$K_a = \frac{\sqrt{3}BX_cI_{dc}^2}{\pi V_l^2} - \frac{\sqrt{6}BI_{dc}V_I \sin(\delta - \alpha - \mu)}{\pi TV_l^2}$$

$$K_b = \frac{\sqrt{3}BM_1}{2\pi X_cT^2} + \frac{\sqrt{6}BI_{dc}V_R \sin(\delta - \alpha - \mu)}{\pi TV_l^2}$$

$$K_c = -\frac{\sqrt{6}B \cos(\delta - \alpha - \mu)}{\pi T}$$

$$K_d = -\frac{\sqrt{6}BI_{dc} \sin(\alpha)\sin(\delta - \alpha - \mu/2)}{\pi T \sin(\alpha + \mu/2)}$$

$$K_e = -\frac{\sqrt{3}BM_1}{2\pi X_cT^2} + \frac{\sqrt{6}BI_{dc}V_I \cos(\delta - \alpha - \mu)}{\pi TV_l^2}$$

$$K_f = -\frac{\sqrt{3}BX_cI_{dc}^2}{\pi V_l^2} - \frac{\sqrt{6}BI_{dc}V_R \cos(\delta - \alpha - \mu)}{\pi TV_l^2}$$

$$K_g = -\frac{\sqrt{6}B \sin(\delta - \alpha - \mu)}{\pi T}$$

$$K_h = \frac{\sqrt{6}BI_{dc} \sin(\alpha)\cos(\delta - \alpha - \mu/2)}{\pi T \sin(\alpha + \mu/2)}$$

$$K_i = \frac{3\sqrt{2}BV_R \cos(\alpha)}{\pi TV_l}$$

$$K_j = \frac{3\sqrt{2}BV_I \cos(\alpha)}{\pi TV_l}$$

$$K_l = -\frac{3X_cB}{\pi}$$

$$K_l = -\frac{3\sqrt{2}BV_I \sin(\alpha)}{\pi T}$$

Note that the signs of K_i, K_j, K_k and K_l should be inverted in order to represent a positive voltage pole inverter.

9.3.3 Phase-locked oscillator

Firing control of valves is one of the major control functions of HVDC converters. The firing moment of a value is determined based on the phase of the AC voltage. Phase-locked oscillators (PLO) based on the phase vector techniques are used to determine the AC voltage phase in HVDC converters [21]. A simplified version of PLO is shown in Figure 9.7. An error signal $[sin(\delta - \delta_m)]$ of the actual phase angle (δ) and the calculated phase angle (δ_m) is given to a PI controller, in order to obtain the angular frequency error. The frequency error is added to the nominal frequency (ω_0) and the resultant frequency is integrated to obtain the phase angle.

The actual phase angle can be obtained in terms of V_R and $V_I(tan(\delta) = V_I/V_R)$. Two state variables: X_{PLO} and δ_m are used to represent the PI controller and the integrator, respectively. The SS PLO model is given in (9.32).

$$\begin{bmatrix} \Delta\dot{X}_{PLO} \\ \Delta\dot{\delta}_m \end{bmatrix} = \begin{bmatrix} 0 & -1 \\ K_{IP} & -K_{PP} \end{bmatrix} \begin{bmatrix} \Delta X_{PLO} \\ \Delta\delta_m \end{bmatrix} + \begin{bmatrix} -\frac{V_I}{V_I^2} & \frac{V_R}{V_I^2} \\ -\frac{K_{PP}V_I}{V_I^2} & \frac{K_{PP}V_R}{V_I^2} \end{bmatrix} \begin{bmatrix} \Delta V_R \\ \Delta V_I \end{bmatrix} \tag{9.32}$$

The phase angle (δ_m) determined by the PLO is added to the desired firing angle (α) to obtain the firing instant.

9.3.4 DC transmission system

A T-model is used to represent the DC system as shown in Figure 9.8. The DC-side inductance and resistance are distributed into two series inductor–resistor units and the line to ground capacitance is concentrated at the middle.

In the steady-state relationships, the derivative parts of the DC current are ignored. In the SS model, these derivative parts are included by considering the average effect of the converter–transformer leakage reactance. The contribution of the transformer leakage inductance is different during the normal conduction and commutation periods. Therefore, an average inductance is obtained and it is added to the half of the DC–line inductance to obtain the effective inductance [22] given in (9.33).

$$L_{eff} = \frac{L_{dc}}{2} + B\left(2 - \frac{3\mu}{2\pi}\right)L_c \tag{9.33}$$

L_{dc} is the DC-side total inductance and L_c is the transformer leakage inductance. The effective inductances in the rectifier side and the inverter side are different because the transformer leakage inductances are different.

The dynamics of the DC line are modeled using three state variables, namely, the rectifier-side DC current (ΔI_{dcr}), the inverter-side DC current (ΔI_{dci}) and the

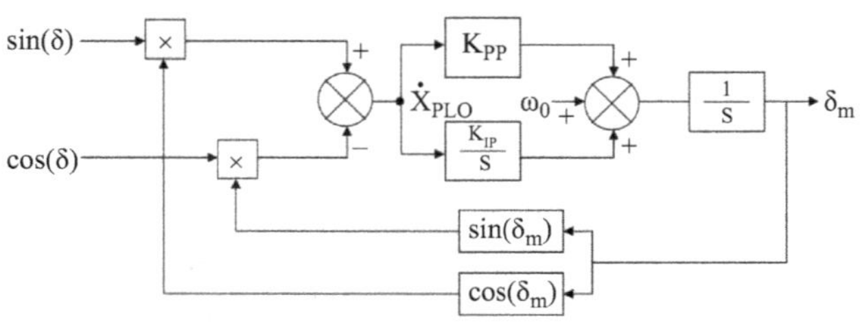

Figure 9.7 Phase-locked oscillator

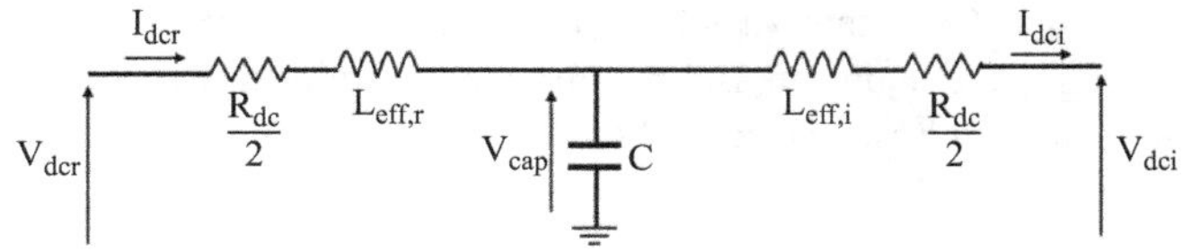

Figure 9.8 A DC transmission system

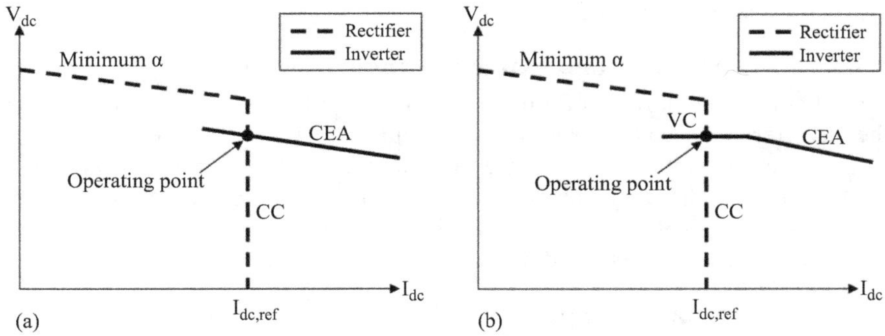

Figure 9.9 HVDC operating points at nominal conditions

mid-point capacitor voltage (ΔV_{cap}). The linearized model of the DC line is given in (9.34).

$$\begin{bmatrix} \Delta \dot{I}_{dcr} \\ \Delta \dot{I}_{dci} \\ \Delta \dot{V}_{cap} \end{bmatrix} = \begin{bmatrix} -\frac{R_{dc}}{2L_{eff,r}} & 0 & -\frac{1}{L_{eff,r}} \\ 0 & -\frac{R_{dc}}{2L_{eff,i}} & \frac{1}{L_{eff,i}} \\ \frac{1}{C} & -\frac{1}{C} & 0 \end{bmatrix} \begin{bmatrix} \Delta I_{dcr} \\ \Delta I_{dci} \\ \Delta V_{cap} \end{bmatrix} + \begin{bmatrix} \frac{1}{L_{eff,r}} & 0 \\ 0 & -\frac{1}{L_{eff,i}} \\ 0 & 0 \end{bmatrix} \begin{bmatrix} \Delta V_{dcr} \\ \Delta V_{dci} \end{bmatrix} \tag{9.34}$$

V_{dcr} and V_{dci} are the rectifier- and the inverter-side DC voltages. $L_{eff,r}$ and $L_{eff,i}$ are the rectifier- and the inverter-side effective inductances. R_{dc} and C are the DC resistance and the capacitance, respectively.

9.3.5 HVDC controllers

Two different operating conditions are considered as shown in Figure 9.9. The rectifier current controller and the inverter extinction angle controller are the active controllers in the operating point shown in Figure 9.9(a). The rectifier current controller and the inverter voltage controller are the active controllers in the operating point shown in Figure 9.9(b). The SS models of the individual controllers are described in the following sections.

9.3.5.1 Rectifier constant current control

The PI controller shown in Figure 9.10(a) is used to control the firing angle of the rectifier. The difference between the measured DC current and the desired current is used as the input. One state variable (X_{ar}) is used to represent the integral controller. The state-space equation and the change in required firing angle (Δar^*) are given by (9.35) and (9.36), respectively.

$$\Delta \dot{X}_{ar} = \Delta I_{dcr} - \Delta I_{dc,order} \tag{9.35}$$

$$\Delta ar^* = K_{Ir}\Delta X_{ar} + K_{Pr}\Delta I_{dcr} - K_{Pr}\Delta I_{dc,order} \tag{9.36}$$

K_{Pr} and K_{Ir} are the proportional and integral gains, respectively, and $I_{dc,order}$ is the rectifier current reference.

9.3.5.2 Inverter constant extinction angle control

The PI controller shown in Figure 9.10(b) is used to control the firing angle of the inverter. One state variable (X_{ai}) is used to represent the integral controller. The difference between the measured extinction angle and the desired angle is used as the input. The measured extinction angle is expressed in terms of the AC-side voltages, the inverter DC current and the controller state variables. The linearized model is given in (9.37) and (9.38).

$$\Delta \dot{X}_{ai} = k_{1a}\Delta X_{ai} + k_{2a}\Delta I_{dci} + k_{3a}\Delta V_{Ri} + k_{4a}\Delta V_{Ii} + k_{5a}\Delta \gamma_{order} \tag{9.37}$$

$$\Delta ai^* = K_{Pi}k_{1a}\Delta X_{ai} + K_{Pi}k_{2a}\Delta X_{dci} + K_{Pi}k_{3a}\Delta V_{Ri} - \frac{\sin(\alpha+\mu)}{\sin(\alpha)}k_{3a}\Delta V_{Ii} + K_{Pi}k_{5a}\Delta\gamma_{order} \tag{9.38}$$

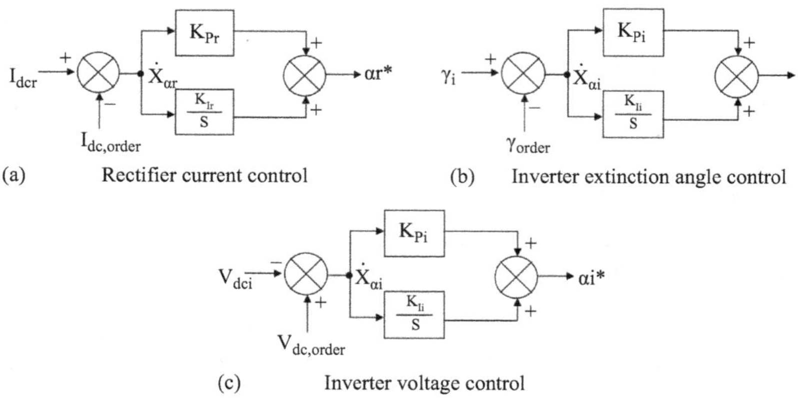

Figure 9.10 HVDC control schemes

In the above equations,

$$k_{1a} = -\frac{K_{Ii}}{\sin(\alpha_i) + \sin(\alpha_i + \mu_i)}$$

$$k_{2a} = -\frac{\sqrt{2}TX_c}{V_{li}(K_{Pi}\sin(\alpha_i) + \sin(\alpha_i + \mu_i))}$$

$$k_{3a} = \frac{\sqrt{2}TX_c I_{dci} V_{Ri}}{V_{li}^3(K_{Pi}\sin(\alpha_i) + \sin(\alpha_i + \mu_i))}$$

$$k_{4a} = \frac{\sqrt{2}TX_c I_{dci} V_{Ii}}{V_{li}^3(K_{Pi}\sin(\alpha_i) + \sin(\alpha_i + \mu_i))}$$

$$k_{5a} = -\frac{\sin(\alpha_i + \mu_i)}{K_{Pi}\sin(\alpha_i) + \sin(\alpha_i + \mu_i)}$$

9.3.5.3 Inverter constant voltage control

The PI controller shown in Figure 9.10(c) is used to control the firing angle of the inverter. Since only one of excitation angle controller or voltage controller is active, the same notation of the state variable used in the excitation controller (X_{ai}) is used to represent the integral controller. The difference between the measured inverter-side DC voltage and the desired DC voltage is used as the input. The inverter-side DC voltage is a function of the AC-side voltages, the inverter DC current and the firing angle as given in the linearized converter model in (9.31) and, therefore, it can be replaced by them. The linearized model is given in (9.39) and (9.40).

$$\Delta\dot{X}_{ai} = k_{1b}\Delta X_{ai} + k_{2b}\Delta I_{dci} + k_{3b}\Delta V_{Ri} + k_{4b}\Delta V_{Ii} + k_{5a}\Delta V_{dc,order} \tag{9.39}$$

$$\begin{aligned}\Delta ai^* = -\frac{k_{1b}}{K_{Li}}\Delta X_{ai} + K_{Pi}k_{2b}\Delta I_{dci} + K_{Pi}k_{3b}\Delta V_{Ri} + K_{Pi}k_{4b}\Delta V_{Ii} \\ + K_{Pi}k_{5b}\Delta V_{dc,order}\end{aligned} \tag{9.40}$$

In the above equations,

$$k_{1b} = -\frac{K_{Ii}Kl_i}{1 + K_{Pi}Kl_i}$$

$$k_{2b} = -\frac{Kk_i}{1 + K_{Pi}Kl_i}$$

$$k_{3b} = -\frac{Ki_i}{1 + K_{Pi}Kl_i}$$

$$k_{4b} = -\frac{Kj_i}{1 + K_{Pi}Kl_i}$$

$$k_{5b} = \frac{1}{1 + K_{Pi}Kl_i}$$

Ki_i, Kj_i, Kk_i and Kl_i can be obtained from the linearized converter model given in (9.31) by substituting the values corresponding to the inverter.

9.3.6 State-space model of HVDC system

The linearized models of the subsystems are combined together to obtain the state-space model of the HVDC system. This combination process is illustrated in the control block diagram shown in Figure 9.11.

The change in desired firing angle ($\Delta\alpha r^*$) given by the rectifier current controller (9.36) is added to the change in phase angle ($\Delta\delta_{mr}$) given by the PLO model (9.32) to obtain the change in firing instant of the rectifier ($\Delta\alpha r$) as shown in (9.41). Similarly, for the inverter, the change in desired firing angle ($\Delta\alpha i^*$) given by the extinction angle controller (9.38) or the voltage controller (9.40) is added to the change in phase angle ($\Delta\delta_{mi}$) to obtain the change in firing instant ($\Delta\alpha i$) as shown in (9.42).

$$\Delta\alpha r = \Delta\alpha r^* + \Delta\delta_{mr} \tag{9.41}$$

$$\Delta\alpha i = \Delta\alpha i^* + \Delta\delta_{mi} \tag{9.42}$$

The firing instants are substituted in the rectifier and inverter models obtained using (9.31). The changes in DC voltages given by the rectifier and inverter models are substituted in the DC transmission system (9.34). This procedure results in the state-space model of the HVDC system given in (9.43). The overall system consists of nine state variables and two control inputs. The real and imaginary components of the rectifier and inverter output currents are obtained from (9.44).

$$\Delta\dot{X}_H = \underset{(9\times9)}{A_H}(\Delta X_H) + \underset{(9\times2)}{B_H}(\Delta U_H) + \underset{(9\times4)}{E_H}(\Delta V_H) \tag{9.43}$$

$$\Delta I_H = \underset{(4\times9)}{C_H}(\Delta X_H) + \underset{(4\times2)}{D_H}(\Delta U_H) + \underset{(4\times4)}{Y_H}(\Delta V_H) \tag{9.44}$$

where

$$\Delta X_H = [\ \Delta X_{ar}\ \ \Delta X_{ai}\ \ \Delta I_{dcr}\ \ \Delta I_{dci}\ \ \Delta V_{cap}\ \ \Delta X_{PLOr}\ \ \Delta\delta_{mr}\ \ \Delta X_{PLOi}\ \ \Delta\delta_{mi}\ ]^T$$

$$\Delta U_H = \begin{bmatrix} \Delta I_{dc,order} \\ \Delta\gamma_{order} \end{bmatrix}, \Delta V_H = \begin{bmatrix} \Delta V_{Rr} \\ \Delta V_{Ir} \\ \Delta V_{Ri} \\ \Delta V_{Ii} \end{bmatrix} \text{ and } \Delta I_H = \begin{bmatrix} \Delta I_{Rr} \\ \Delta I_{Ir} \\ \Delta I_{Ri} \\ \Delta I_{Ii} \end{bmatrix}.$$

A_H, B_H, E_H, C_H, D_H and Y_H matrices are obtained using the procedure described above.

9.4 Accuracy of AC network models

The conventional SS model is accurate enough to analyze electromechanical oscillations of a power system. However, a better model, which includes the AC network dynamics, is required to analyze higher-frequency oscillations such as HVDC interactions.

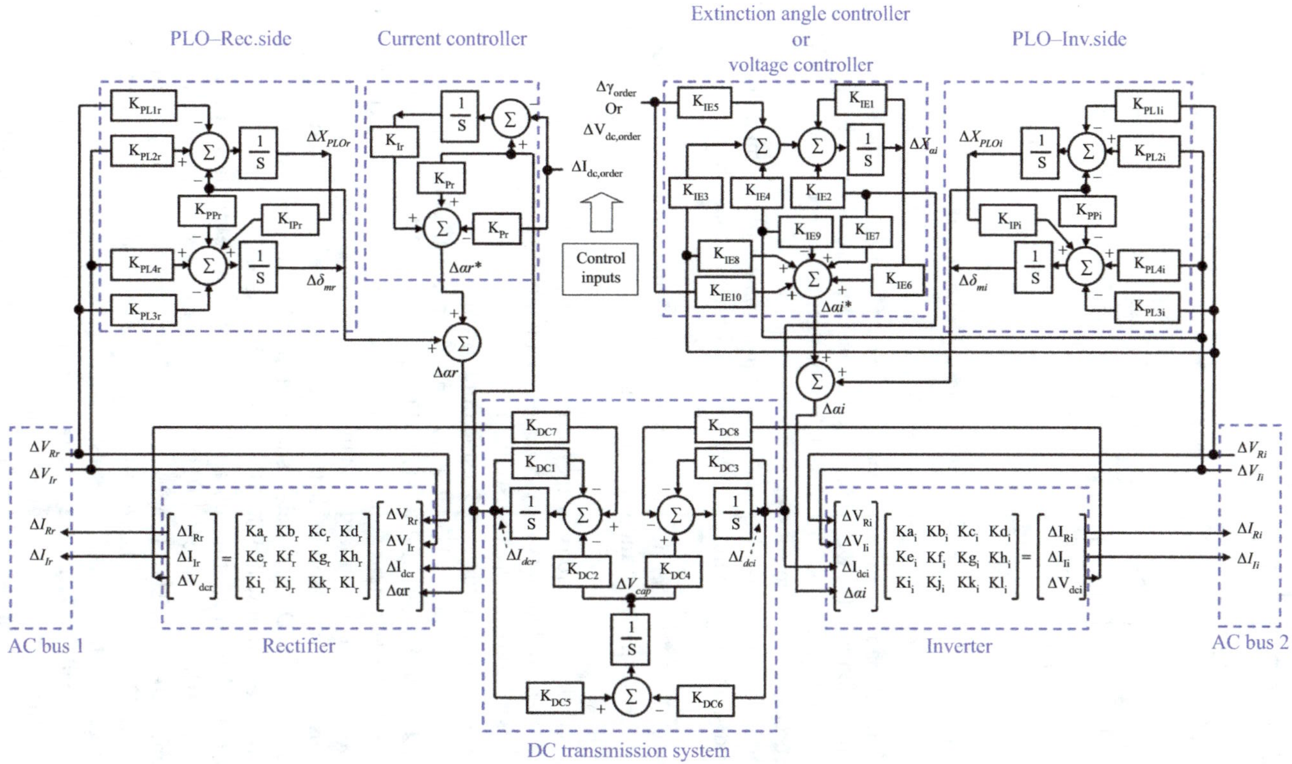

Figure 9.11 Linearized model of HVDC system

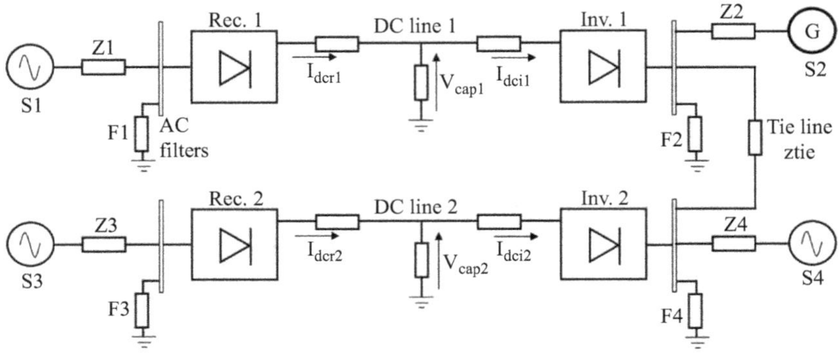

Figure 9.12 Multi-in-feed HVDC test system

In order to analyze the adequacy of the AC network models, a simple test system, in which two HVDC in-feeds are connected through a tie-line, is used. The circuit is shown in Figure 9.12. A synchronous generator is connected at S2 and all other sources are voltage sources.

Model-1: Conventional model (admittance matrix representation for AC network and standard dynamic model of synchronous machine used in stability studies).

Model-2: SS model including network dynamics (dynamic phasor representation for AC network and synchronous machine with stator winding differential equations).

For small perturbations in the control inputs, the system responses obtained using the linearized models were compared with the responses of the detailed EMT simulations obtained using PSCAD/EMTDC.

A pulse of magnitude of +5% and duration of 0.3 s was applied to the rectifier current controller input in HVDC1. The initial transients obtained with SS models are compared with those of PSCAD/EMTDC in Figure 9.13. Note that the scales of the *y*-axes of the sub-figures (a) and (b) are different. Since the perturbation was applied at HVDC1, much larger changes in the HVDC1 rectifier-side DC current can be observed compared to that of HVDC2. It is seen that Model-2 agrees well with PSCAD/EMTDC. Model-1 shows a poorly damped oscillatory frequency which does not agree with PSCAD/EMTDC results.

An extended simulation up to 1.5 s for the same perturbation is shown in Figure 9.14. The DC current of HVDC1 increases by 5% (0.05 kA) during the disturbance (0–0.3 s). The changes in the AC network voltages cause small changes in the HVDC2 currents. The results of Model-2 show a very good agreement with the average variation of the time response. It is noticed that the effect of electromechanical oscillations of the generator is also embedded in the current waveforms. Although Model-1 gives quite different results for high-frequency transients, the waveforms coincide with those of Model-2 and PSCAD/EMTDC in the latter part, where only the low-frequency electromechanical oscillations are

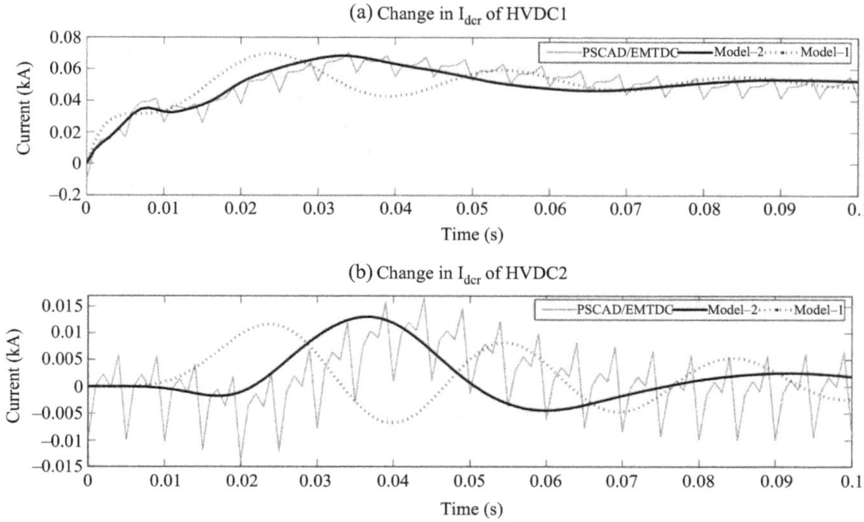

Figure 9.13 Accuracy of different models (initial changes in rectifier-side DC currents for a 5%, 0.3 s pulse on the current controller input in HVDC1)

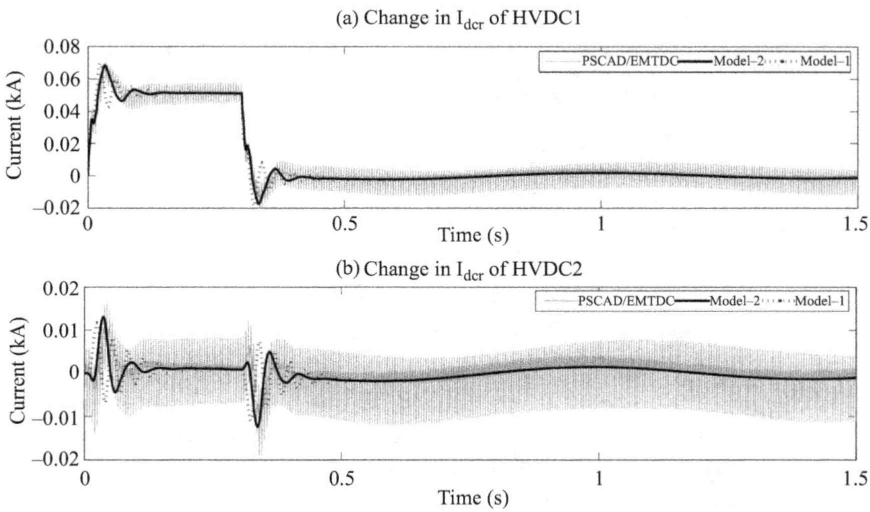

Figure 9.14 Changes in rectifier side DC currents for a 5%, 0.3 s pulse on the current controller input in HVDC1 (an extended simulation)

present. Furthermore, the comparison given in Figure 9.15 for the rotor speed of the generator confirms that both of the models are good for low-frequency electromechanical oscillation studies.

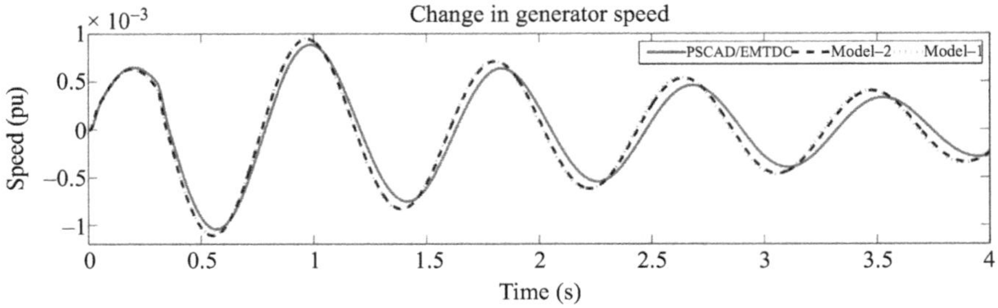

Figure 9.15 Change in generator speed for a 5%, 0.3 s pulse on the current controller input in HVDC1

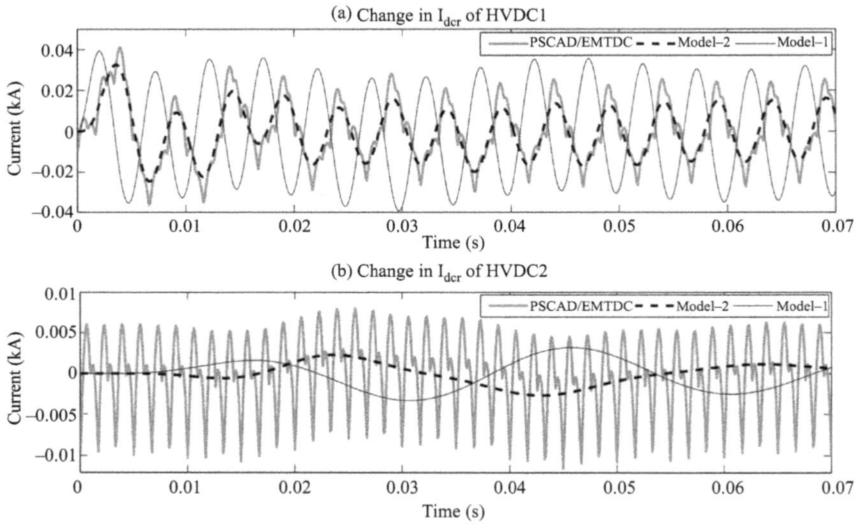

Figure 9.16 Changes in rectifier-side DC currents for a 5%, 200-Hz sinusoidal change of the HVDC1 rectifier-side AC source voltage (V_{S1})

Accuracy of the models at high frequencies was further verified by applying a 5%, 200-Hz sinusoidal change to the HVDC1 rectifier-side AC source voltage (V_{S1}). The changes in rectifier-side DC currents of HVDC1 and HVDC2 are compared in Figure 9.16. The disturbance applied at S1 propagates through the AC network (Z1) to the HVDC1. The 200-Hz sinusoidal signal can be observed in the rectifier DC current of HVDC1 obtained using PSCAD/EMTDC. The rectifier-side DC current obtained using Model-2 follows PSCAD/EMTDC results very well. However, the results of the conventional model (Model-1) are inaccurate at this frequency. The high-frequency disturbance dies out at HVDC2 while propagating through Z1, HVDC1 and the tie line (Ztie). Therefore, the 200-Hz signal cannot be observed in the rectifier-side DC current of HVDC2. PSCAD/EMTDC and Model-

2 results agree well for the low-frequency changes observed in that current and Model-1 does not show these changes accurately.

9.5 VSC-HVDC systems

In order to perform detailed studies such as controller interaction, it is essential to model VSC-HVDC systems considering dynamics of the AC system, DC system and controllers. Importantly, the measuring delays have to be modeled to assess the stability accurately [23]. Typically, VSCs are controlled using the $d-q$ decoupled control based on synchronous $(d-q)$ reference frame. The main advantage of using d-q quantities is that at the steady state, the time-varying AC quantities will become DC quantities in d-q reference frame and hence, simple PI-controllers can be used to control these quantities. A generic two-terminal symmetrical monopole modular multilevel converter (MMC) based VSC system shown in Figure 9.17 is considered in this section. The power system is modeled using real and imaginary (R-I) quantities and the control system is modeled in d-q quantities. The data pertaining to the system are given in Table 9.1. The following vectors are used to represent each sub-systems in state-space form.

$$
\begin{aligned}
\Delta X_{ac} &= [\Delta I_{sR}, \Delta I_{sI}, \Delta I_{tR}, \Delta I_{tI}]^T\\
\Delta X_{dc} &= [\Delta v_{dc1}, \Delta v_{dc2}, \Delta i_{dc}]^T\\
\Delta X_{ac12} &= [I_{t1R}, \Delta I_{t1I}, \Delta I_{t2R}, \Delta I_{t2I}]^T\\
\Delta X_{c1} &= [\Delta X_{v_{dc1}}, \Delta X_{v_{dc1}}, \Delta X_{i_{t1d}}, \Delta X_{i_{t1q}}]^T\\
\Delta X_{c2} &= [\Delta X_{P_{t2}}, \Delta X_{v_{t2}}, \Delta X_{i_{t2d}}, \Delta X_{i_{t2q}}]^T\\
\Delta X_{ms1} &= [\Delta v_{dc1(m)}, \Delta v_{t1(m)}]^T\\
\Delta X_{ms2} &= [\Delta P_{t2(m)}, \Delta v_{t2(m)}]^T\\
\Delta X_{dv} &= [\Delta v_{td(m)}, \Delta v_{tq(m)}]^T\\
\Delta X_{di} &= [\Delta i_{td(m)}, \Delta i_{tq(m)}]^T
\end{aligned}
$$

$$
\begin{aligned}
\Delta \mathbf{U}_{ac} &= [\Delta V_{sR}, \Delta V_{sI}]^T\\
\Delta \mathbf{U}_{c1} &= [\Delta v_{dc1(r)}, \Delta V_{t1(r)}]^T\\
\Delta \mathbf{U}_{c1} &= [\Delta P_{t2(r)}, \Delta V_{t2(r)}]^T\\
\Delta \mathbf{I}_t &= [\Delta I_{tR}, \Delta V_{tI}]^T\\
\Delta \mathbf{I}_t^{dq} &= [\Delta i_{td}, \Delta i_{tq}]^T\\
\Delta \mathbf{V}_t &= [\Delta V_{tR}, \Delta V_{tI}]^T\\
\Delta \mathbf{V}_i &= [\Delta V_{iR}, \Delta V_{iI}]^T\\
\Delta \mathbf{V}_t^{dq} &= [\Delta v_{td}, \Delta v_{tq}]^T\\
\Delta \mathbf{V}_i^{dq} &= [\Delta v_{id}, \Delta v_{iq}]^T\\
\Delta \mathbf{V}_o^{dq} &= [\Delta v_{od}, \Delta v_{oq}]^T
\end{aligned}
$$

9.5.1 Representation of AC system

The AC network is represented using a Thévenin equivalent model that comprises a voltage source in series with an R-R//L type impedance as shown in Figure 9.17.

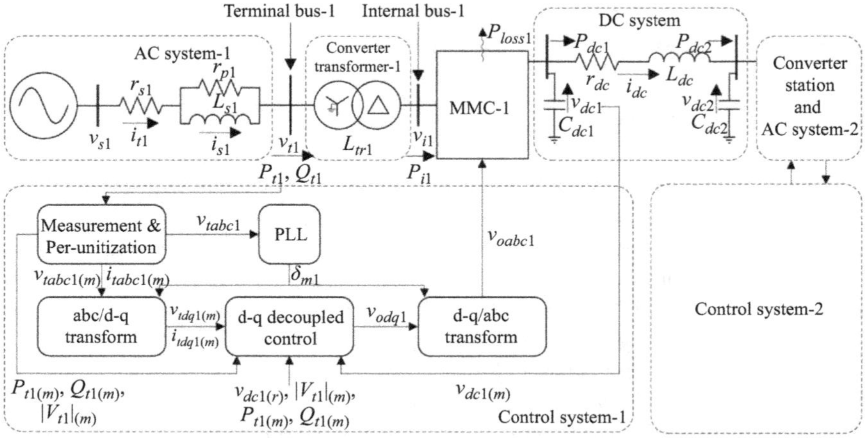

Figure 9.17 Two-terminal MMC-VSC system (m-measured and r-reference)

Table 9.1 Parameters of MMC-VSC system

Parameter	Value
Rating	$V_{ac} = 230$ kV, SCR $= 2.0/84°$, $V_{dc} = 500$ kV, $P_{dc} = 500$ MW
Transformer	230 kV:250 kV, $L_{tr} = 0.18\,p.u.$ on 575 MVA
Converter	250 sub-modules (SM)/arm, $C_{SM} = 10000$ μF, $L_{arm} = 50$ mH
DC Line	400 km, $R_{dc} = 7.17\ \Omega$, $L_{dc} = 0.74$ H, $C_{dc} = 5$ μF

For easiness, the arm inductance is considered together with the transformer reactance and, hence, the voltage v_i represents the internal AC voltage of the converter. Dynamic phasors are used to write the analytical equations of the AC system as shown in (9.15). The linearized model of the AC system in the state-space form is given in (9.45).

$$\Delta\dot{\mathbf{X}}_{ac} = \mathbf{A}_{ac}\Delta\mathbf{X}_{ac} + \mathbf{B}_{ac}\Delta\mathbf{U}_{ac} + \mathbf{E}_{ac}\Delta\mathbf{V}_i \tag{9.45}$$

9.5.2 Representation of DC system

The DC transmission network is represented using a π-section. Capacitors C_{dc1} and C_{dc2} in Figure 9.17 are modeled to reflect the combined effects of the shunt capacitances of DC line and sub-module capacitances of converter. The dynamics of capacitors are modeled by writing energy-balance relationships shown in (9.46) and (9.47). The dynamics of DC line inductor is given in (9.48). The average power loss in MMC sub-modules is modeled by considering the average flows of both AC and DC currents through sub-modules. Equation (9.49) shows the state-space form of the DC system model.

$$C_{dc1}\frac{d}{dt}v_{dc1} = P_{i1} - P_{dc1} - P_{loss1} \tag{9.46}$$

$$C_{dc2}\frac{d}{dt}v_{dc2} = P_{i2} + P_{dc2} - P_{loss2} \tag{9.47}$$

$$L_{dc}\frac{d}{dt}i_{dc} = -r_{dc}i_{dc} + v_{dc1} - v_{dc2} \tag{9.48}$$

$$\Delta\dot{\mathbf{X}}_{dc} = \mathbf{A}_{dc}\Delta\mathbf{X}_{dc} + \mathbf{A}^*{}_{dc}\Delta\mathbf{X}_{ac12} + \mathbf{E}_{dc}\Delta\mathbf{V}_{i12} \tag{9.49}$$

9.5.3 Representation of MMC

The block diagram of MMC is shown in Figure 9.18. Relationships between the voltage orders (v_o) from the control system and the internal voltage (v_i) of converter are given by (9.50) and (9.51). The algebraic model of the MMC is given in (9.52).

$$v_{id} = \frac{1}{2}\left(\frac{\sqrt{3}v_{od}}{\sqrt{2}}\right)v_{dc} \tag{9.50}$$

$$v_{iq} = \frac{1}{2}\left(\frac{\sqrt{3}v_{oq}}{\sqrt{2}}\right)v_{dc} \tag{9.51}$$

$$\Delta\mathbf{V}_i^{dq} = \mathbf{C}_m\Delta v_{dc} + \mathbf{D}_m\Delta\mathbf{V}_o^{dq} \tag{9.52}$$

9.5.4 Representation of control system

The functional block diagram of control system is shown in Figure 9.19. This control scheme consists of cascaded PI-control loops referred to as outer-loop (voltage/power) controllers and inner-loop (current) controllers. At the outer-loop, either the DC voltage or real power is selected as the control variable to produce *d*-axis current reference. Similarly, either the AC voltage or reactive power is selected as the *q*-axis control variable. The control mode discussed below is where the MMC1 controls the DC voltage and AC voltage, while MMC2 controls real power and AC voltage. In inner-loop controllers, the measured d-q currents are regulated to meet the reference currents produced by outer-loop controllers.

Dynamic equations of each PI-controller are written for the model shown in Figure 9.20. The error between reference value $(\alpha_{(r)})$ and measured value $(\alpha_{(m)})$ is regulated through the PI-controller. The dynamic and linear equations of PI-controller are given by (9.53) and (9.54).

$$\frac{d}{dt}X_a = \alpha_{(r)} - \alpha_{(m)} \tag{9.53}$$

$$Z = K_{Pa}\left(\alpha_{(r)} - \alpha_{(m)}\right) + K_{Ia}X_a \tag{9.54}$$

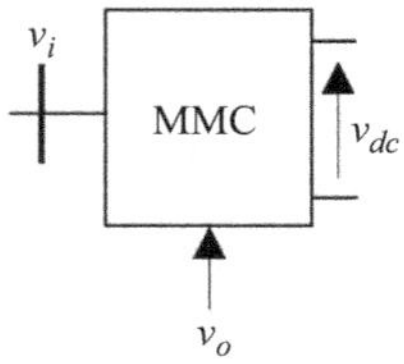

Figure 9.18 Block diagram of MMC

The dynamic equations for outer-loop controllers, given through (9.55–9.58), are obtained by substituting in (9.53) with v_{dc1}, P_{t2}, V_{t1} and V_{t2}.

$$\frac{d}{dt}X_{vdc1} = \left(v_{dc1(r)} - v_{dc1(m)}\right) \tag{9.55}$$

$$\frac{d}{dt}X_{P_{t2}} = \left(P_{t2(r)} - P_{t2(m)}\right) \tag{9.56}$$

$$\frac{d}{dt}X_{V_{t1}} = \left(V_{t1(r)} - V_{t1(m)}\right) \tag{9.57}$$

$$\frac{d}{dt}X_{V_{t2}} = \left(V_{t2(r)} - V_{t2(m)}\right) \tag{9.58}$$

The outputs of outer-loop controller are the reference currents for inner-loop current controllers. The output relationships given by (9.59–9.62) are written by substituting the parameter Z in (9.54) with $i_{t1d(r)}$, $i_{t2d(r)}$, $i_{t1q(r)}$ and $i_{t2q(r)}$.

$$i_{t1d(r)} = K_{P(v_{dc1})}\left(v_{dc1(r)} - v_{dc1(m)}\right) + K_{I(v_{dc1})}X_{v_{dc1}} \tag{9.59}$$

$$i_{t2d(r)} = K_{P(P_{t1})}\left(P_{t1(r)} - P_{t1(m)}\right) + K_{I(P_{t1})}X_{P_{t1}} \tag{9.60}$$

$$i_{t1q(r)} = K_{P(V_{t1})}\left(V_{t1(r)} - V_{t1(m)}\right) + K_{I(V_{t1})}X_{V_{t1}} \tag{9.61}$$

$$i_{t2q(r)} = K_{P(V_{t2})}\left(V_{t2(r)} - V_{t2(m)}\right) + K_{I(V_{t2})}X_{V_{t2}} \tag{9.62}$$

The state equations correspond to inner-loop controllers on both sides MMCs are given by (9.63–9.66).

$$\frac{d}{dt}X_{it1d} = i_{t1d(r)} - i_{t1d(m)} \tag{9.63}$$

$$\frac{d}{dt}X_{it2d} = i_{t2d(r)} - i_{t2d(m)} \tag{9.64}$$

$$\frac{d}{dt}X_{it1q} = i_{t1q(r)} - i_{t1q(m)} \tag{9.65}$$

$$\frac{d}{dt}X_{it2q} = i_{t2q(r)} - i_{t2q(m)} \tag{9.66}$$

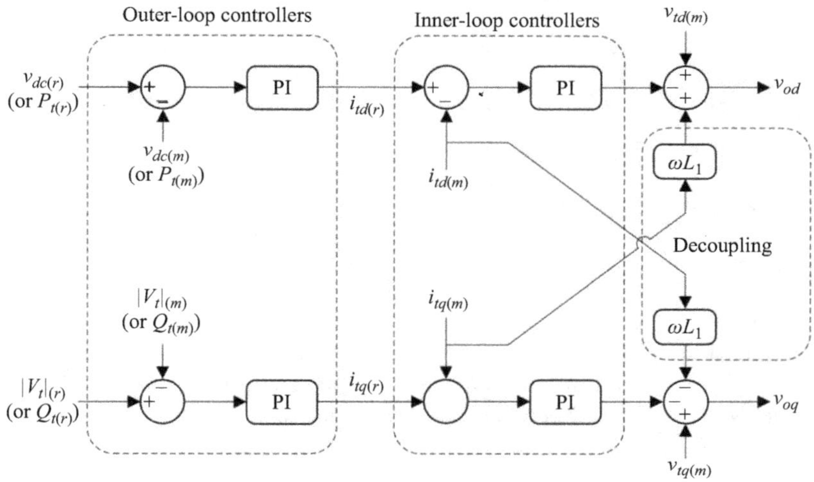

Figure 9.19 Block diagram of the d-q decoupled control system (m-measured and r-reference)

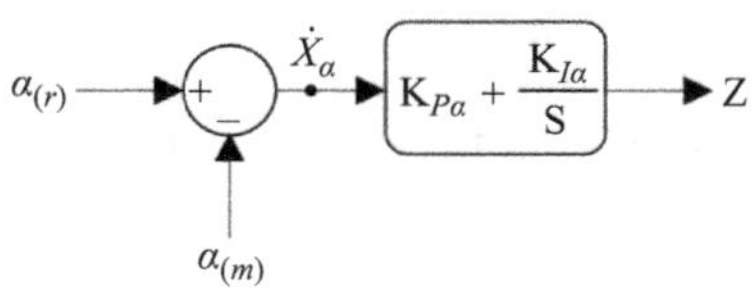

Figure 9.20 Block diagram of the PI-controller

A first-order measurement filter model shown in Figure 9.21 is used to represent measuring delays of quantities, which are regulated at inner-loop and outer-loop controllers. The dynamics of filter model is given by (9.67).

$$\frac{d}{dt}\alpha_{(m)} = \frac{1}{T_\alpha}(\alpha - \alpha_{(m)}) \tag{9.67}$$

Measurement filter dynamics of outer-loop quantities are given through (9.68–9.71).

$$\frac{d}{dt}v_{dc1(m)} = \frac{1}{T_{vdc1}}\left(v_{dc1} - v_{dc1(m)}\right) \tag{9.68}$$

$$\frac{d}{dt}P_{t2(m)} = \frac{1}{T_{Pt2}}\left(P_{t2} - P_{t2(m)}\right) \tag{9.69}$$

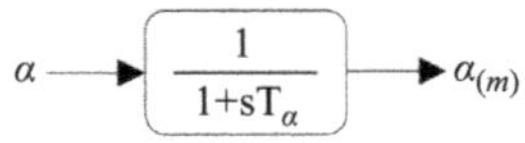

Figure 9.21 Block diagram of the measurement filter

$$\frac{d}{dt}V_{t1(m)} = \frac{1}{T_{V_{t1}}}\left(V_{t1} - V_{t1(m)}\right) \tag{9.70}$$

$$\frac{d}{dt}V_{t1(m)} = \frac{1}{T_{V_{t2}}}\left(V_{t2} - V_{t2(m)}\right) \tag{9.71}$$

By substituting P_{t2}, V_{t1} and V_{t2} in-terms of the R-I frame quantities of voltages and currents and by linearizing the equations, the state-space models of measurement delays given by (9.72) and (9.73) are obtained.

For MMC1,

$$\Delta\dot{\mathbf{X}}_{ms1} = \mathbf{A}_{ms1}\Delta\mathbf{X}_{ms1} + \mathbf{A}^*{}_{ms1}\Delta v_{dc1} + \mathbf{F}_{ms1}\Delta\mathbf{I}_{t1} \tag{9.72}$$

For MMC2,

$$\Delta\dot{\mathbf{X}}_{ms2} = \mathbf{A}_{ms2}\Delta\mathbf{X}_{ms2} + \mathbf{A}^*{}_{ms2}\Delta\mathbf{I}_{t2} + \mathbf{F}_{ms2}\Delta\mathbf{V}_{t2} \tag{9.73}$$

The measurement filter dynamics of inner-loop quantities are given by (9.74–9.77) and the corresponding state-space forms are given in (9.78) and (9.79).

$$\frac{d}{dt}i_{td(m)} = \frac{1}{T_{itd}}\left(i_{td} - i_{td(m)}\right) \tag{9.74}$$

$$\frac{d}{dt}i_{tq(m)} = \frac{1}{T_{itq}}\left(i_{tq} - i_{tq(m)}\right) \tag{9.75}$$

$$\frac{d}{dt}v_{td(m)} = \frac{1}{T_{vtd}}\left(v_{td} - v_{td(m)}\right) \tag{9.76}$$

$$\frac{d}{dt}v_{tq(m)} = \frac{1}{T_{vtq}}\left(v_{tq} - v_{tq(m)}\right) \tag{9.77}$$

$$\Delta\dot{\mathbf{X}}_{di} = \mathbf{A}_{di}\Delta\mathbf{X}_{di} + \mathbf{F}_{di}\Delta\mathbf{I}_t^{dq} \tag{9.78}$$

$$\Delta\dot{\mathbf{X}}_{dv} = \mathbf{A}_{dv}\Delta\mathbf{X}_{dv} + \mathbf{E}_{dv}\Delta\mathbf{V}_t^{dq} \tag{9.79}$$

The linear equations from outer-loop controllers are used to replace reference currents in the above equations to obtain the dynamic equations. The state-space models of control system can be obtained by linearizing the controller dynamic equations and by combining them into the state-space form as shown in (9.80) and (9.81).

For MMC1,

$$\Delta\dot{\mathbf{X}}_{c1} = \mathbf{A}_{c1}\Delta\mathbf{X}_{c1} + \mathbf{A}^{*}{}_{c1}\Delta\mathbf{X}_{ms1} + \mathbf{A}^{\dagger}{}_{c1}\Delta\mathbf{X}_{di1} + \mathbf{B}_{c1}\Delta\mathbf{U}_{c1} \tag{9.80}$$

For MMC2,

$$\Delta\dot{\mathbf{X}}_{c2} = \mathbf{A}_{c2}\Delta\mathbf{X}_{c2} + \mathbf{A}^{*}{}_{c2}\Delta\mathbf{X}_{ms2} + \mathbf{A}^{\dagger}{}_{c2}\Delta\mathbf{X}_{di2} + \mathbf{B}_{c2}\Delta\mathbf{U}_{c2} \tag{9.81}$$

The outputs of inner-loop controllers are combined with the measured terminal voltages and decoupling terms to produce the voltage orders (v_o) as shown in Figure 9.19. The voltage-order equations for terminal1 and terminal2 are given by (9.82) and (9.83), respectively.

For MMC1,

$$\Delta\mathbf{V}_{o1}^{dq} = \mathbf{C}_{o1}\Delta\mathbf{X}_{c1} + \mathbf{C}^{*}{}_{o1}\Delta\mathbf{X}_{ms1} + \mathbf{C}^{\dagger}{}_{o1}\Delta\mathbf{X}_{di1} + \mathbf{C}^{\ddagger}{}_{o1}\Delta\mathbf{X}_{dv1} + \mathbf{D}_{o1}\Delta\mathbf{U}_{c1} \tag{9.82}$$

For MMC2,

$$\Delta\mathbf{V}_{o2}^{dq} = \mathbf{C}_{o2}\Delta\mathbf{X}_{c2} + \mathbf{C}^{*}{}_{o2}\Delta\mathbf{X}_{ms2} + \mathbf{C}^{\dagger}{}_{o2}\Delta\mathbf{X}_{di2} + \mathbf{C}^{\ddagger}{}_{o2}\Delta\mathbf{X}_{dv2} + \mathbf{D}_{o2}\Delta\mathbf{U}_{c2} \tag{9.83}$$

9.5.5 Overall state-space model

The state-space models of control system modeled in synchronous reference $(d-q)$ frame and power system modeled in real-imaginary (R-I) frame are combined using the $d-q$ to R-I transformation. The terminal voltage phase angle δ_m used in the above transformation is obtained using phase locked loop (PLL). The state-space model of the PLL used in this section is identical to the model given in (9.32). The complete state-space model of the system in the form given by (9.84) is obtained by combining the linearized models of AC system, DC system, control system, PLL and measurement filters. Signal flows between the above sub-systems are shown in Figure 9.22 [24]. This complete system consists 35 state variables and 8 control inputs.

$$\Delta\dot{\mathbf{X}} = \mathbf{A}\Delta\mathbf{X} + \mathbf{B}\Delta\mathbf{U} \tag{9.84}$$

The model can be excited by applying perturbations to its control inputs. The outputs of the system can be written as linear combinations of state variables and control inputs as given by (9.85), where C and D are user-defined matrices to generate required outputs.

$$\Delta\mathbf{Y} = \mathbf{C}\Delta\mathbf{X} + \mathbf{D}\Delta\mathbf{U} \tag{9.85}$$

9.5.6 Validation of the linearized model

The linearized model is validated against the non-linear EMT simulation on the real-time digital simulator (RTDS) [25]. The controller parameters given in Table 9.2 are used for the validation purpose. A small-pulse with a magnitude of

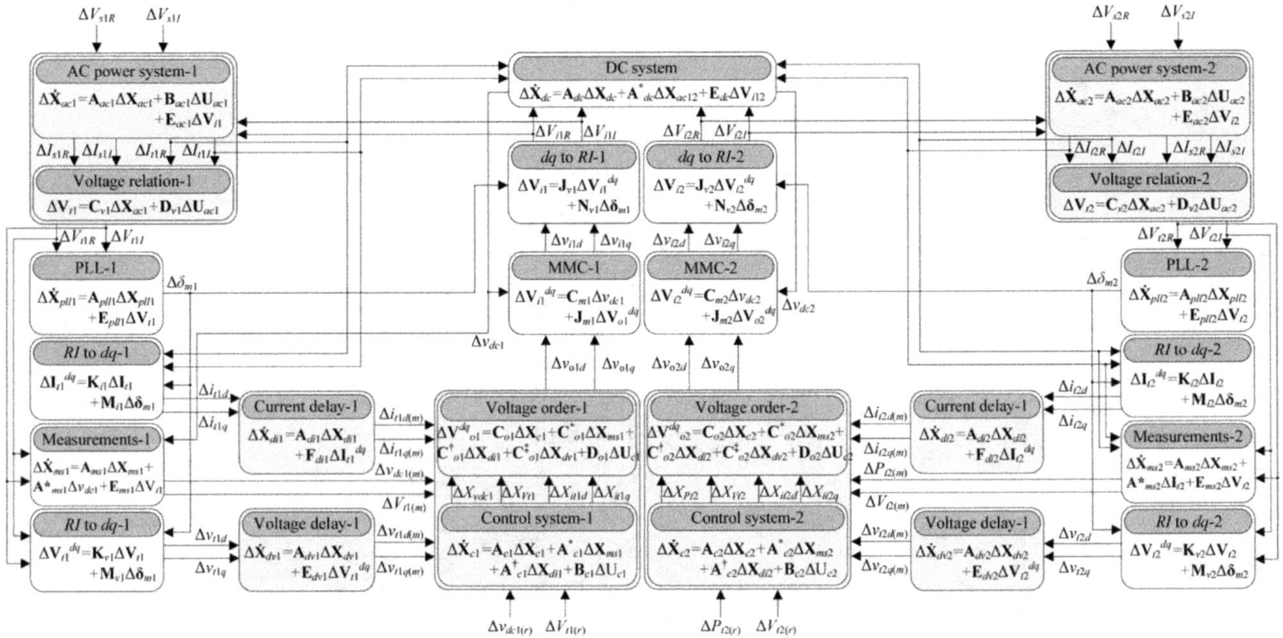

Figure 9.22 Overall state-space model of the point-to-point MMC-VSC system

Table 9.2 Control system parameters

Controller	**Gain K_P , K_I (p.u.)**	**Filter T_a(ms)**
PLL (MMC1, MMC2)	50, 250	—
DC voltage (MMC1)	2, 20	0.5
Real Power (MMC2)	1, 20	20
AC voltage (MMC1, MMC2)	1, 20	20
d-q current (MMC1, MMC2)	1, 100	1
d-q voltage (MMC1, MMC2)	—	5

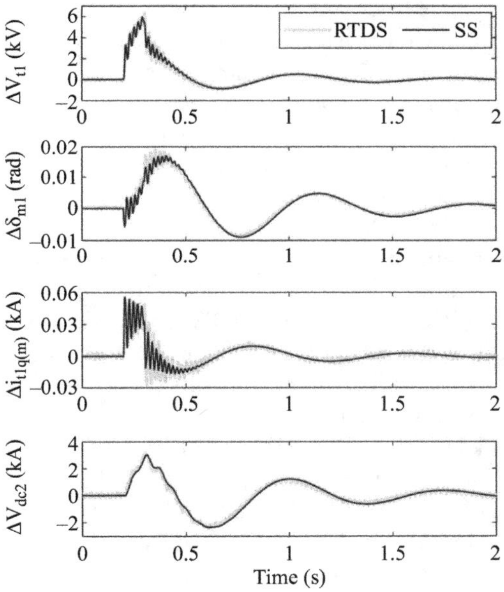

Figure 9.23 Variation of changes in AC voltage-v_{t1}, PLL angle- δ_{m1}, measured q-axis current-$i_{t1q(m)}$ and DC voltage-v_{dc2} for a small pulse (0.03 p.u. with 100 ms duration) applied to the AC voltage reference of MMC1 at 0.2 s

0.03 p.u. and a duration of 100 ms is applied to MMC1 AC voltage reference. The simulation results from RTDS and SS models are compared in Figure 9.23. It can be observed that the low frequency as well as high-frequency oscillations are captured by the linearized model with an acceptable accuracy.

9.6 Summary

The application of dynamic phasors for developing SS models of HVDC systems was presented in this chapter. The chapter started with an introduction to dynamic phasors. Then, the modeling of transmission lines using dynamic phasors was presented. This was followed by modeling of individual components of the HVDC systems. How to combine the individual components together was explained with the aid of block diagrams to illustrate the signal flow between components. The developed SS models have been validated against detailed EMT simulation models.

References

[1] P. Kundur, J. Paserba, and V. Ajjarapu. 'Definition and classification of power system stability ieee/cigre joint task force on stability terms and definitions'. *IEEE Transactions on Power Systems*. 2004, vol. 19, pp. 1387–1401.

[2] V. Venkatasubramanian, H. Schattler, and J. Zaborszky. 'Fast time-varying phasor analysis in the balanced three-phase large electric power system'. *IEEE Transactions on Automatic Control*. 1995, vol. 40, pp. 1975–1982.

[3] H. S. J. Zaborszky, and V. Venkatasubramanian. 'Limitations of the quasistationary dynamics and error due to phasor computation in network equations'. Power Systems Computations Conference; Le Touquet, France, Oct 1993, pp. 721–729.

[4] V. Venkatasubramanian, H. Schattler, and J. Zaborszky. 'Dynamics of large constrained nonlinear systems-a taxonomy theory [power system stability]'. *Proceedings of the IEEE*. 1995, vol. 83, pp. 1530–1561.

[5] S. Henschel. 'Analysis of electromagnetic and electromechanical power system transients with dynamic phasors'. PhD thesis, The University of British Columbia, Canada, 1999.

[6] U. D. Annakkage, N. K. C. Nair, Y. Liang, et al. 'Dynamic system equivalents: A survey of available techniques'. *IEEE Transactions on Power Delivery*. 2012, vol. 27, pp. 411–420.

[7] V. Venkatasubramanian. 'Tools for dynamic analysis of the general large power system using time-varying phasors'. *International Journal of Electrical Power & Energy Systems*. 1994, vol. 16(6), pp. 365–376.

[8] P. Zhang, J. R. Marti, and H. W. Dommel. 'Synchronous machine modeling based on shifted frequency analysis'. *IEEE Transactions on Power Systems*. 2007, vol. 22, pp. 1139–1147.

[9] P. Zhang, J. R. Marti, and H. W. Dommel. 'Shifted-frequency analysis for emtp simulation of power-system dynamics'. *IEEE Transactions on Circuits and Systems I: Regular Papers*. 2010, vol. 57, pp. 2564–2574.

[10] C. Karawita. 'HVDC interaction studies using small signal stability assessment'. PhD thesis, University of Manitoba, Canada, April 2009.

[11] T. H. Demiray. 'Simulation of power system dynamics using dynamic phasor models'. PhD thesis, Swiss Federal Institute of Technology Zurich, 2008.

[12] C. Karawita, and U. D. Annakkage. 'Multi-infeed hvdc interaction studies using small-signal stability assessment'. *IEEE Transactions on Power Delivery*. 2009, vol. 24, pp. 910–918.

[13] C. Karawita, and U. D. Annakkage. 'A hybrid network model for small signal stability analysis of power systems'. *IEEE Transactions on Power Systems*. 2010, vol. 25, pp. 443–451.

[14] S. R. Sanders, J. M. Noworolski, X. Z. Liu, and G. C. Verghese. 'Generalized averaging method for power conversion circuits'. *IEEE Transactions on Power Electronics*. 1991, vol. 6, pp. 251–259.

[15] A. M. Stankovic, S. R. Sanders, and T. Aydin. 'Dynamic phasors in modeling and analysis of unbalanced polyphase ac machines'. *IEEE Transactions on Energy Conversion*. 2002, vol. 17, pp. 107–113.

[16] A. M. Stankovic, and T. Aydin. 'Analysis of asymmetrical faults in power systems using dynamic phasors'. *IEEE Transactions on Power Systems*. 2000, vol. 15, pp. 1062–1068.

[17] J. Mahdavi, A. Emaadi, M. D. Bellar, and M. Ehsani. 'Analysis of power electronic converters using the generalized state-space averaging approach'. *IEEE Transactions on Circuits and Systems I: Fundamental Theory and Applications*. 1997, vol. 44, pp. 767–770.

[18] V. A. Caliskan, O. C. Verghese, and A. M. Stankovic. 'Multifrequency averaging of dc/dc converters'. *IEEE Transactions on Power Electronics*. 1999, vol. 14, pp. 124–133.

[19] M. A. Kulasza. 'Generalized dynamic phasor-based simulation for power systems'. Master's thesis, University of Manitoba, Canada, 2014.

[20] C. L. DeMarco, and G. Verghese. 'Bringing phasor dynamics into the power system load flow'. Proceedings of the 25th North American Power Symposium, Washington, D.C., 1993.

[21] V. Sood, V. Khatri, H. Jin, D. A. N. Jacobson, and P. Wang. 'Performance assessment using emtp of two gate firing units for HVDC converters operating with weak ac systems'. International Conference on Power System Transients (IPST 1995); Sept 1995, pp. 517–522.

[22] W. Hammer. 'Dynamic modeling of line and capacitor commutated converters for HVDC power transmission'. PhD thesis, Swiss Federal Institute of Technology, Zurich, 2003.

[23] S. Arunprasanth, R. Kuffel, U. D. Annakkage, and C. Karawita. 'Dynamic behaviour of VSC-HVDC systems under different AC system strengths'. Proceedings of IEEE/PES Transmission and Distribution Conference and Exposition (T&D) Apr 2018, pp. 1–9.

[24] S. Arunprasanth. 'A systematic procedure to determine controller parameters for MMC-VSC systems'. PhD thesis, University of Manitoba, Canada, 2016.

[25] RTDS Simulator. Available: https://www.rtds.com/contact/RTDS Technologies Inc.

Chapter 10
Small-signal modeling of HVDC systems

Lennart Harnefors[1], Lidong Zhang[2] and Alan Wood[3]

10.1 Introduction

Analytical modeling of high-voltage DC (HVDC) systems has traditionally been regarded as a very difficult task, and its importance has been very often underestimated in the presence of a large number of available electromagnetic simulation packages.

However, an approach based on analytical modeling, for reaching generalized conclusions, or performing in-depth analysis, cannot be matched by the trial and error type of studies carried out using simulation software.

The small-signal analysis of power electronic systems has been an active area of research since Ainsworth identified a mechanism of converter instability related to individual firing angle control in 1967 [1]. Although there have been a significant number of contributions to the subject in the intervening years, still there are many aspects of system modeling and dynamics that are not clearly defined in the literature. With the increasing use of power electronic controllers, the availability of dynamic models, which provide insights into system behavior and facilitate analysis and design, will be of considerable benefit. An opportunity currently exists for the establishment of a framework, based on small-signal analysis, which is capable of accepting the modeling challenges of existing and future systems.

It is useful to analyze the interaction between a HVDC converter and the AC grid to which the converter is connected. The higher the rated power of the converter is, the more important such studies are, since the converter will have a greater dynamic impact on the grid. HVDC terminals are among the most powerful grid-connected converters.

HVDC converters contribute to dynamic phenomena in various frequency ranges. Among the slowest – with time scale in the tens-of-seconds range – we find the grid-frequency dynamics, determined by the total inertia of all rotating machines directly connected to the grid together with the synchronous machine which perform primary frequency control (PFC) [2]. Next there are the interarea

[1]ABB Corporate Research, Västerås, Sweden
[2]ABB Power Grid Research, Västerås, Sweden
[3]University of Canterbury, Christchurch, New Zealand

modes, with eigenfrequencies of about 0.5 Hz. HVDC installations are often powerful enough to affect both, and schemes for PFC and interarea-mode damping have been employed for decades [3–6]. For the study of these phenomena, it is often sufficient to consider the HVDC converter as a controllable active-power source.

In the range from approximately 10 Hz to the fundamental frequency (50 Hz or 60 Hz), we find subsynchronous interaction phenomena [7]. Here, an HVDC converter interacts with nearby synchronous machines and/or series-compensated power lines. It is important to ensure that the interaction adds to the damping of the critical subsynchronous modes, as otherwise oscillations of high amplitude may result [8–13]. Frequency variations and interarea oscillations are global within one synchronous area, whereas subsynchronous oscillations are local in the electrical region where the interaction occurs. For their study, merely the adjacent synchronous machines and transmission lines need to be modeled. On the other hand, a more detailed voltage-source converter (VSC) model is necessary. All control loops – not only that for the active power – need to be modeled.

From the fundamental frequency up to a few times thereof, there is the risk for destabilization of electrical resonances, or even an unstable interconnection without the presence of a resonance [14–20]. This is a problem mainly for weak grids, i.e., grids that have a low short-circuit ratio (SCR). The SCR is given as

$$\mathrm{SCR} = \frac{V_{\mathrm{base}}^2}{X_{\mathrm{sc}} S_{\mathrm{base}}} \tag{10.1}$$

where X_{sc} is the short-circuit reactance seen from the converter terminals, and V_{base} S_{base} are the base (i.e., rated) line-to-line root mean square voltage and apparent power of the converter, respectively. An HVDC link of high-power rating thus “sees” a weaker grid than an HVDC link of low-power rating, demonstrating clearly that high-power HVDC terminals are particularly critical. SCR < 3 is generally considered to constitute a weak grid. As for subsynchronous interaction, near-synchronous stability studies require a complete HVDC model.

In the highest frequency range, typically above 1 kHz, there is a risk of destabilization of electrical resonances as well [21]. This is caused by the time delay in the execution of the current control and the pulse-width modulation (PWM) [22]. Here, it is sufficient to consider only the current control loop, because the outer loops have a negligible impact at higher frequencies.

It can be concluded that subsynchronous and near-synchronous interaction phenomena are the most difficult to study, since a full line-commutated-converter (LCC) or VSC model is needed. To facilitate such studies, we review how such a full model can be constructed. This chapter is structured into nine sections, which include the Introduction and Conclusion sections (Sections 5 and 9). Section 2 introduces the LCC structure and small-signal modeling technique. Thereafter, the small-signal models of LCC subsystems are developed in Section 3, and applied and verified in Section 4. Section 6 presents the system model and the structure of the VSC-HVDC control system. This lays the foundation on which the small-signal modeling and analysis in Section 7 builds. The latter allows the construction of a converter input admittance as seen from the point of common coupling (PCC). In Section 8, this input admittance is expressed in matrix form and it is exemplified how the generalized Nyquist criterion can be used to analyze the stability of the converter–grid interconnection.

10.2 System model and control structure of LCC

The purpose of this section is the presentation of modeling techniques that allow for the accurate representation of flexible AC transmission systems (FACTS) and HVDC systems using linearized state models. Aspects of system modeling covered include the use of a structured subsystems approach for model formation, modeling of the HVDC converter, accounting for the frequency conversion process in a time-invariant manner, alternative representations of the AC voltage and current variables, and the modeling of the firing angle control system phase locked loop (PLL). While the dynamics of the systems under consideration are the result of complex interactions, the use of a subsystems modeling approach greatly simplifies the formation of the system models.

The modeling techniques described are demonstrated using a mono-polar HVDC transmission system. The small-signal analysis of HVDC converter systems has received significant attention in the literature, and the model described here has similarities with those previously presented by Persson [23], Sucena-Paiva [24], Todd [25], and Jovcic [26]. In the context of the generalization of the methods described to FACTS, the models of the STATCOM by Padiyar [27] and Schauder [28] are also of relevance.

10.2.1 General description of the model

The configuration of the HVDC transmission system model is shown in Figure 10.1 and is of the same form as the CIGRE benchmark HVDC transmission system [29]. The method used to form a linearized model of the system is a three-step process, which involves the division of the system into a number of smaller subsystems, the description of each subsystem using a linearized model, and finally the interconnection of the subsystems. This approach relies on the principle of superposition

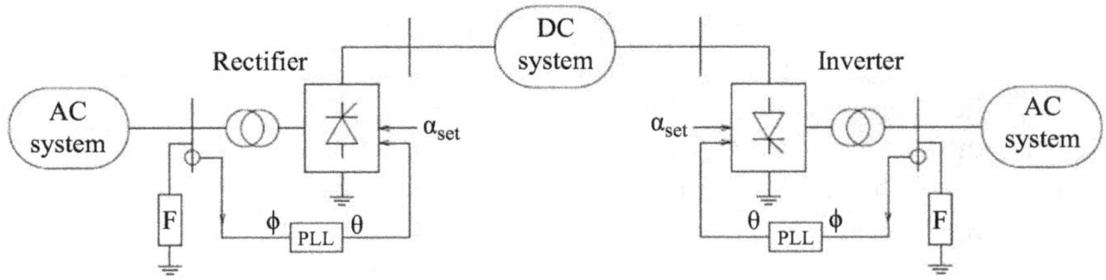

Figure 10.1 HVDC transmission system model

for linear or linearized systems and is a simpler approach than linearizing the nonlinear equations that describe the system directly.

The HVDC system is modeled using nine subsystems, which are the rectifier and inverter AC systems, the rectifier and inverter AC filters and shunt capacitors (F), the DC system, the rectifier and inverter HVDC converters, and the rectifier and inverter PLLs. Although the firing angle control inputs are left uncontrolled in the system shown, the addition of controller subsystems is a straightforward process.

The purpose of the model is the representation of the general dynamics of an HVDC system in the frequency range between 2 and 200 Hz on the DC side. Analysis of more specialized phenomena, such as subsynchronous resonance [7] and transformer core saturation instability [30], requires the inclusion of additional subsystems relevant to the phenomena under consideration in the model. For subsynchronous resonance the dynamics of the generator-turbine units are of importance, whereas for core saturation instability modeling the coupling between the slowly varying zero frequency and positive-sequence second harmonic components of the transformer magnetizing current is necessary.

10.2.2 State model formation

In order for the HVDC system to be represented in state model form, availability of a linear state model of each subsystem is necessary. A linear time-invariant state model dynamically relates the subsystem inputs $\boldsymbol{u}$, outputs $\boldsymbol{y}$, and states $\boldsymbol{x}$ using a state equation (10.2) and an output equation (10.3), which are specified by the constant matrices $\boldsymbol{A}$, $\boldsymbol{B}$, $\boldsymbol{C}$, and $\boldsymbol{D}$. In the situation where a state model is obtained by linearizing a system around an operating point, the input, output, and state variables represent the deviation of the system variables from their operating point values.

$$\dot{x} = \boldsymbol{Ax} + \boldsymbol{Bu} \tag{10.2}$$

$$\boldsymbol{y} = \boldsymbol{Cx} + \boldsymbol{Du} \tag{10.3}$$

The inputs and outputs of the state model are either signal or electrical variables. Signal variables are associated with the measurement of electrical variables and control subsystems, whereas electrical variables occur as voltage–current pairs and are associated with the electrical terminals of the subsystems. When connecting electrical subsystems together at a bus bar, only one subsystem can be represented in current-input voltage-output (impedance) form, while all others must be represented in voltage-input current-output (admittance) form. The subsystem in impedance form provides the voltage input for all of the subsystems in admittance form, whereas the addition of the current outputs of the subsystems in admittance form provides the current input to the subsystem in impedance form. A convention is adopted where the flow of current into an electrical terminal is assumed to be of positive sign.

It is important to correctly choose the inputs and outputs of electrical subsystems such that first, the inputs and outputs that are to be connected together are compatible, and second, the subsystems are proper. Only those systems that are proper, meaning the transfer functions between the inputs and outputs of the system have at least as many poles as zeros, are able to be described in the above state model form. If it is not possible to accommodate an improper electrical subsystem (more zeros than poles) by interchanging inputs and outputs, then a proper subsystem can be formed by adding extra poles above the frequency range of interest to the system. Since standard methods are available to convert between proper s-domain transfer functions and state model representations, with respect to the input and output relationships of the system, these two representations are equivalent. In cases where subsystems are defined in terms of frequency response data, it is necessary to fit an s-domain transfer function to the frequency response data. For simple frequency responses, an s-domain representation can be obtained by inspection, whereas for more complicated frequency responses transfer function fitting algorithms are available [31,32].

When forming system models, it is likely that a particular type of component, such as the PLL and HVDC converter subsystems in the case described, will occur in multiplicity. In this situation, it is of significant advantage to adopt a modular approach where the state model of the component is specified as the output of a function. The subsystem state model functions collectively form a library of components, which may be called repeatedly during the formation of system models. Once the subsystems have been defined, they are connected together to form a state model of the overall system using the algorithm described in Appendix A. The availability of a linearized state model facilitates the use of standard classical and modern control theory for steady-state stability analysis and controller design.

10.2.3 Modeling the converter frequency conversion process

The conversion of electrical energy between AC and DC frequency is achieved by the periodic firing of the HVDC converter thyristors. The switching action is the direct cause of HVDC system nonlinearity, and the linearized representation of this process is of significant importance to the system model.

Although the behavior of the converter is that of a complex single-frequency input multiple-frequency output modulator, all but the most significant frequency interactions are above the frequency range of interest and can be ignored in the analysis. Due to the frequency conversion process, a number of these interactions occur between different frequencies on the AC and DC sides. An arbitrary frequency $k\omega_0$ on the DC side of the converter is related to two frequencies on the AC side separated by twice the fundamental frequency, positive-sequence frequency $(k+1)\omega_0$ and negative-sequence frequency $(k-1)\omega_0$. A model that considers only these interactions is referred as a three-port model [33] and is essentially of the describing function type. The zero sequence component of the AC system waveforms is omitted from the system models as they are neither generated by nor affect the operation of three-phase power electronic devices such as the HVDC converter.

Frequency conversion exhibits time-variance and cannot be directly represented in the required state model form. In order to model the system in a time-invariant manner, it is necessary to decouple the frequency conversion process from the model of the converter. The effect of the frequency conversion process is accounted for by frequency shifting the equations that describe the dynamics of the subsystems on the AC side of the converter. Even though frequency conversion does not appear explicitly in the analysis, the interactions are correctly represented through the altered subsystem dynamics. The dynamics of the subsystems on the AC side of the converter can be frequency shifted using Park's transformation, or the transfer function zero-pole shifting approach described in this chapter.

10.2.4 AC system variable representation

Three distinct representations of three-phase AC system voltage and current variables are useful for the purposes of system modeling and control. The representations are positive and negative-sequence (*pn*) components, direct and quadrature (*dq*) components, and magnitude and angle (*ma*) components. The transformations between the three AC variable representations are described in Appendix C. The *dq* and *ma* representations are with respect to a synchronously rotating frame of reference, whereas sequence components may be at their actual frequency or referred to their equivalent DC-side frequency depending on the context in which they are used.

10.3 Representation of the subsystems of LCC

10.3.1 AC system and filters

The rectifier and inverter AC systems and filters are represented using s-domain equivalents that describe their electrical characteristics as seen from the AC terminals of the converters. The HVDC converter uses AC voltage as an input, whereas the nature of the AC system (series inductance) and the AC filters (shunt capacitance) means that these subsystems are most appropriately represented in admittance and impedance forms, respectively. To enable the connection of the subsystems at the AC terminals of the converters without the need for variable transformations, it is chosen to use a common *pn* sequence components representation of the AC variables. This has a natural relationship with the frequencies coupled together by LCC-HVDC converters.

The relationships between the inputs and outputs of the AC system model are described in (10.4). It is assumed that the positive and negative-sequence admittances of the AC system Y_{ac} are equal, and that there is no coupling between the sequences.

$$\begin{bmatrix} I_{acp} \\ I_{acn} \end{bmatrix} = \begin{bmatrix} Y_{ac} & \\ & Y_{ac} \end{bmatrix} \begin{bmatrix} V_{acp} \\ V_{acn} \end{bmatrix} \tag{10.4}$$

The frequency conversion process of the converter is accounted for by frequency shifting the AC system equations. This is achieved by representing the admittance Y_{ac} in zero-pole transfer function form, and then adding $\pm j\omega_0$ to the values of the zeroes and poles, as described in (10.5) and (10.6). As the transfer function zeroes and poles have been shifted in opposite directions, the poles of the subsystem still form complex conjugate pairs. The AC filters are modeled in impedance form using the same process described. In this way, the DC-side frequency is chosen as the single frequency represented by s, and the two AC-side frequencies and admittances that it is coupled to are referred to that same frequency.

$$Y_{acp}(s) = Y_{ac}(s + j\omega_0) \tag{10.5}$$

$$Y_{acn}(s) = Y_{ac}(s - j\omega_0) \tag{10.6}$$

10.3.2 DC system

The DC system has two electrical terminals that are connected to the DC terminals of the rectifier and inverter. The series inductive nature of the DC system (smoothing reactor and DC transmission line) means that the system is best described in admittance form as described in (10.7). This representation is directly compatible with the converter DC terminal current input and voltage output variables.

$$\begin{bmatrix} I_1 \\ I_2 \end{bmatrix} = \begin{bmatrix} Y_{11} & Y_{12} \\ Y_{21} & Y_{22} \end{bmatrix} \begin{bmatrix} V_1 \\ V_2 \end{bmatrix} \tag{10.7}$$

10.3.3 HVDC converter

The state model of the HVDC converter used in this paper is obtained from the frequency domain model derived in Reference 34. The analysis described in the reference uses a piecewise linear state variable representation of the converter voltage and current waveforms to model all interactions associated with the operation of the converter exactly in the small-signal sense. While the frequency domain model is more accurate than that is required for the analysis of HVDC system dynamics, its availability has allowed the effect of converter model accuracy on the dynamics of the system model to be assessed.

The small-signal linearized relationships between the converter input and output variables are described in (10.8). The quantities a to l are referred to as the transfers and are described as analytic functions of the converter operating point in Appendix B.

$$\begin{bmatrix} \Delta I_{acp} \\ \Delta I_{acn} \\ \Delta V_{dc} \end{bmatrix} = \begin{bmatrix} a & b & c & d \\ e & f & g & h \\ i & j & k & l \end{bmatrix} \begin{bmatrix} \Delta V_{acp} \\ \Delta V_{acn} \\ \Delta I_{dc} \\ \Delta\alpha \end{bmatrix} \tag{10.8}$$

The transfers in Appendix B have had the weak frequency dependencies removed. This is possible as the frequency range of interest is low. The change in the HVDC system dynamics, resulting from the use of converter models that take into account varying degrees of frequency dependence in the transfers, indicates that a model where the transfers are approximated as constants is of sufficient accuracy. The constants are naturally chosen to be the value of the frequency domain transfers at zero frequency on the DC side and are consistent with the differentiation of the standard steady-state converter equations. Transfer k, from ΔI_{dc} to ΔV_{dc}, has the form of a zero and is the only case where a constant approximation is inappropriate. The constant component of this transfer is the value that is obtained by differentiating the converter steady-state equations, whereas the frequency-dependent component is the time averaged value of the commutation reactance seen from the DC side.

The frequency domain HVDC converter model was derived assuming an ideal equidistant firing angle control system, where the firing angle ramp references are fixed in time. This is not the case in reality as the ramp references are made to track the changes in the converter terminal AC voltage angle using a PLL system. The firing angle required by the converter model $\Delta\alpha$ is the angle such that the correct firing instants are obtained using ramp references that are fixed in time. This is given by subtracting the firing angle desired by the controller $\Delta\alpha_{\text{set}}$ from the PLL output reference angle

$$\Delta\alpha = \Delta\alpha_{\text{set}} - \Delta\theta \tag{10.9}$$

10.3.4 Phase locked loop

The PLL is a negative feedback control system that tracks the changes in the phase angle of the positive-sequence fundamental frequency component of the converter AC bus voltage. The PLL generates a ramp reference function that is synchronized to the AC voltage. This output is used to define the ramp reference associated with each of the converter thyristors and ensures that the firing instants are synchronized to the AC voltage.

The PLL system modeled is of the DQZ type [35], the three major components of which are an error signal calculator, proportional–integral (PI) controller, and voltage-controlled oscillator (VCO). The error signal is calculated as the q component of the AC voltage with respect to a sinusoidal representation of the PLL output ramp reference. This signal, which in the small-signal case is proportional to the phase difference between the AC voltage and output reference, is used to slow down or speed up the VCO such that the q component and hence phase difference between the AC voltage and PLL output becomes zero.

The small-signal dynamics of the PLL system are represented by the block diagram in Figure 10.2. The input to the model is the angle component of the AC bus voltage in the synchronous reference frame, which is obtained from a sequence or dq component representation of the AC voltage using the transforms described in Appendix C. The open loop transfer function consists of the series combination of a gain K, which is the operating point magnitude component of the AC voltage, a PI controller, and an integrator that represents the operation of the VCO. Controller integral action is required for the PLL to track changes in the frequency of the AC bus voltage with zero steady-state error.

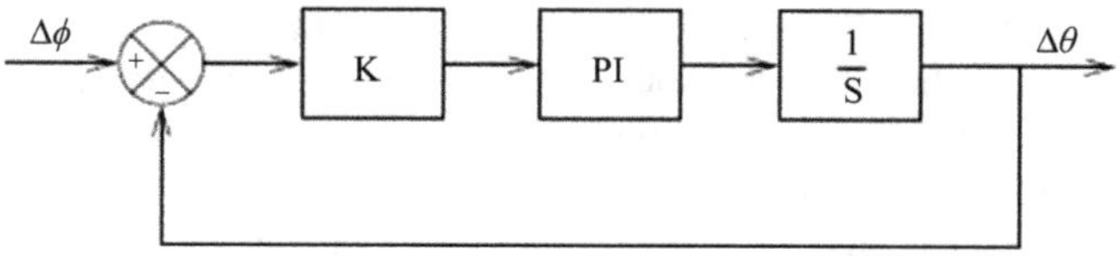

Figure 10.2 Small-signal model of the phase locked loop

Table 10.1 Eigenvalues of the uncontrolled CIGRE HVDC system

#	Eigenvalue	PLL	ELEC	REC	INV
1	-0.0181	×		×	
2	-0.1206	×		×	
3,4	$-0.0204 \pm j0.0154$	×			×
5	-0.4389		×	×	×
6,7	$-0.2726 \pm j0.5882$		×	×	×
8,9	$-0.3136 \pm j1.2051$		×	×	×
10,11	$-0.1685 \pm j1.3844$		×	×	×
12,13	$-0.3231 \pm j3.0378$		×	×	
14,15	$-0.6052 \pm j3.0689$		×		×

ELEC: electrical network; REC: rectifier; INV: inverter.

The parameters of the PI controller are normally chosen such that the output reference angle is only able to follow changes in the AC voltage angle which are slower than about 5 Hz. Since the modes of oscillation resulting from the interconnection of the electrical subsystems are usually at significantly higher frequencies than 5 Hz, the inclusion of the PLL has only a very limited effect on these modes. Despite this the representation of the PLL is still of importance, particularly at the inverter where a low frequency instability arises when the inverter AC system has a very low SCR [36].

10.4 Application and validation

The formation of a small-signal model of the CIGRE benchmark HVDC transmission system using the methods described in the previous sections results in a state model that is of 39th order. The majority of the eigenvalues of this system are associated with the AC filter subsystems and are generally well damped and/or above the frequency range of interest. The dynamics of importance are characterized by the 15 eigenvalues listed in Table 10.1, which are in p.u. with respect to 50 Hz. Of these eigenvalues, 11 is associated with the natural interaction of the electrical subsystems, whereas the further 4 is introduced by the PLL subsystems. The right-hand columns of the table indicate whether the eigenvalues are associated with the PLL subsystems, electrical subsystems, and rectifier and/or inverter ends of the HVDC system.

Eigenvalue 5 is the time constant of the equivalent series resistance and series inductance seen on the DC side, whereas the complex conjugate eigenvalues 6 and 7 are associated with a series resonance between the equivalent inductance and the mid-point capacitor of the T-network DC system model. Above this series resonance, the rectifier and inverter ends of the system become increasingly decoupled as the impedance of the mid-point capacitor decreases to a short circuit. The eigenvalues above the frequency of the series resonance are associated with the parallel resonances that exist between the inductance of the AC system models and the equivalent shunt capacitance of the AC filter subsystems.

The accuracy of the small-signal model was investigated by comparing the disturbance response with that obtained from PSCAD/EMTDC simulations of the CIGRE system at a 2 kA DC current operating point. Constant firing angle control was assumed to be the control strategy at the inverter, whereas at the rectifier the PSCAD/EMTDC CIGRE system model PI DC current controller (CC), $(1.1 + 92/s)$, was implemented. An extra signal input, representing a small-signal disturbance change in the angle of the remote source voltage, was added to the model of the rectifier AC system. The s-domain transfer function between the disturbance input and the rectifier DC current was determined and the corresponding step response calculated.

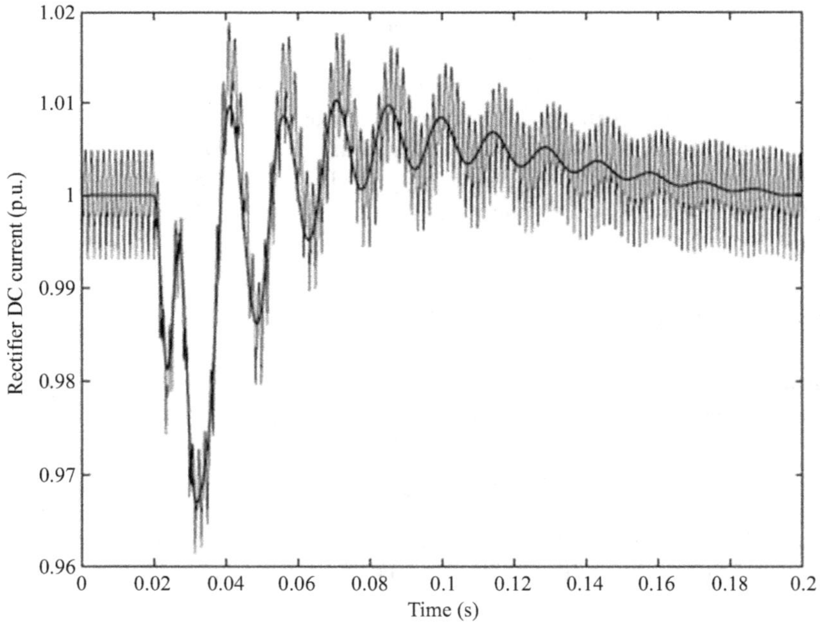

Figure 10.3 Rectifier DC current disturbance response : noisy line – PSCAD/EMTDC, smooth line – small-signal state model

Figure 10.3 shows the comparison of the rectifier DC current response to a 5° advance of the remote AC voltage (equivalent to a 10% increase in a quadrature voltage) at 0.02 s. The ripple on the PSCAD/EMTDC trace is related to the DC-side characteristic harmonics. The agreement between the state model and PSCAD/EMTDC responses is very good and suggests that the discrete switching nature of the converter can be modeled using a continuous-time model in the frequency range of interest. The response of other system variables such as the DC voltage and current, and the AC voltage and current, at both the rectifier and inverter have also been found to be in excellent agreement with time domain simulations.

A comparison of the eigenvalues of the system with and without the DC CC indicated that the 70 Hz mode of oscillation (eigenvalues 10 and 11 in Table 10.1) seen in the disturbance response is destabilized by the DC CC used. This was confirmed by simulating the disturbance response without the DC CC and clearly highlights the applicability of small-signal dynamic models to control system design. Of relevance to the design of a rectifier DC CC is the open loop frequency response $\Delta\alpha$ to ΔI_{dc} shown in Figure 10.4, the frequency scale being with respect to 50 Hz.

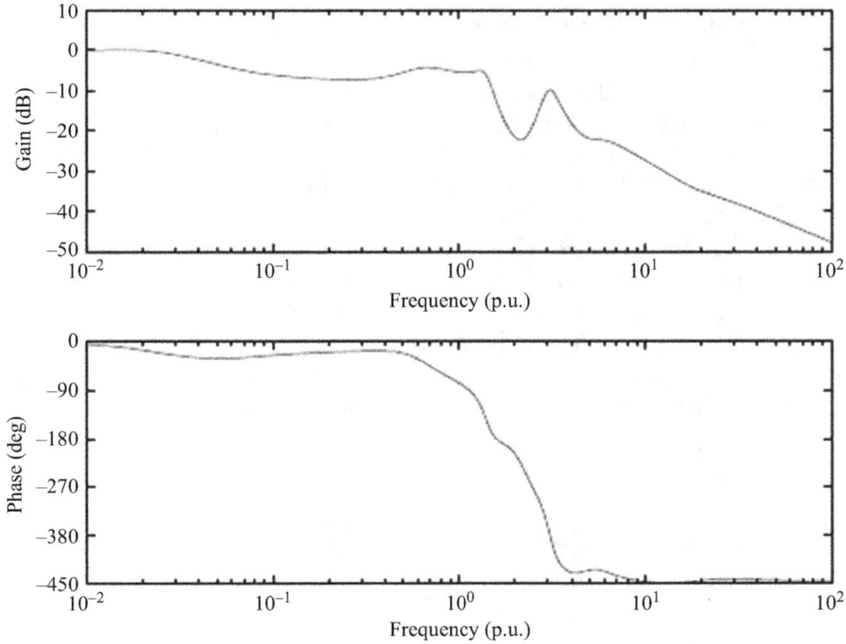

Figure 10.4 Bode diagram of the open loop transfer function from the rectifier firing angle (rad) to the rectifier DC current (p.u.)

10.5 Conclusion of LCC modeling

A structured subsystems modeling approach has been described for the formation of small-signal dynamic state models of FACTS and HVDC systems. The approach involves the division of the system into a number of smaller subsystems, the representation of each subsystem using a linear state model, and the interconnection of the subsystems. Application of the analysis techniques to a mono-polar HVDC transmission system has shown that small-signal models can be accurate and of benefit for the analysis of system dynamics. The HVDC system model is currently being used to investigate the variation of system dynamics with changes in the parameters of the AC systems, DC systems, and operating point of the converters, and will be used to investigate HVDC/FACTS control interactions.

10.6 System model and control structure of VSC

In this section, after giving overviews of VSC-HVDC technology in general and the VSC-HVDC control system in particular, we consider in detail the current control loop. This can be considered as the core of the converter control system. Feeding into the current control loop are signals from outer, and slower, loops, which are considered as well.

10.6.1 VSC-HVDC overview

Although classical LCC-HVDC became available commercially already in the 1950s, VSC-HVDC in comparison is a relatively recent development, with the first installations commissioned in the late 1990s. Although there are some multi-terminal transmissions in operation and at the planning stage, most LCC and VSC-HVDC transmissions have two terminals, as illustrated in Figure 10.5.

The applications of VSC-HVDC are partly overlapping with those of LCC-HVDC, e.g., to interconnect two asynchronous AC networks or to connect remote generation to load centers. However, LCC-HVDC requires fairly strong grids at the terminals to allow stable operation, whereas VSC-HVDC can interface with very weak grids (although maintaining stability then becomes more challenging). Since a VSC – as the name suggests – is voltage stiff, it can form its own terminal voltage. It is thus not even reliant on an active grid. This can be used to re-energize

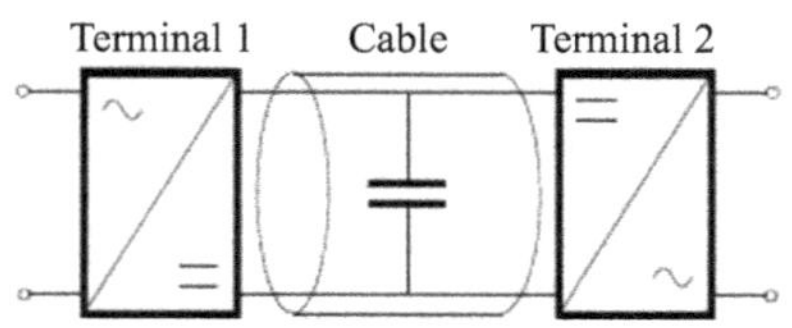

Figure 10.5 HVDC transmission system model

a grid after a blackout (called black start) or to connect isolated networks, such as offshore wind farms.

VSC-HVDC has its origins in variable-speed VSC drives – a technology that has approximately two more decades of maturity – and more precisely back-to-back VSC drives. A two-terminal VSC-HVDC transmission can be considered as an upscaling of a back-to-back drive system in terms of the voltage and power ratings. The successful series connection of numerous power transistors (particularly insulated-gate bipolar transistors) was the enabler of DC-bus voltages up to several hundred kilovolts.

During the first decade, all commissioned VSC-HVDC installations utilized two or three-level converters. Since 2007, the prevalent VSC-HVDC topology has been the modular multilevel converter (MMC). Although the MMC requires twice the number of semiconductor devices as a two-level VSC of similar voltage rating, the topology shift brings several benefits. One is that series connection of power transistors no longer is necessary. However, more important is that tens or even hundreds of levels are available, allowing an output voltage with very low harmonic content. This reduces the need for passive filters at the AC terminals to a minimum.

10.6.2 Control-system overview

When analyzing the dynamic interaction of a VSC-HVDC transmission with the AC grid(s) to which it is connected, it is mostly sufficient to consider each terminal as an independent system. Although the DC bus – which in HVDC transmissions in effect may be a cable several hundred kilometers long – interconnects the terminals, the dynamic impact of this interconnection is usually relatively weak. In addition, the cable can be approximated as purely capacitive, as shown in Figure 10.5. Consequently, the DC bus as seen from one VSC-HVDC terminal can be considered as a lumped capacitance to which a constant power load or source is connected, representing the other terminal(s).

Furthermore, provided that the control system is designed well, the converter topology is transparent. That is, the VSC-HVDC terminal acts as a controllable voltage source regardless whether it uses a two or three-level converter or a MMC. In the latter case, though, care is needed, since the MMC has complex internal dynamics. However, by properly compensating for the sum-cell-voltage ripples [37–39], even a MMC can be made to behave as a controllable voltage source as seen from the AC terminals. In the sequel, it is assumed that such internal MMC control is used, and we therefore do not restrict the focus to a certain converter topology.

Before discussing the theory, let us make a few key definitions.

- The positive- and negative-sequence synchronous components are referred to as $+1$ and -1, respectively. Harmonics are referred to with their signed order in a similar fashion.
- Upright boldface letters are used to denote complex space vector, e.g., $\mathbf{v}$, and transfer functions that operate on complex space vectors, e.g., $\mathbf{Z}(s)$.

- Vectors and transfer functions referred to the stationary $\alpha\beta$ frame are denoted with the superscript s, whereas such referred to the synchronous dq frame aligned with +1 are denoted without a superscript.
- The derivative operator is denoted as $s = d/dt$ (which shall be considered as the complex Laplace variable, where appropriate).
- The angular synchronous frequency is denoted by ω_1.

In most of the VSC control systems – including that of a VSC-HVDC terminal – the CC is the innermost and fastest, closed loop; that for the converter current $\mathbf{i}^s$ (Figure 10.6). If overmodulation is assumed not to occur and switching harmonics are disregarded, the PWM process can be modeled as lumped with the computational time delay into the total time delay T_d. The converter is assumed to be equipped with an inductive input filter, with inductance L and negligible resistance. Hence, the converter-current dynamics are in the $\alpha\beta$ frame governed by

$$\mathbf{i}^s = \frac{\mathbf{E}^s - \mathbf{v}^s}{sL}, \qquad \mathbf{v}^s = e^{-sT_d}\mathbf{v}^s_{\mathrm{ref}} \tag{10.10}$$

where $\mathbf{E}^s$ is the PCC voltage and $\mathbf{v}^s_{\mathrm{ref}}$ is the reference vector to the PWM, from which the converter voltage $\mathbf{v}^s$ is generated. The dq-frame correspondence is obtained simply by substituting $s \to s + j\omega_1$ [40]

$$\mathbf{i} = \frac{\mathbf{E} - \mathbf{v}}{(s + j\omega_1)L}, \qquad \mathbf{v} = e^{-(s+j\omega_1)T_d}\mathbf{v}_{\mathrm{ref}} \tag{10.11}$$

The grid impedance $\mathbf{Z}^s(s)$, which for modeling using complex space vectors needs to be balanced (also known as symmetric [41]), adds the following relations in the $\alpha\beta$ and dq frames, respectively:

$$\mathbf{v}^s_g - \mathbf{Z}^s(s)\mathbf{i}^s = \mathbf{E}^s \qquad \mathbf{v}_g - \mathbf{Z}(s)\mathbf{i} = \mathbf{E} \tag{10.12}$$

where $\mathbf{Z}(s) = \mathbf{Z}^s(s + j\omega_1)$ and $\mathbf{v}^s_g$ is the grid voltage.

A number of outer control loops feed into the CC.

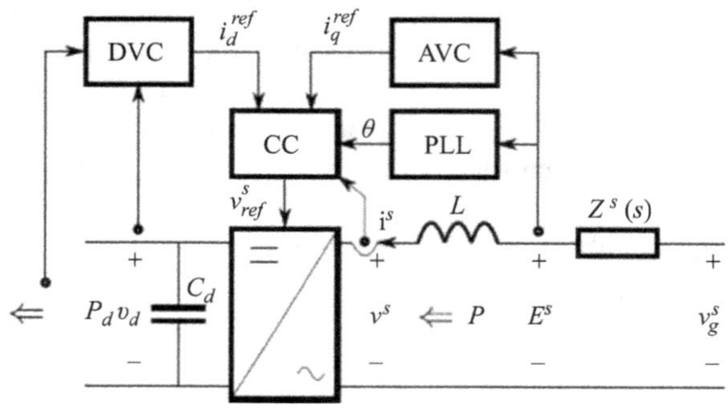

Figure 10.6 VSC circuit and control-system block diagram

- The PLL synchronizes the converter by aligning the dq frame used by the control system with the PCC voltage by computing the transformation angle θ.
- The DC-bus voltage v_d is controlled by the DC-bus-voltage controller (DVC), which sets the d component of the dq-frame current reference $\mathbf{i}_{\text{ref}} = i_d^{\text{ref}} + ji_q^{\text{ref}}$. Because the dq frame is aligned with the PCC voltage, i_d^{ref} produces active power. An increase of i_d^{ref} thus increases the power flow into the converter, boosting the DC-bus voltage.
- The reference component i_q^{ref} is perpendicular to the PCC voltage and thus determines the reactive-power exchange between the converter and the grid. It can be set by an AC-bus-voltage controller (AVC), which controls the modulus of the PCC voltage to its desired value. Incorporating an AVC in the control system is particularly important for connection to a weak grid.

10.6.3 Current controller

From the previous discussion, it is clear that controlling the input current of the converter is important, not the least since its d and q components determine the active and reactive-power exchanges, respectively. Other reasons are to reduce the harmonic content of the current and to prevent overcurrent during transients, particularly faults in the grid. Current control can be made either in the dq frame or in the $\alpha\beta$ frame (possibly even on the actual phase quantities). Often, dq-frame control is preferable, since the quantities then ideally are constant in the steady state, allowing the control problem to be solved using (variants of) the standard PI controller. We therefore approach the CC design from this end and thereafter consider how the dq-frame controller can be transformed to the $\alpha\beta$ frame for an equivalent implementation there.

1. *dq-frame CC*: The CC is in this case given by

$$\mathbf{v}_{\text{ref}} = e^{j\omega_1 T_d}[-\mathbf{F}_c(s)(\mathbf{i}_{\text{ref}} - \mathbf{i}) - j\omega_1 L\mathbf{i} + \mathbf{H}(s)\mathbf{E}] \tag{10.13}$$

where $\mathbf{F}_c(s)$ is the controller transfer function and the angle-adjustment factor $e^{j\omega_1 T_d}$ compensates the displacement factor $e^{-j\omega_1 T_d}$ in (10.11). PCC-voltage feedforward via a low-pass filter $\mathbf{H}(s)$ is included [40], as is a dq decoupling term $j\omega_1 L\mathbf{i}$. $\mathbf{F}_c(s)$ is typically selected as a proportional (P)–resonant (R) controller, which includes a number of parallel-connected R parts, also called generalized integrators. If reduced-order generalized integrators [42] are used as R parts, then

$$\mathbf{F}_c(s) = \alpha_c L\left(1 + \sum_h \frac{\alpha_h e^{j\phi_h}}{s - jh\omega_1}\right) \tag{10.14}$$

where α_c is the ideal closed-loop system bandwidth, α_h (with dimension angular frequency) is the individual gain factor of the R part for the harmonic order h, and ϕ_h is the compensation angle of that R part [43,44]. Characteristically, R parts are included for $+1$ and -1 as well as for balanced harmonics, i.e., orders $-5, +7, -11, +13, \ldots$ [42]. These translate to

$h = 0, -2, \pm 6, \pm 12, \ldots$ in the dq frame. The R part at $h = 0$ reduces to a pure integrator

$$\frac{\alpha_0}{s} \tag{10.15}$$

Selecting $\alpha_h \ll \alpha_c$ is recommended [45].

2. *αβ-frame CC*: Since dq transformation implies substituting $s \to s + j\omega_1$, $\alpha\beta$ transformation implies substituting $s \to s - j\omega_1$. Consequently, an $\alpha\beta$-frame control law that is equivalent to (10.13) can be obtained as

$$\mathbf{V}^s_{\mathrm{ref}} = e^{j\omega_1 T_\mathrm{d}}\left[-\mathbf{F}_c(s - j\omega_1)(\mathbf{i}^s_{\mathrm{ref}} - \mathbf{i}^s) - j\omega_1 L\mathbf{i}^s + \mathbf{H}(s - j\omega_1)\mathbf{E}^s\right] \tag{10.16}$$

The compensation factor $e^{j\omega_1 T_\mathrm{d}}$, the dq decoupling, and the P part of $\mathbf{F}_c(s)$ remain unchanged. However, the I part is now a generalized integrator, since $\alpha\beta$ transformation of (10.15) yields

$$\frac{\alpha_0}{s - j\omega_1} \tag{10.17}$$

Similarly, the low-pass filter in (10.13) is transformed to a complex-coefficient band-pass filter. Implementation in the $\alpha\beta$ is more difficult than dq-frame implementation in the sense that the complex-coefficient transfer functions need to be implemented. This requires greater care in the discretization for implementation, e.g., on a digital signal processor. On the other hand, only one coordinate transformation is needed – the $\alpha\beta$ transformation of the dq-frame reference $\mathbf{i}_{\mathrm{ref}}$ (whose components are set by the outer loops) to get $\mathbf{i}^s_{\mathrm{ref}}$ – whereas dq-frame implementation requires three: dq transformations of $\mathbf{i}^s$ and $\mathbf{E}^s$ and $\alpha\beta$ transformation of $\mathbf{v}_{\mathrm{ref}}$.

10.6.4 Ideal current control loop

Since the $\alpha\beta$-frame CC can be designed to give performance identical to the dq-frame CC, we henceforth consider the latter and do all modeling in the dq frame. Combining (10.11) and (10.13) yields

$$\mathbf{i} = \mathbf{G}_{\mathrm{ci}}(s)\mathbf{i}_{\mathrm{ref}} + \mathbf{Y}_\mathrm{i}(s)\mathbf{E} \tag{10.18}$$

where $\mathbf{G}_{\mathrm{ci}}(s)$ and $\mathbf{Y}_\mathrm{i}(s)$ are called the inner closed-loop system and the inner input admittance, respectively. By "inner" it is meant that the outer control loops, i.e., the PLL, DVC, and AVC, are not yet taken into account. If T_d is neglected – which is a valid approximation for the study of interaction phenomena from subsynchronous to a few times the fundamental frequency – the transfer functions in (10.18) are, respectively, given by

$$\mathbf{G}_{\mathrm{ci}}(s) = \frac{\mathbf{F}_c(s)}{sL + \mathbf{F}_c(s)} \qquad \mathbf{Y}_\mathrm{i}(s) = \frac{1 - \mathbf{H}(s)}{sL + \mathbf{F}_c(s)} \tag{10.19}$$

So as long as $\mathbf{F}_c(s)$ and $\mathbf{H}(s)$ have purely real coefficients, $\mathbf{G}_{\mathrm{ci}}(s)$ and $\mathbf{Y}_{\mathrm{i}}(s)$ are real as well, meaning that there is no coupling between the d and q signal paths. Current components i_d and i_q are controlled independently. Although it can be observed in (10.14) that $\mathbf{F}_c(s)$ well may be complex due to the R parts, this effect is generally beyond the frequency range that is studied here, i.e., from sub-synchronous to a few times the fundamental. Thus, we assume that the controller and feedforward-filter transfer functions are real (for s real) and henceforth use

$$\mathbf{F}_c(s) \rightarrow F_c(s) \qquad \mathbf{H}(s) \rightarrow H(s) \tag{10.20}$$

giving also

$$\mathbf{i} = G_{\mathrm{ci}}(s)\mathbf{i}_{\mathrm{ref}} + Y_{\mathrm{i}}(s)\mathbf{E} \tag{10.21}$$

in place of (10.18).

The control of the fundamental component, i.e., $h = 0$ in the dq frame, is by far the most important task. Suppose that, for this reason, pure P control is used (i.e., $\alpha_h = 0$) together with a first-order feedforward filter

$$F_c(s) = \alpha_c L \qquad H(s) = \frac{\alpha_f}{s + \alpha_f} \tag{10.22}$$

Then

$$G_{ci}(s) = \frac{\alpha_c}{s + \alpha_c} \qquad Y_i(s) = \frac{s}{(s + \alpha_c)(s + \alpha_f)L} \tag{10.23}$$

We find that the inner closed-loop system has bandwidth α_c (as desired) and unity static gain. The inner input admittance has a zero at $s = 0$, implying that there is no steady-state impact of the PCC voltage. If instead PI control without voltage feedforward is used

$$F_c(s) = \alpha_c L\left(1 + \frac{\alpha_0}{s}\right) \qquad H(s) = 0 \tag{10.24}$$

then

$$G_{ci}(s) = \frac{\alpha_c(s + \alpha_c)}{s^2 + \alpha_c s + \alpha_c \alpha_0} \qquad Y_i(s) = \frac{s}{(s^2 + \alpha_c s + \alpha_c \alpha_0)L} \tag{10.25}$$

Since $\alpha_0 \ll \alpha_c$ normally, $\alpha_c \approx \alpha_c + \alpha_0$, which allows the denominator polynomials in (10.25) to be approximately factorized as $(s + \alpha_c)(s + \alpha_0)$, reducing the transfer functions to

$$G_{ci}(s) \approx \frac{\alpha_c}{s + \alpha_c} \qquad Y_i(s) = \frac{s}{(s + \alpha_c)(s + \alpha_0)L} \tag{10.26}$$

It is found that (10.26) for $\alpha_0 = \alpha_f$ is approximately identical to (10.23), showing that integral action and PCC-voltage feedforward via a low-pass filter are complementary; if one is used, the other need not be. There is a slight difference, though. Had the resistance of the inductive filter not been neglected, we would have obtained $G_{ci}(0) < 1$ (but yet close to 1) in (10.23). That is, a small static control error in practice results due to the resistance if integral action is not used. However, since the outer controllers generally employ integral action, this is usually of minor importance.

10.6.5 Outer control loops

The aforementioned outer control loops PLL, DVC, and AVC are here described in more detail. All have in common that a PI controller is typically used. In addition, although it is not elaborated on further, a low-pass filter may be cascaded with the PI controller in order to suppress parasitic higher-frequency signal components.

1. *PLL*: The purpose of the PLL is to track the rotation of the PCC voltage vector, thereby aligning the dq frame (given by the angle θ relative the $\alpha\beta$ frame) with the +1 component of $\mathbf{E}^s$, whose magnitude is controlled to E_0, see below. The PLL uses the imaginary part of $\mathbf{E} = e^{-j\theta}\mathbf{E}^s$ as input signal, which is fed to the PLL controller $F_p(s)$. This is typically a PI controller, which can be parametrized as

$$F_p(s) = \frac{\alpha_p}{E_0}\left(1 + \frac{\alpha_{ip}}{s}\right) \tag{10.27}$$

where normalization of the input signal is obtained by dividing it by E_0, and where the gains (with dimension angular frequency) typically are selected as $\alpha_{ip} < \alpha_p \approx \alpha_c$. To the PLL-controller output, ω_1 is added, and the sum signal is then integrated to form the transformation angle as

$$\theta = \frac{1}{s}\left[F_p(s)\mathrm{Im}\{\mathbf{E}\} + \omega_1\right] \tag{10.28}$$

The PLL thus forces $\mathrm{Im}\{\mathbf{E}\}$ to zero (disturbances disregarded), leaving $\mathbf{E} = E_0$ in the steady state.

2. *DVC*: The purpose of the DVC is to make v_d track its reference v_d^{ref}. Introducing $W_d = C_d v_d^2/2$ as the energy stored in the DC bus, the energy balance of the DC link can be expressed as

$$\frac{dW_d}{dt} = \frac{C_d}{2}\frac{dv_d^2}{dt} = P - P_d \tag{10.29}$$

where P is the AC-side input power and P_d is the DC-side output power (i.e., to the other terminal or terminals in the HVDC transmission), including the converter losses ($P < 0$ and $P_d < 0$ for inverter operation). Since i_d is the active-power-producing current component, the following control law can be used:

$$i_d^{ref} = F_d(s)(W_d^{ref} - W_d) + P_d^f \qquad W_d^{ref} = \frac{C_d(v_d^{ref})^2}{2} \tag{10.30}$$

where term P_d^f is an optional feedforward of P_d (possibly low-pass filtered) as measured on the DC bus. The DVC is structurally similar to the PLL controller (10.27)

$$F_d(s) = \frac{\alpha_d}{E_0}\left(1 + \frac{\alpha_{id}}{s}\right) \tag{10.31}$$

and a similar parameter selection recommendation applies, i.e., $\alpha_{id} < \alpha_d \approx \alpha_c$. If the feedforward term P_d^f is included, weak integral action, i.e., $\alpha_{id} \approx \alpha_d$ can be used, or none at all. (A deviation of the value of C_d used in the control system from the actual value does not give a static control error, but effectively alters α_d.)

3. *AVC*: The AVC control law is given by

$$i_q^{ref} = F_a(s)(E_0 - |\mathrm{E}|) \tag{10.32}$$

where $F_a(s)$ typically is a PI controller. Unlike the PLL and the DVC, there is no analytical recommendation for its parameter selection; often trial and error tuning is used.

10.7 Small-signal modeling and impact of the outer control loops of VSC

The purpose of this section is to extend the closed-loop system model (10.21) to include the effect of the outer loops as well. Since they all involve nonlinearities, small-signal modeling and analysis need to be relied upon.

10.7.1 PLL impact for αβ-frame CC

Unlike Section 10.6.3, we begin by analyzing the PLL impact for $\alpha\beta$-frame CC, because the analysis is simpler than for dq-frame CC. To do so by linearization, a perturbation ΔE about the operating point E_0 of the PCC voltage is considered. For a constant ω_1, this yields the $\alpha\beta$-frame vector

$$\mathbf{E}^s = e^{j\omega_1 t}(E_0 + \Delta\mathbf{E}) \tag{10.33}$$

Introducing a perturbation also in the dq-frame angle, as $\theta = \omega_1 t + \Delta\theta$, gives $\mathbf{E} = e^{-j\theta}\mathbf{E}^s = e^{-j\Delta\theta}(E_0 + \Delta\mathbf{E})$. This relation can be linearized by approximating $e^{-j\Delta\theta} \approx 1 - j\Delta\theta$ and by neglecting cross terms between the perturbation quantities, yielding

$$\mathrm{E} = E_0 + \Delta\mathrm{E} - jE_0\Delta\theta \tag{10.34}$$

Substitution of (10.34) in (10.28) results in

$$\Delta\theta = \frac{F_\mathrm{p}(s)}{s}\mathrm{Im}\{\mathbf{E}\} - E_0\frac{F_\mathrm{p}(s)}{s}\Delta\theta \Rightarrow \Delta\theta = \underbrace{\frac{F_\mathrm{p}(s)}{s + E_0F_\mathrm{p}(s)}}_{G_\mathrm{P}(s)}\mathrm{Im}\{\mathbf{E}\} \tag{10.35}$$

If the integral term of (10.27) is neglected, then

$$G_\mathrm{p}(s) = \frac{\alpha_\mathrm{p}}{(s+\alpha_\mathrm{p})E_0} \tag{10.36}$$

i.e., α_p is the closed-loop PLL bandwidth.

The *dq*- and $\alpha\beta$-frame CCs (10.13) and (10.16) are equivalent under the design premises stated, concerning the current control loop only. However, they differ concerning their PLL impact [20]. For an $\alpha\beta$-frame CC, the *dq*-frame reference $\mathbf{i}_\mathrm{ref}$ is transformed into the $\alpha\beta$ frame as

$$\mathbf{i}^s_\mathrm{ref} = e^{j\theta}\mathbf{i}_\mathrm{ref} = e^{j(\omega_1 t+\Delta\theta)}(\mathbf{i}_0 + \Delta\mathbf{i}_\mathrm{ref}) \tag{10.37}$$

where $\Delta\mathbf{i}_\mathrm{ref}$ is the perturbation about the operating point $\mathbf{i}_0 = i_{d0} + ji_{q0}$. Equation (10.37) can be linearized as

$$\mathbf{i}^s_\mathrm{ref} \approx e^{j\omega_1 t}[(1+j\Delta\theta)\mathbf{i}_0 + \Delta\mathbf{i}_\mathrm{ref}] \tag{10.38}$$

so to take the PLL effect into account, we should substitute

$$\mathbf{i}_\mathrm{ref} \to (1+j\Delta\theta)\mathbf{i}_0 + \Delta\mathbf{i}_\mathrm{ref} \tag{10.39}$$

in (10.13). Resolving the closed-loop system, this implies that (10.39) should be substituted in (10.21).

10.7.2 PLL impact for dq-frame CC

In this case, there are three coordinate transformations. Starting with the *dq* transformation of the measured current, we have

$$\mathbf{i} = e^{-j\theta}\mathbf{i}^s = e^{-j(\omega_1 t+\Delta\theta)}e^{j\omega_1 t}(\mathbf{i}_0 + \Delta\mathbf{i}) \approx (1-j\Delta\theta)\mathbf{i}_0 + \Delta\mathbf{i} \tag{10.40}$$

so taking the PLL impact into account implies substituting in (10.13)

$$\mathbf{i} \to (1-j\Delta\theta)\mathbf{i}_0 + \Delta\mathbf{i} \tag{10.41}$$

The *dq* transformation of $\mathbf{E}^s$ gives in a similar way a PLL impact that is taken into account by substituting in (10.13)

$$\mathbf{E} \to (1-j\Delta\theta)E_0 + \Delta\mathbf{E} \tag{10.42}$$

Finally, there is the $\alpha\beta$ transformation of the converter voltage, which implies that $j\mathbf{v}_0\Delta\theta$, where $\mathbf{v}_0$ is the operating point of the converter voltage, needs to be added to (10.13), in a similar fashion as (10.39). With T_d neglected and with real transfer functions, (10.13) then transforms to

$$\begin{aligned}\mathbf{v}_{\mathrm{ref}} &= -F_{\mathrm{c}}(s)(\Delta\mathbf{i}_{\mathrm{ref}} - \Delta\mathbf{i} + j\mathbf{i}_0\Delta\theta) - j\omega_1 L[(1 - j\Delta\theta)\mathbf{i}_0 + \Delta\mathbf{i}] \\ &\quad + H(s)[(1 - j\Delta\theta)E_0 + \Delta\mathbf{E}] + j\mathbf{v}_0\Delta\theta \\ &= H(s)E_0 - j\omega_1 L\mathbf{i}_0 - F_{\mathrm{c}}(s)(\Delta\mathbf{i}_{\mathrm{ref}} - \Delta\mathbf{i} + j\mathbf{i}_0\Delta\theta) - j\omega_1 L\Delta\mathbf{i} \\ &\quad + H(s)\Delta\mathbf{E} + j[\mathbf{v}_0 - H(s)E_0 + j\omega_1 L\mathbf{i}_0]\Delta\theta\end{aligned} \tag{10.43}$$

Putting $s = 0$ in (10.11) gives the static relation $\mathbf{v}_0 = E_0 - j\mathbf{i}_0\Delta\theta$, which allows (10.43) to be simplified as

$$\begin{aligned}\Delta\mathbf{v}_{\mathrm{ref}} &= -F_{\mathrm{c}}(s)(\Delta\mathbf{i}_{\mathrm{ref}} - \Delta\mathbf{i} + j\mathbf{i}_0\Delta\theta) \\ &\quad - j\omega_1 L\Delta\mathbf{i} + H(s)\Delta\mathbf{E} + j[1 - H(s)]E_0\Delta\theta\end{aligned} \tag{10.44}$$

Because the measured current enters (10.44) with the opposite sign to its reference, the PLL impact obtained from dq transformation of $\mathbf{i}^s$ is identical to that obtained from $\alpha\beta$ transformation of i_{ref} for $\alpha\beta$-frame CC. In addition, dq-frame CC adds the term $j[1 - H(s)]E_0\Delta\theta$. Combining (10.44) with (10.11), it is found that the PLL impact can be accounted for by the small-signal relation

$$\begin{aligned}\Delta\mathbf{i} &= G_{\mathrm{ci}}(s)(\Delta\mathbf{i}_{\mathrm{ref}} + j\mathbf{i}_0) + Y_{\mathrm{i}}(s)(\Delta\mathbf{E} - j\xi E_0\Delta\theta) \\ &= G_{\mathrm{ci}}(s)\Delta\mathbf{i}_{\mathrm{ref}} + Y_{\mathrm{i}}(s)\Delta\mathbf{E} + j[\mathbf{i}_0 G_{\mathrm{ci}}(s) - \xi E_0 Y_{\mathrm{i}}(s)]\Delta\theta\end{aligned} \tag{10.45}$$

where

$$\xi = \begin{cases} 0, & \text{for } \alpha\beta - \text{frame CC} \\ 1, & \text{for } dq - \text{frame CC} \end{cases} \tag{10.46}$$

10.7.3 DVC impact

We now proceed to determine the impact of the DVC. Assuming power-invariant space-vector scaling or per-unit (p.u.) values, the complex input power to the converter is given by $\mathbf{Ei}^*$. The active input power is thus given by $P = \mathrm{Re}\{\mathbf{Ei}^*\} = \mathrm{Re}\{\mathbf{E}^*\mathbf{i}\}$. Combining this relation with (10.29) and (10.30) yields

$$i_d^{\mathrm{ref}} = F_{\mathrm{d}}(s)[W_{\mathrm{d}}^{\mathrm{ref}} - \frac{\mathrm{Re}\{(E_0 + \Delta\mathbf{E})^*(\mathbf{i}_0 + \Delta\mathbf{i})\} - P_{\mathrm{d}} + P_{\mathrm{d}}^{\mathrm{f}}}{s}] \tag{10.47}$$

which, for a constant P_{d}, can be linearized as

$$\Delta i_d^{\mathrm{ref}} = -\frac{F_{\mathrm{d}}(s)}{s}\mathrm{Re}\{E_0\Delta\mathbf{i}^* + \mathbf{i}_0^*\Delta\mathbf{E}^*\} = -\frac{F_{\mathrm{d}}(s)}{s}\mathrm{Re}\{E_0\Delta\mathbf{i}^* + i_0^*\Delta\mathbf{E}^*\} \tag{10.48}$$

Substitution of (10.45) with $\Delta\mathbf{i}_{\mathrm{ref}} = \Delta i_d^{\mathrm{ref}} + j\Delta i_q^{\mathrm{ref}}$ in (10.48) yields

$$\begin{aligned}\Delta i_d^{\mathrm{ref}} &= -\frac{F_{\mathrm{d}}(s)}{s}\mathrm{Re}\Big\{E_0, G_{\mathrm{ci}}(s)(\Delta i_d^{\mathrm{ref}} + j\Delta i_q^{\mathrm{ref}}) +, E_0, Y_{\mathrm{i}}(s)\Delta \mathrm{E} \\ &\quad + jE_0[, \mathbf{i}_0, G_{\mathrm{ci}}(s) - \xi, E_0, Y_{\mathrm{i}}(s)]\Delta\theta + \mathbf{i}_0^*\Delta\mathbf{E}\Big\}\end{aligned} \tag{10.49}$$

This equation can be simplified to

$$\begin{aligned}\Delta i_d^{\text{ref}} = & -\frac{F_{\text{d}}(s)}{s}\left[E_0 G_{\text{ci}}(s)\Delta i_d^{\text{ref}} + E_0 \underbrace{\text{Re}\{j\mathbf{i}_0\}}_{-\text{Im}\{\mathbf{i}_0\}} \Delta\theta\right. \\ & \left. + E_0, Y_{\text{i}}(s)\text{Re}\{\Delta\mathbf{E}\} + \text{Re}\{\mathbf{i}_0^*\Delta\mathbf{E}\}\right]\end{aligned} \tag{10.50}$$

Now, Δi_d^{ref} can be solved as

$$\begin{aligned}\Delta i_d^{\text{ref}} = & \underbrace{\frac{F_{\text{d}}(s)}{s + E_0 F_{\text{d}}(s) G_{\text{ci}}(s)}}_{G_{\text{d}}(s)}[E_0 Y_{\text{i}}(s)\text{Re}\{\Delta\mathbf{E}\} - E_0\text{Im}\{\mathbf{i}_0\}\Delta\theta \\ & + \text{Re}\{i_0^*\Delta E\}]\end{aligned} \tag{10.51}$$

It can be observed that there is an implicit PLL impact in (10.51), owing to the term proportional to $\Delta\theta$. It is significant only for high reactive-power exchange, due to the factor $\text{Im}\{\mathbf{i}_0\} = i_{q0}$.

10.7.4 AVC impact

Control law (10.32) involves the nonlinearity $|\mathbf{E}| = \sqrt{E_d^2 + E_q^2}$. Since the dq frame is oriented along the PCC voltage, E_q is small (ideally zero). This allows us to approximate $|\mathbf{E}| \approx E_d$, so (10.32) can be linearized as

$$\Delta i_q^{\text{ref}} = -F_{\text{a}}(s)\Delta E_d = -F_{\text{a}}(s)\text{Re}\{\Delta\mathbf{E}\} \tag{10.52}$$

10.7.5 Total impact

The combined DVC and AVC impact is obtained by multiplying (10.52) by j and adding it to (10.51), giving

$$\begin{aligned}\Delta i_{\text{ref}} & = -G_{\text{d}}(s)[E_0 Y_{\text{i}}(s)\text{Re}\{\Delta\mathbf{E}\} - E_0\text{Im}\{\mathbf{i}_0\}\Delta\theta + \text{Re}\{\mathbf{i}_0^*\Delta\mathbf{E}\}] \\ & \quad - jF_{\text{a}}(s)\text{Re}\{\Delta\mathbf{E}\} \\ & = -[G_{\text{d}}(s)E_0 Y_{\text{i}}(s) + jF_{\text{a}}(s)]\text{Re}\{\Delta\mathbf{E}\} - G_{\text{d}}(s)\text{Re}\{\mathbf{i}_0^*\Delta\mathbf{E}\} \\ & \quad + G_{\text{d}}(s)E_0\text{Im}\{\mathbf{i}_0\}\Delta\theta\end{aligned} \tag{10.53}$$

Remaining now is to take the PLL impact into account by substituting (10.53) in (10.45). We obtain

$$\begin{aligned}\Delta\mathbf{i} = & \; Y_{\text{i}}(s)\Delta\mathbf{E} - G_{\text{ci}}(s)[G_{\text{d}}(s)E_0 Y_{\text{i}}(s) + jF_{\text{a}}(s)]\text{Re}\{\Delta\mathbf{E}\} \\ & - G_{\text{ci}}(s)G_{\text{d}}(s)\text{Re}\{\mathbf{i}_0^*\Delta\mathbf{E}\} + [G_{\text{ci}}(s)G_{\text{d}}(s)E_0\text{Im}\{i_0\} \\ & + j\mathbf{i}_0 G_{\text{ci}}(s) - j\xi E_0 Y_{\text{i}}(s)]G_{\text{p}}(s)\text{Im}\{\Delta\mathbf{E}\}\end{aligned} \tag{10.54}$$

where, according to (10.35), $\Delta\theta = G_p(s)\text{Im}\{\Delta\mathbf{E}\}$ has been used.

10.8 Input-admittance matrix and closed-loop stability analysis of VSC

All terms on the right-hand side of (10.54) are functions of $\Delta\mathbf{E}$. The equation thus, in a sense, represents an admittance. Yet, it cannot immediately be used for analyzing the dynamics and stability of the converter–grid interconnection. In Reference 22, how a similar relation can be rewritten is shown, so that the interconnection stability can be analyzed (with some approximations) for the special case of a balanced grid impedance. This restriction prevents analyzing, e.g., the subsynchronous torsional interaction with a synchronous machine [13].

To lift this restriction, (10.54) needs to be rewritten using real-space vectors in place of the complex-space vectors. We use notations of the real vectors which are identical to those of their complex counterparts but with slanted boldface letters, e.g.,

$$\mathbf{i} = i_d + ji_q \Leftrightarrow \boldsymbol{i} = \begin{bmatrix} i_d \\ i_q \end{bmatrix} \tag{10.55}$$

Moreover, the complex transfer functions need to be replaced by their corresponding transfer (function) matrices [41]. Since (10.54) involves real transfer functions which are scaled by complex coefficients, this is relatively straightforward. The first step is to introduce the matrix correspondences to 1 and j, which are the identity matrix $\boldsymbol{I}$ and the skew-symmetric identity matrix $\boldsymbol{J}$, respectively

$$1 \Leftrightarrow \boldsymbol{I} = \begin{bmatrix} 1 & 0 \\ 0 & 1 \end{bmatrix} \qquad j \Leftrightarrow \boldsymbol{J} = \begin{bmatrix} 0 & -1 \\ 1 & 0 \end{bmatrix} \tag{10.56}$$

To illustrate their usage, the matrix representation of a purely inductive grid impedance is given by

$$\boldsymbol{Z}(s) = (s + j\omega_1)L_g \Leftrightarrow \mathbf{Z}(s) = (s\boldsymbol{I} + \boldsymbol{J}\omega_1)L_g = \begin{bmatrix} sL_g & -\omega_1 L_g \\ \omega_1 L_g & sL_g \end{bmatrix} \tag{10.57}$$

Unlike complex transfer functions, transfer matrices allow imbalanced models, e.g., different d- and q-direction inductances

$$\boldsymbol{L} = \begin{bmatrix} L_d & 0 \\ 0 & L_q \end{bmatrix} \Rightarrow \boldsymbol{Z}(s) = (s\boldsymbol{I} + \boldsymbol{J}\omega_1)\boldsymbol{L} = \begin{bmatrix} sL_d & -\omega_1 L_q \\ \omega_1 L_d & sL_q \end{bmatrix} \tag{10.58}$$

This can be utilized, e.g., to model the imbalanced dynamics of a synchronous machine as seen from its terminals [13]. In addition to (10.56), the operations that extract the real and imaginary parts need to be put in matrix form. Taking the real part maps real to real and puts a zero in the imaginary. Taking the imaginary part maps imaginary to real and puts a zero in the imaginary as well. Consequently, the matrices that correspond to these operations are

$$\boldsymbol{Re} = \begin{bmatrix} 1 & 0 \\ 0 & 0 \end{bmatrix} \qquad \boldsymbol{Im} = \begin{bmatrix} 0 & 1 \\ 0 & 0 \end{bmatrix} \tag{10.59}$$

Finally, it is useful to introduce the matrix that corresponds to multiplication by $\mathbf{i}_0 = i_d + ji_{q0}$. This matrix is obviously given by

$$i_0 = \boldsymbol{I}i_{d0} + \boldsymbol{J}i_{q0} + = \begin{bmatrix} i_{d0} & -i_{q0} \\ i_{q0} & i_{d0} \end{bmatrix} \tag{10.60}$$

The matrix correspondence to the conjugate $\mathbf{i}_0^*$ is obtained by substituting $i_{q0} \rightarrow -i_{q0}$ and thus corresponds to transposing i_0

$$i_0^{\mathrm{T}} = \boldsymbol{I}i_{d0} - \boldsymbol{J}i_{q0} + = \begin{bmatrix} i_{d0} & i_{q0} \\ -i_{q0} & i_{d0} \end{bmatrix} \tag{10.61}$$

Equation (10.54) can now be converted to matrix and real-vector form term by term as

$$\begin{aligned} \Delta\boldsymbol{i} = \; & \boldsymbol{I}Y_{\mathrm{i}}(s)\Delta\boldsymbol{E} - [\boldsymbol{I}G_{\mathrm{ci}}(s)G_{\mathrm{d}}(s)E_0Y_{\mathrm{i}}(s) + \boldsymbol{J}G_{\mathrm{ci}}(s)F_{\mathrm{a}}(s)]\boldsymbol{Re}\ \Delta\boldsymbol{E} \\ & -G_{\mathrm{ci}}(s)G_{\mathrm{d}}(s)\boldsymbol{Re}\ \boldsymbol{i}_0^{\mathrm{T}}\Delta\boldsymbol{E} + [G_{\mathrm{ci}}(s)G_{\mathrm{d}}(s)E_0\boldsymbol{Im}\ \boldsymbol{i}_0 + \boldsymbol{J}\boldsymbol{i}_0G_{\mathrm{ci}}(s) \\ & -\boldsymbol{J}\xi E_0Y_{\mathrm{i}}(s)]G_{\mathrm{p}}(s)\boldsymbol{Im}\ \Delta\boldsymbol{E} \end{aligned} \tag{10.62}$$

Unlike (10.54), all terms on the right-hand side of (10.54) are multiplied (from the right) by ΔE, allowing the compact representation

$$\Delta\boldsymbol{i} = \boldsymbol{Y}(s)\Delta\boldsymbol{E} \tag{10.63}$$

where $\boldsymbol{Y}(s)$ is the total input-admittance matrix, given by

$$\begin{aligned} \boldsymbol{Y}(s) = \; & \boldsymbol{I}Y_{\mathrm{i}}(s) - [\boldsymbol{I}G_{\mathrm{ci}}(s)G_{\mathrm{d}}(s)E_0Y_{\mathrm{i}}(s) + \boldsymbol{J}G_{\mathrm{ci}}(s)F_{\mathrm{a}}(s)]\boldsymbol{Re} \\ & -G_{\mathrm{ci}}(s)G_{\mathrm{d}}(s)\boldsymbol{Re}\ \boldsymbol{i}_0^{\mathrm{T}} + [G_{\mathrm{ci}}(s)G_{\mathrm{d}}(s)E_0\boldsymbol{Im}\ \boldsymbol{i}_0 + \boldsymbol{J}\boldsymbol{i}_0G_{\mathrm{ci}}(s) \\ & -\boldsymbol{J}\xi E_0Y_{\mathrm{i}}(s)]G_{\mathrm{p}}(s)\boldsymbol{Im} \end{aligned} \tag{10.64}$$

Equation (10.63) can now be combined with the corresponding relation for the grid, obtained from (10.12) as

$$\Delta\boldsymbol{E} = \Delta\boldsymbol{v}_g - \boldsymbol{Z}(s)\Delta\boldsymbol{i} = \Delta\boldsymbol{v}_g - \boldsymbol{Z}(s)\boldsymbol{Y}(s)\Delta\boldsymbol{E} \tag{10.65}$$

giving the following closed-loop system:

$$\Delta\boldsymbol{E} = [\boldsymbol{I} + \boldsymbol{Z}(s)\boldsymbol{Y}(s)]^{-1}\Delta\boldsymbol{v}_g \tag{10.66}$$

In Reference 14, a method for assessing the interconnection stability is presented. The dissipative (passivity) properties of $\boldsymbol{Y}(j\omega)$ are studied. It is concluded that, as long as all poorly damped grid resonances fall in frequency regions where $\boldsymbol{Y}(j\omega)$ dissipates power, the interconnection remains stable. The method outlined in Reference 14 is intuitive and provides a good "feel" for the behavior of the VSC. Principally, it conveys the message that the PLL and the

DVC degrade the stability properties mainly for inverter and rectifier operation, respectively. Yet, the method is qualitative rather than quantitative.

The Nyquist criterion, on the other hand, provides a quantitative way of analyzing the interconnection stability. Since $\boldsymbol{Z}(s)$ and $\boldsymbol{Y}(s)$ both are matrices, the generalized Nyquist criterion has to be applied to the open-loop system $\boldsymbol{Z}(s)\boldsymbol{Y}(s)$. This implies evaluating the eigenvalues $\lambda\{\boldsymbol{Z}(j\omega)\boldsymbol{Y}(j\omega)\}$ for $-\infty < \omega < \infty$ and plotting the so obtained characteristic loci (i.e., Nyquist curves for both eigenvalues) [40].

Example

To illustrate the suggested way of analyzing the stability of a converter–grid interconnection, a VSC whose control system uses the following p.u. controller parameters is considered:

$$\alpha_c = 5 \quad \alpha_h = 0 \quad \alpha_f = 1 \quad \alpha_p = 0.2 \quad \alpha_{ip} = 0 \quad \alpha_d = 0.15 \quad \alpha_{id} = 0.03 \tag{10.67}$$

A PI AVC is used. Rather than feeding back **E**, the output signal of the feed-forward filter in the CC is used. Thus, we have

$$F_a(s) = \left(k_{pa} + \frac{k_{ia}}{s}\right)\frac{\alpha_f}{s + \alpha_f} \tag{10.68}$$

with $k_{pa} = k_{ia} = 1$ p.u. (All given parameters shall be considered as examples, not necessarily as recommendations.) Finally, $L = 0.1$ p.u. and (10.57) with $L_g = 0.5$ p.u. is used for the grid impedance (thus, the SCR seen from the PCC is 2, i.e., a relatively weak grid). The base frequency is 50 Hz.

The operating point $i_{d0} = -0.5$ p.u., i.e., inverter operation at half the rated power, is considered. To maintain the PCC voltage at $E_0 = 1$ p.u. for $|\mathbf{v}_g| = 1$ p.u., a reactive-current injection of $i_{q0} = 0.07$ p.u. is needed. Figure 10.7(a) shows that a

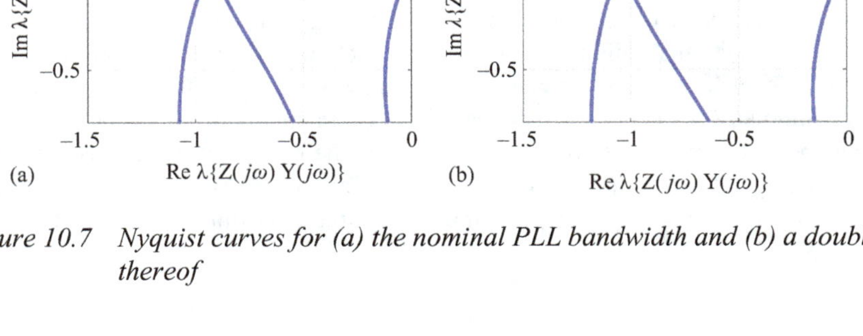

Figure 10.7 Nyquist curves for (a) the nominal PLL bandwidth and (b) a doubling thereof

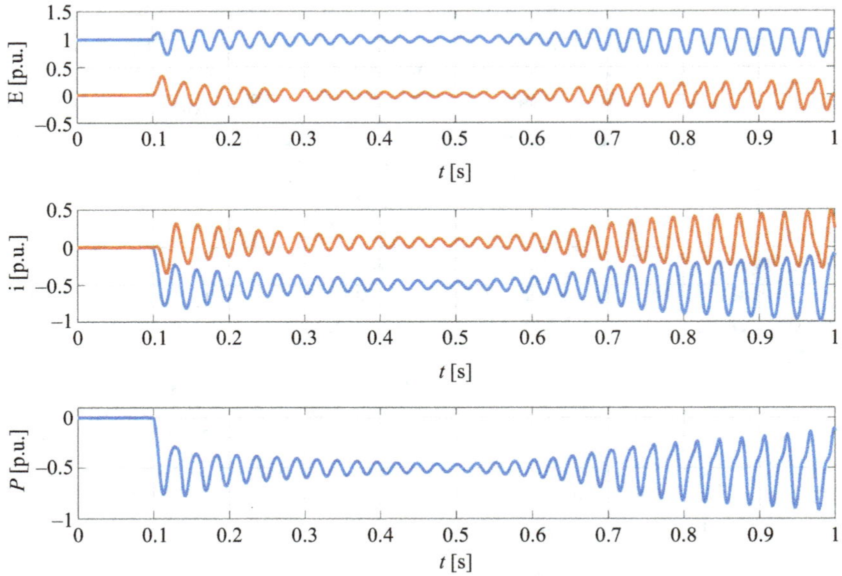

Figure 10.8 Simulation results. In the graphs for **E** *and* **i**, *the q components are those with (near) zero mean. PLL bandwidth α_p is doubled at $t = 0.5$ s*

stable closed-loop system is obtained, but with very poor stability margins. Doubling the PLL bandwidth yields an unstable system, as shown in Figure 10.7 (b). (An excerpt of the MATLAB® code used for obtaining the Nyquist plots is shown in Appendix D.) To verify these results in the time domain, a simulation is made where the converter is started with $i_d^{\text{ref}} = 0$. At $t = 0.1$ s, i_d^{ref} is stepped to −0.5 p.u. A very oscillatory response results, but the system is stable. At $t = 0.5$ s, α_p is doubled. As predicted by theory, this results in growing oscillations, i.e., instability. Simulation results are shown in Figure 10.8.

10.9 Conclusion of VSC modeling

In this chapter, how the current control loop and the outer loops that feed to it can be analyzed via small-signal modeling was shown. Thereby, an input-admittance matrix is formed, which characterizes the behavior of the VSC-HVDC terminal as seen from the PCC. This matrix is operating-point dependent. By modeling the grid impedance as a matrix as well, the stability of the converter–grid interconnection can be analyzed using the generalized Nyquist criterion. Although this was exemplified for a purely inductive grid impedance, the grid-impedance matrix can be expanded to include, e.g., models of other converters and synchronous machines that are located in the electrical vicinity of the VSC-HVDC terminal.

Appendix A Algorithm for the interconnection of the subsystems

The first step of the algorithm requires that the state models are diagonally appended. The appended system, represented using capital letters for the inputs, outputs, and states, is then rearranged such that all inputs/outputs that are to be left unconnected are grouped together (indicated by subscript 1), and all inputs/outputs that are to be connected are grouped together (indicated by subscript 2)

$$\dot{\boldsymbol{X}} = \boldsymbol{A}\boldsymbol{X} + \begin{bmatrix} \boldsymbol{B}_1 & \boldsymbol{B}_2 \end{bmatrix} \begin{bmatrix} \boldsymbol{U}_1 \\ \hline \boldsymbol{U}_2 \end{bmatrix} \tag{A.1}$$

$$\begin{bmatrix} \boldsymbol{Y}_1 \\ \hline \boldsymbol{Y}_2 \end{bmatrix} = \begin{bmatrix} \boldsymbol{C}_1 \\ \hline \boldsymbol{C}_2 \end{bmatrix} \boldsymbol{X} + \left[\begin{array}{c|c} \boldsymbol{D}_{11} & \boldsymbol{D}_{12} \\ \hline \boldsymbol{D}_{21} & \boldsymbol{D}_{22} \end{array}\right] \begin{bmatrix} \boldsymbol{U}_1 \\ \hline \boldsymbol{U}_2 \end{bmatrix} \tag{A.2}$$

Using the matrix $\boldsymbol{H}$ that specifies the connections between the outputs and the inputs

$$\boldsymbol{U}_2 = \boldsymbol{H}\boldsymbol{Y}_2 \tag{A.3}$$

the variables $\boldsymbol{U}_2$ and $\boldsymbol{Y}_2$ are eliminated from (A.1) and (A.2), resulting in a state model of the form

$$\dot{\boldsymbol{X}} = \left[\boldsymbol{A} + \boldsymbol{B}_2\boldsymbol{H}(\boldsymbol{I} - \boldsymbol{D}_{22}\boldsymbol{H})^{-1}\boldsymbol{C}_2\right]\boldsymbol{X} + \left[\boldsymbol{B}_1 + \boldsymbol{B}_2\boldsymbol{H}(\boldsymbol{I} - \boldsymbol{D}_{22}\boldsymbol{H})^{-1}\boldsymbol{D}_{21}\right]\boldsymbol{U}_1 \tag{A.4}$$

$$\begin{aligned} \boldsymbol{Y}_1 = {} & \left[\boldsymbol{C}_1 + \boldsymbol{D}_{12}\boldsymbol{H}(\boldsymbol{I} - \boldsymbol{D}_{22}\boldsymbol{H})^{-1}\boldsymbol{C}_2\right]\boldsymbol{X} \\ & + \left[\boldsymbol{D}_{11} + \boldsymbol{D}_{12}\boldsymbol{H}(\boldsymbol{I} - \boldsymbol{D}_{22}\boldsymbol{H})^{-1}\boldsymbol{D}_{21}\right]\boldsymbol{U}_1 \end{aligned} \tag{A.5}$$

where $\boldsymbol{I}$ is the identity matrix.

Appendix B HVDC converter model

The six-pulse HVDC converter is described by the transfers

$$
\begin{aligned}
a &= -j\frac{3\mu_0}{2\pi X_L}\left[1 - \mathrm{sinc}\left(\frac{\mu_0}{2\pi}\right)\angle - \mu_0/2\right] \\
f &= +j\frac{3\mu_0}{2\pi X_L}\left[1 - \mathrm{sinc}\left(\frac{\mu_0}{2\pi}\right)\angle + \mu_0/2\right] \\
b &= \frac{3}{\pi X_L}\sin^2\frac{\mu_0}{2}\angle - 2(\alpha_0 - \phi) - \mu_0 \\
e &= \frac{3}{\pi X_L}\sin^2\frac{\mu_0}{2}\angle + 2(\alpha_0 - \phi) + \mu_0 \\
c &= -\frac{\sqrt{3}}{\pi}\angle - (\alpha_0 - \phi)\left[-j - 2\sin\left(\frac{\mu_0}{2}\right)\angle - \mu_0/2\right] \\
g &= -\frac{\sqrt{3}}{\pi}\angle + (\alpha_0 - \phi)\left[+j - 2\sin\left(\frac{\mu_0}{2}\right)\angle + \mu_0/2\right] \\
d &= -\frac{3V_1}{\pi X_L}\sin(\alpha_0)\sin\left(\frac{\mu_0}{2}\right)\angle - (\alpha_0 - \varphi) - \mu_0/2 \\
h &= -\frac{3V_1}{\pi X_L}\sin(\alpha_0)\sin\left(\frac{\mu_0}{2}\right)\angle + (\alpha_0 - \varphi) + \mu_0/2 \\
i &= +j\frac{3\sqrt{3}}{\pi}\angle + (\alpha_0 - \varphi) \\
j &= -j\frac{3\sqrt{3}}{\pi}\angle - (\alpha_0 - \varphi) \\
k &= \frac{3X_L}{\pi} + j\omega\frac{X_L}{\omega_0}\left(2 - \frac{3\mu_0}{2\pi}\right) \\
l &= -\frac{3\sqrt{3}V_1}{2\pi}\sin(\alpha_0)
\end{aligned}
$$

where the AC-side variables are written in terms of their equivalent DC-side frequencies. To model higher pulse converters, it is necessary to scale the transfers in accordance with the series and parallel connections on the AC and DC sides. All AC-side variables are referred to the valve side of the converter transformer. The transfers are dependent on the operating point of the converter, which is specified by the following parameters: V_1 and φ are the peak magnitude and angle of the fundamental frequency positive-sequence component of the AC phase voltages, α_0 and μ_0 are the firing and commutation angles, and X_L is the commutation reactance in ohms. In this form, the transfers describe the operation of a positive pole rectifier where the current flow into the converter has been assigned a positive value. In order to model a positive pole inverter, it is necessary to change the signs of the transfers c, g, i, j, l.

Appendix C AC system variable transformations

The transform between *pn* and *dq* components (C.1) is obtained by applying Park's transformation to positive and negative-sequence distortions. In this case, the direct axis is referenced to a phase angle of zero, and the quadrature axis has been chosen to lead the direct axis. If the frequency conversions involved with the transform are assumed to be implicit, then the transform is considered linear.

$$\begin{bmatrix} X_d \\ X_q \end{bmatrix} = \begin{bmatrix} i & -i \\ 1 & 1 \end{bmatrix} \begin{bmatrix} X_p \\ X_n \end{bmatrix} \tag{C.1}$$

The linearization of the transform between *ma* and *dq* components in the synchronous reference frame is

$$\begin{bmatrix} \Delta X_d \\ \Delta X_q \end{bmatrix} = \begin{bmatrix} \cos(X_{a0}) & -X_{m0}\sin(X_{a0}) \\ \sin(X_{a0}) & X_{m0}\cos(X_{a0}) \end{bmatrix} \begin{bmatrix} \Delta X_m \\ \Delta X_a \end{bmatrix}$$

where X_{m0} and X_{a0} are the operating point magnitude and angle of the AC variable.

Appendix D Excerpt of MATLAB® code

```
w = -20:0.0011:20; s = j*w;
I = eye(2); J = [0 -1; 1 0];
i0 = I*id0+J*iq0; i0T = I*id0-J*iq0;
Re = [1 0; 0 0]; Im = [0 1; 0 0];
Gci = ac./(s+ac);
Yi = s./((s+ac).*(s+af).*L);
Gp = ap./((s+ap)*E0);
Fa = (kpa+kia./s).*af./(s+af);
Fd = ad*(1+aid./s)/E0;
Gd = Fd./(s+E0*Fd.*Gci);
for n = 1:length(w)
Y = I*Yi(n)-(I*Gci(n)*Gd(n)*E0*Yi(n)+J*Gci(n)*Fa(n))*Re-Gci(n)*Gd(n)
*Re*i0T;
Y = Y+(Gci(n)*Gd(n)*E0*Im*i0+J*i0*Gci(n)-J*xi*E0*Yi(n))*Gp(n)*Im;
Z = (I*s(n)+J*w1)*Lg;
e = eig(Z*Y);
eig1(n) = e(1); eig2(n) = e(2);
end
```

References

[1] Ainsworth J. Harmonic instability between controlled static convertors and ac networks. In: Proceedings of the Institution of Electrical Engineers. Vol. 114. IET; 1967. pp. 949–957.

[2] Hingorani NG and Gyugyi L. Understanding FACTS. IEEE Press, New York; 2000.
[3] Bakken B and Faanes H. Technical and economic aspects of using a long submarine HVDC connection for frequency control. IEEE Transactions on Power Systems. 1997;12(3):1252–1258.
[4] Cresap RL and Mittelstadt WA. Small-signal modulation of the Pacific HVDC intertie. IEEE Transactions on Power Apparatus and Systems. 1976;95(2):536–541.
[5] Cresap R, Mittelstadt W, Scott D, *et al.* Operating experience with modulation of the Pacific HVDC intertie. IEEE Transactions on Power Apparatus and Systems. 1978;(4):1053–1059.
[6] Hauer JF. Robust damping controls for large power systems. IEEE Control Systems Magazine. 1989;9(1):12–18.
[7] Anderson PM, Agrawal BL, and Van Ness JE. Subsynchronous resonance in power systems (vol. 9). Wiley–IEEE Press, New York; 1999.
[8] Bahrman M, Larsen E, Piwko R, *et al.* Experience with HVDC-turbine-generator torsional interaction at Square Butte. IEEE Transactions on Power Apparatus and Systems. 1980;(3):966–975.
[9] Piwko R and Larsen E. HVDC system control for damping of subsynchronous oscillations. IEEE Transactions on power apparatus and systems. 1982;(7):2203–2211.
[10] Larsen E and McIntyre I. Sub synchronous torsional interaction with voltage-source converter HVDC systems-reference-frame study. Final Report. 2001.
[11] Jiang-Hafner Y, Duchen H, Linden K, *et al.* Improvement of subsynchronous torsional damping using VSC HVDC. In: Proceedings. International Conference on Power System Technology (vol. 2). IEEE; 2002. pp. 998–1003.
[12] Dickmander D, Pourbeik P, Tulkiewicz T, *et al.* SSTI characteristics of HVDC Light. Electric systems consulting. ABB Inc, Raleigh, USA; 2003.
[13] Harnefors L. Analysis of subsynchronous torsional interaction with power electronic converters. IEEE Transactions on power systems. 2007;22 (1):305–313.
[14] Harnefors L, Bongiorno M, and Lundberg S. Input-admittance calculation and shaping for controlled voltage-source converters. IEEE transactions on industrial electronics. 2007;54(6):3323–3334.
[15] Sun J. Impedance-based stability criterion for grid-connected inverters. IEEE Transactions on Power Electronics. 2011;26(11):3075–3078.
[16] Céspedes M and Sun J. Modeling and mitigation of harmonic resonance between wind turbines and the grid. In: 2011 IEEE Energy Conversion Congress and Exposition. IEEE; 2011. pp. 2109–2116.
[17] Cespedes M and Sun J. Impedance shaping of three-phase grid-parallel voltage-source converters. In: 2012 Twenty-Seventh Annual IEEE Applied Power Electronics Conference and Exposition (APEC). IEEE; 2012. pp. 754–760.

[18] Vesti S, Suntio T, Oliver J, *et al.* Impedance-based stability and transient-performance assessment applying maximum peak criteria. IEEE Transactions on Power Electronics. 2012;28(5):2099–2104.
[19] Turner R, Walton S, and Duke R. A case study on the application of the Nyquist stability criterion as applied to interconnected loads and sources on grids. IEEE Transactions on Industrial Electronics. 2012;60(7):2740–2749.
[20] Cespedes M and Sun J. Impedance modeling and analysis of grid-connected voltage-source converters. IEEE Transactions on Power Electronics. 2013;29(3):1254–1261.
[21] Zou C, Rao H, Xu S, *et al.* Analysis of Resonance between a VSC-HVDC converter and the AC grid. IEEE Transactions on Power Electronics. 2018;33(12):10157–10168.
[22] Harnefors L, Wang X, Yepes AG, *et al.* Passivity-based stability assessment of grid-connected VSCs—An overview. IEEE Journal of emerging and selected topics in Power Electronics. 2015;4(1):116–125.
[23] Persson EV. Calculation of transfer functions in grid-controlled convertor systems. With special reference to HVDC transmissions. In: Proceedings of the Institution of Electrical Engineers. vol. 117. IET; 1970. pp. 989–997.
[24] Sucena-Paiva J and Freris L. Stability of a dc transmission link between weak ac systems. In: Proceedings of the Institution of Electrical Engineers (vol. 121). IET; 1974. pp. 508–515.
[25] Todd S, Wood A, and Bodger P. An s-domain model of an HVDC converter. IEEE transactions on power delivery. 1997;12(4):1723–1729.
[26] Jovcic D, Pahalawaththa N, and Zavahir M. Analytical modelling of HVDC-HVAC systems. IEEE Transactions on Power Delivery. 1999;14(2):506–511.
[27] Padiyar K and Kulkarni A. Design of reactive current and voltage controller of static condenser. International Journal of Electrical Power & Energy Systems. 1997;19(6):397–410.
[28] Schauder C and Mehta H. Vector analysis and control of advanced static VAR compensators. In: IEE Proceedings C (Generation, Transmission and Distribution). vol. 140. IET; 1993. pp. 299–306.
[29] Szechtman M. First benchmark model for HVDC control studies. Electra. 1991;135:55–73.
[30] Chen S, Wood A, and Arrillaga J. HVDC converter transformer core saturation instability: a frequency domain analysis. IEE Proceedings—Generation, Transmission and Distribution. 1996;143(1):75–81.
[31] Todd S, Wood A, Bodger P, *et al.* Rational functions as frequency dependent equivalents for transient studies. In: Proceedings of the 1997 International Conference on Power systems transients; 1997. pp. 137–142.
[32] Gustavsen B and Semlyen A. Rational approximation of frequency domain responses by vector fitting. IEEE Transactions on power delivery. 1999;14 (3):1052–1061.
[33] Larsen E, Baker D, and McIver J. Low-order harmonic interactions on AC/DC systems. IEEE Transactions on Power Delivery. 1989;4(1):493–501.

[34] Osauskas C, Hume D, and Wood A. Small signal frequency domain model of an HVDC converter. IEE Proceedings—Generation, Transmission and Distribution. 2001;148(6):573–578.
[35] Manitoba H. Research Centre, PSCAD/EMTDC Users Manual 1994. Tutorial Manual.
[36] Larsen EV, Clark K, and Lorden DJ. Stability analysis of a digitally based HVDC firing-pulse synchronization control. IEEE Transactions on Power Delivery. 1992;7(3):1415–1424.
[37] Jacobson B, Karlsson P, Asplund G, *et al.* VSC-HVDC transmission with cascaded two-level converters. In: Cigré session. 2010. pp. B4–B110.
[38] Antonopoulos A, Angquist L, and Nee HP. On dynamics and voltage control of the modular multilevel converter. In: 2009 13th European Conference on Power Electronics and Applications. IEEE; 2009. pp. 1–10.
[39] Harnefors L, Antonopoulos A, Norrga S, *et al.* Dynamic analysis of modular multilevel converters. IEEE Transactions on Industrial Electronics. 2012;60 (7):2526–2537.
[40] Harnefors L, Zhang L, and Bongiorno M. Frequency-domain passivity-based current controller design. IET Power Electronics. 2008;1(4):455–465.
[41] Harnefors L. Modeling of three-phase dynamic systems using complex transfer functions and transfer matrices. IEEE Transactions on Industrial Electronics. 2007;54(4):2239–2248.
[42] Busada CA, Jorge SG, Leon AE, *et al.* Current controller based on reduced order generalized integrators for distributed generation systems. IEEE Transactions on Industrial Electronics. 2011;59(7):2898–2909.
[43] Buso S and Mattavelli P. Digital control in power electronics. Lectures on Power Electronics. 2006;1(1):1–158.
[44] Mattavelli P. Synchronous-frame harmonic control for high-performance AC power supplies. IEEE Transactions on Industry Applications. 2001;37 (3):864–872.
[45] Holmes DG, Lipo TA, Mcgrath BP, *et al.* Optimized design of stationary frame three phase AC current regulators. IEEE transactions on power electronics. 2009;24(11):2417–2426.

Chapter 11

Hybrid simulation for HVDC

Wenchuan Wu[1] *and Yizhong Hu*[1]

11.1 Background

11.1.1 Large-scale AC–DC hybrid power system

Modern electric power girds are gradually getting converted from traditional pure AC systems to AC–DC hybrid systems due to the rapid development/deployment of high-voltage DC (HVDC) projects all over the world. One typical example is the China Southern Power Grid (CSG) where "8 AC and 10 DC" – totaling 18 channels – have been deployed in the West-to-East Power Transmission Project with a maximum transmission capacity of 47.5 GW (according to the data of 2018) [1].

11.1.2 Power system transient simulation

Power system transient simulation is an important approach to study the system's dynamic behaviors and guide its daily operation schedule. It can be classified into the following two categories based on the different scales of time step: electromagnetic transient (EMT) simulation and transient stability analysis (TSA).

EMT has time step in microseconds, uses instantaneous value of electrical variables, and studies the transient phenomena caused by the interaction between electrical and magnetic fields. EMT is able to present the detailed switch process of power electronic devices in HVDC, achieving very accurate simulation. However, generally it is not suitable for large-scale system simulation because of its large computational burden due to complicated model and small time step. EMT is commonly used in detailed study of HVDC system in a relatively small scale. Currently there are several commercial EMT tools, such as PSCAD/EMTDC, EMTP, MATLAB®/Simulink®, Real Time Digital Simulator (RTDS), HYPERSIM, and RT-LAB. Particularly, some EMT tools can achieve the so-called real-time simulation, such as RTDS, by parallel computation hardware. This capacity makes it possible to connect real control/protect devices via power amplifiers, implementing hardware in loop simulation, which is quite valuable for in-factory tests.

[1]Tsinghua University, Beijing, China

TSA has time step in milliseconds, uses phasor to present all electrical variables, and studies the transient caused by the interaction between electrical field and machine. Since the dynamic behavior of traditional pure AC system comes from the conversion of mechanical and electromagnetic power in generators, TSA can appropriately describe it just using millisecond time step. However, TSA is unable to model HVDC in detail. TSA is commonly used in the analysis of stability of generators in large-scale AC system after a disturbance. Currently there are several commercial TSA tools, such as PSS/E, PSASP, and BPA.

In summary, EMT is precise but less efficient (large computational burden) and suitable for small-scale DC system; TSA is efficient but not accurate enough to model HVDC (switching device) and suitable for large-scale AC system.

11.1.3 Simulation of large-scale AC–DC hybrid system

The AC and DC parts in large-scale AC–DC hybrid systems are closely connected and interacted. This statement has been verified by the fact that fault in the AC system near terminal of HVDC can cause commutation failure of HVDC. Therefore, the large-scale AC–DC hybrid system should be simulated as integrity, rather than separately, which brings new challenge to the system simulation.

It is obvious that TSA is not suitable for simulating large-scale AC–DC hybrid system, since it cannot model DC accurately. On the other hand, although EMT can accurately capture the electromagnetic process of DC, the extremely large computational cost makes it impractical. For non-real-time EMT tools, such as PSCAD/EMTDC, the simulation time approximately grows exponentially with the size of the system to be simulated. For real-time EMT tools, such as RTDS, the required hardware grows linearly with the size of the system to be simulated. So EMT is not quite suitable for large-scale AC–DC hybrid system simulation either.

To appropriately simulate large-scale AC–DC hybrid systems, both the precision of the DC parts and the efficiency of the whole system should be considered. Then, it is straightforward to introduce the idea of EMT–TSA hybrid simulation.

11.2 Review of hybrid simulation

11.2.1 Introduction

As its name implies, the so-called EMT–TSA hybrid simulation combines EMT and TSA together and simulates different parts of the system with different precision. NETOMAC [2], the software developed by SIEMENS at 1976, is generally considered the first EMT–TSA hybrid simulation tool. In the next 40+ years, several works in this area were performed and reported.

The basic framework of EMT–TSA hybrid simulation is illustrated in Figure 11.1. According to the characteristics of the two simulation tools, usually DC part is modeled in EMT to achieve accurate simulation and large-scale AC part is modeled in TSA to achieve efficient simulation. It is important to point out that, to implement the interaction between EMT network and TSA network, network

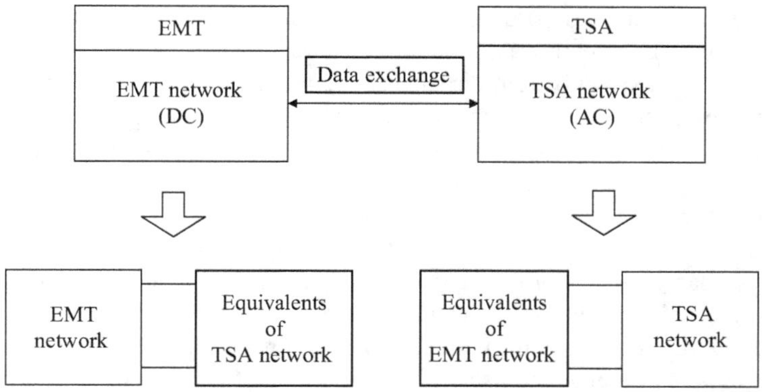

Figure 11.1 Basic framework of EMT–TSA hybrid simulation

equivalents are needed in the boundary (the interface between TSA and EMT). Taking the example of the part simulated in EMT, besides the detailed model of DC network (EMT network), equivalents of AC network (TSA network) should be established in the boundary; so is the part simulated in TSA. In the specific moment, EMT and TSA should exchange data to update the network equivalents.

Therefore, the network equivalent is a critical content in hybrid simulation, especially the equivalent of TSA network in EMT. Classic network equivalents are Norton/Thevenin circuits, whose parameters are calculated in fundamental frequency. Strictly, the Norton/Thevenin equivalent circuits are only valid in fundamental frequency, and they just represent the network's characteristics in fundamental frequency. Directly put these equivalents into EMT to present TSA network, cannot count in the influence brought by harmonics, or more accurate, non-fundamental-frequency components, which cannot be neglected in DC system. As for the equivalent of EMT in TSA, since all the electrical variables in EMT are instantaneous value and all the electrical variables in TSA are fundamental phasors, the main challenge is how to extract fundamental frequency phasor from instantaneous values.

Below we summarize the development of network equivalents of both sides.

11.2.2 Development of TSA network's equivalents

In 1981, Heffernan and Turner [3–5] realized that the traditional TSA tools use quasi-steady-state model to represent HVDC system, making it impossible to accurately study AC–DC hybrid system's transient characteristics. Therefore, they tried to develop EMT–TSA hybrid simulation. They divided the system at the AC bus where the converter is located, and put the HVDC into Transient Converter Simulation (TCS, a kind of EMT tool) and AC system into TSA. Thevenin equivalent circuit was used to present TSA network in TCS. Their work also considered the harmonics generated by HVDC, by introducing several important harmonic impedances into Thevenin equivalent circuit [6,7].

In the Reeve and Adapa's work on EMT–TSA hybrid simulation in 1988 [8,9], Norton equivalent circuit was adopted as the equivalent of TSA network. They pointed that Norton equivalent circuit is only valid at fundamental frequency and cannot represent the TSA network in other frequency. Thus waveform distortion in the boundary has a remarkable influence on the hybrid simulation accuracy. To avoid this, they suggested dividing the system in AC buses away from the DC part, rather than the AC buses where the converters located.

In 1995, Anderson [10,11] reiterated that it was necessary to consider the frequency characteristics of the equivalent of TSA network in EMT. Moreover, his work introduced frequency-dependent network equivalents (FDNEs) into this area. He used FDNE with Thevenin voltage source to present the TSA network in EMT. The FDNE was implemented as an RLC circuit, and the parameters of R, L, and C were carefully chosen so that RLC circuit would have the same frequency characteristics as the original network [12].

The following work of Wang [13] (Manitoba HVDC Research Centre, Canada), Yue [14] (China EPRI), and Liu [15] (Tsinghua University, China) still use Norton/Thevenin equivalent circuits, since the RLC-circuit-based FDNE is complicated and not quite practical.

In 2009, due to the development of rational modeling, Lin [16] (University of Manitoba, Canada) applied rational-model-based FDNE into EMT–TSA hybrid simulation, using FDNE parallel with Norton current source as the equivalents of TSA network. Then Liang [17] (University of Manitoba, Canada), Zhang [18] (Tsinghua University, China), and Hu [19] (Tsinghua University, China) used the same method in their own work. Notable precision improvements are achieved by this strategy.

11.2.3 Development of EMT network's equivalents

Turner and Heffernan [3–5] used frequency spectrum and window filtering to extract the fundamental frequency phasor of the power injected by EMT network to TSA, as well as the voltage, according to the instantaneous information in EMT. Then they used power and voltage phasor to represent the EMT network in TSA.

Reeve and Adapa [8,9] used curve fitting algorithm (CFA) to get the fundamental frequency current and voltage in the boundary and then calculated active/reactive power injected by EMT based on the previous current/voltage phasor. A load was used to present the EMT network in TSA.

Anderson [10,11] applied fast Fourier transform (FFT) to get the fundamental current and voltage in the boundary and then calculated an equivalent current source to represent the EMT network in TSA.

In the following work of Wang [13], Yue [14], and Liu [15], similar strategy was adopted. The idea was to use CFA/FFT to extract the fundamental frequency information.

In the method of both CFA and FFT, sampling is an obligatory process and this will introduce additional delay. To overcome this issue, Lin [16] proposed a power-balanced method in his work. He estimated the instantaneous power injected by

EMT network and calculated an equivalent current injection to represent the EMT network in TSA, based on power balance. Liang [17], Zhang [18], and Hu [19] followed this method.

11.3 Implementation of EMT–TSA hybrid simulation

Here, we theoretically discuss the possible ways to implement EMT–TSA hybrid simulation. We focus on the network equivalents of EMT and TSA network.

11.3.1 AC–DC hybrid system to be simulated

Figure 11.2 shows the hybrid system to be simulated. We want to model the DC part (left dashed box) in EMT tool and model the AC part (right dashed box) in the TSA tool. Thus the question is how to respectively model each part in EMT and TSA.

11.3.2 Part of system simulated in EMT

The DC part (EMT network) and the network equivalent of AC part (TSA network) should be put into the EMT tool.

A straightforward idea is to use current/voltage source to represent the boundary. Here we take the example of current source to discuss, as shown in Figure 11.3 (the concept is same as voltage source). Obviously, the current source should inject the same amount of current as the boundary. In hybrid simulation, the EMT time step is smaller than TSA time step. For example, EMT time step is 50 μs and TSA time step is 2 ms, i.e., one TSA time step contains 40 EMT time steps. The data exchange between EMT and TSA occurs in every TSA time step, which is used to update the network equivalents on both sides. Now let us come back to Figure 11.3. Using a current source to present the TSA network means, for the entire TSA time step (40 EMT time steps), the boundary is constant for EMT tool and no dynamic characteristics of TSA network will be brought into EMT

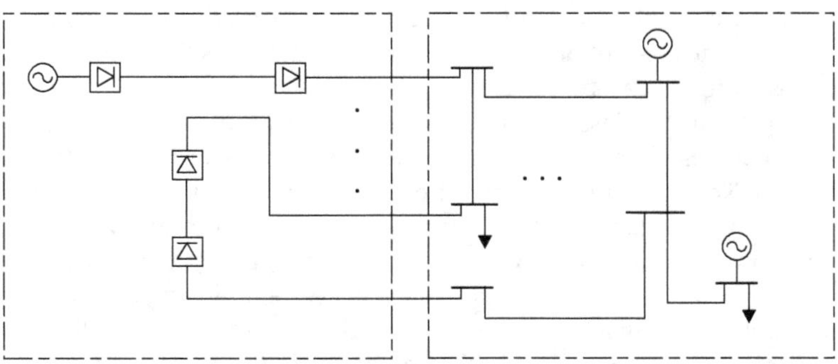

Figure 11.2 The AC–DC hybrid system to be simulated

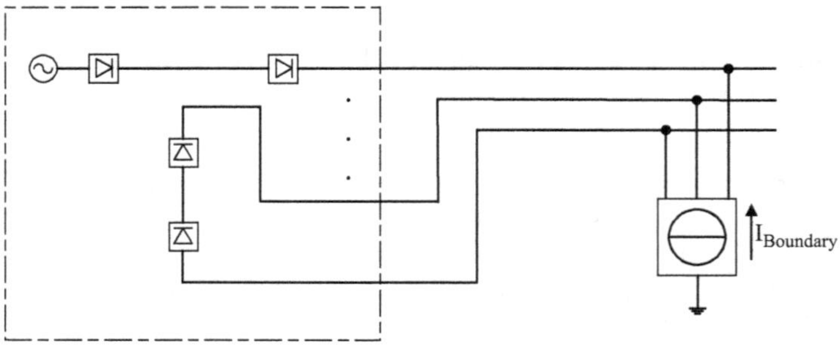

Figure 11.3 The part of system simulated in EMT. This figure represents the TSA network as current injection

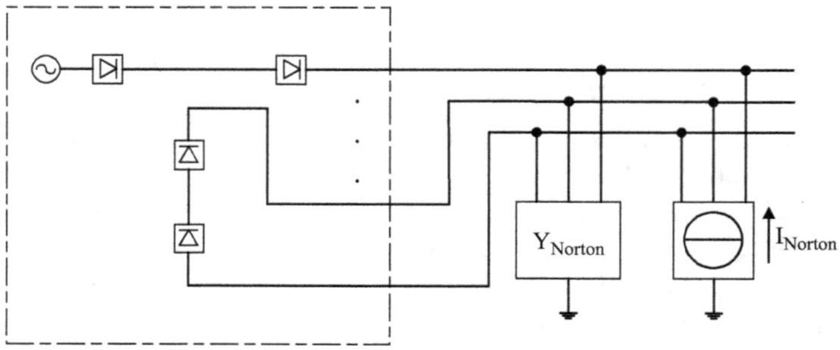

Figure 11.4 The part of system simulated in EMT. This figure represents the TSA network using multi-port Norton equivalent circuit

simulation. This is oversimplified and has problems of both accuracy and stability. None of the work mentioned in the previous section uses this strategy.

Compared with solely boundary current injection, multi-port Norton Equivalent circuit is a better idea, as shown in Figure 11.4. For the entire TSA time step (40 EMT time steps), although the Norton currents are constant, the admittance matrix can interact with EMT network and represent some dynamic characteristics of TSA network. The concept of Thevenin equivalent circuit is the same, and these two are the most commonly used strategies.

Starting from the classic multi-port Norton/Thevenin equivalent circuit, there are two derivatives. One is toward more practical use. Since the admittance matrix in multi-port Norton equivalent circuit can be complicated to implement in EMT tool, especially with large port number, single-port Norton equivalent circuit for each port is an alternative solution, as illustrated in Figure 11.5. It is worthy to point out that, in this case, the single-port Norton equivalent circuit for one port is seen from that port with other ports open-circuited. Theoretically, this form will lose some accuracy compared with multi-port form. However, as long as there is no dramatic change in the boundary (this is achieved by leaving buffer zone, i.e., the

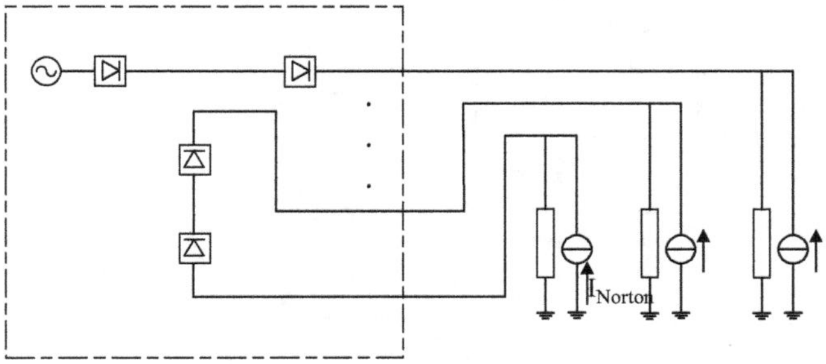

Figure 11.5 The part of system simulated in EMT. This figure represents the TSA network using single-port Norton equivalent circuits

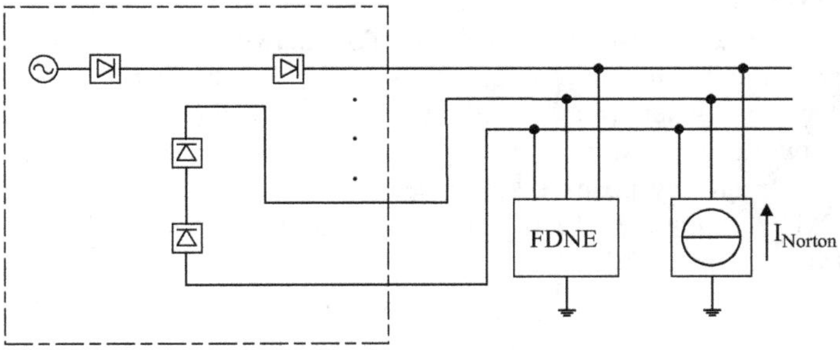

Figure 11.6 The part of system simulated in EMT. This figure represents the TSA network using FDNE techniques

boundary is chosen in AC system with certain distance to DC part), which actually is also required by the multi-port Norton equivalent circuit, this single-port form works fine. (Single-port Thevenin equivalent circuit is the same.)

As mentioned before, the validity of using multi-port Norton equivalent circuit to represent TSA network is that there should be no dramatic change in the boundary. Otherwise, the admittance matrix, which is derived on fundamental frequency, cannot describe the TSA network's characteristics in other frequency, leading to non-negligible accuracy loss. To overcome this problem, the FDNE is brought in and used to replace the admittance matrix in traditional Norton equivalent circuit, as shown in Figure 11.6.

More details of the FDNE techniques will be presented in the following sections. This chapter will focus on FDNE-based EMT–TSA hybrid simulation, since it is the main theoretical improvement in the hybrid simulation area in recent years.

11.3.3 Part of system simulated in TSA

The AC part (TSA network) and the network equivalent of DC part (EMT network) should be put into the TSA tool.

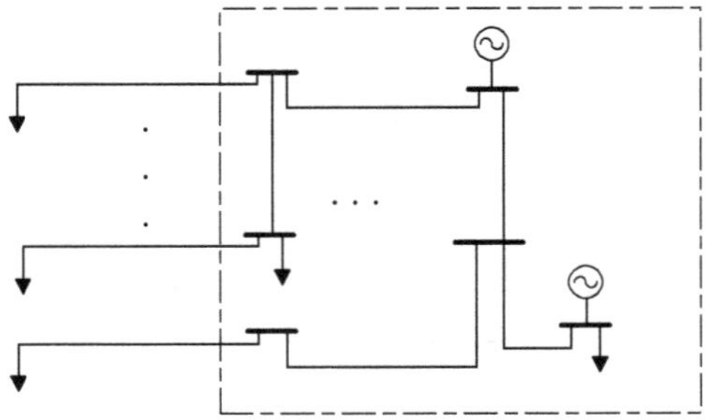

Figure 11.7 The part of system simulated in TSA. This figure represents the EMT network with loads

The EMT network is usually presented in TSA tool with loads (Figure 11.7). This is reasonable because the dynamic of TSA network comes from the swing of generators and that is mainly influenced by active power. Therefore, as long as the EMT network equivalent (loads) can inject the same amount of active power as the original network, it works fine. To be more accurate, the active power should be fundamental frequency positive sequence. It can be extracted from instantaneous value in EMT through CFA/FFT-based method or power-balance-based method, as reviewed in Section 11.2.3.

11.4 Frequency-dependent network equivalent

11.4.1 Introduction to FDNE

FDNE is proposed to replace the admittance matrix in traditional Norton equivalent. Unlike the conventional admittance matrix, which is constant, it varies with frequency and thus able to describe the frequency-dependent characteristics of the network. Figure 11.8 shows the concept of FDNE.

RLC-circuit-based FDNE is illustrated in Figure 11.9 (an example of two-port). If the parameters of R, L, and C are tuned appropriately, the circuit's terminal frequency characteristics can be quite close to the original network. A classic way to solve the RLC parameters is based on resonance, as described in Reference 12. In general, RLC-circuit-based FDNE is complicated and not flexible. It has been gradually replaced by the rational model form. FDNE in the following context refers to the rational model form.

FDNE's rational model is a special admittance matrix, whose entries are rational functions of frequency. It can be expressed as

$$\mathbf{Y}(s) = \begin{bmatrix} y_{11}(s) & y_{12}(s) & \cdots & y_{1N}(s) \\ y_{21}(s) & y_{22}(s) & \cdots & y_{2N}(s) \\ \vdots & \vdots & \ddots & \vdots \\ y_{N1}(s) & y_{N2}(s) & \cdots & y_{NN}(s) \end{bmatrix} \tag{11.1}$$

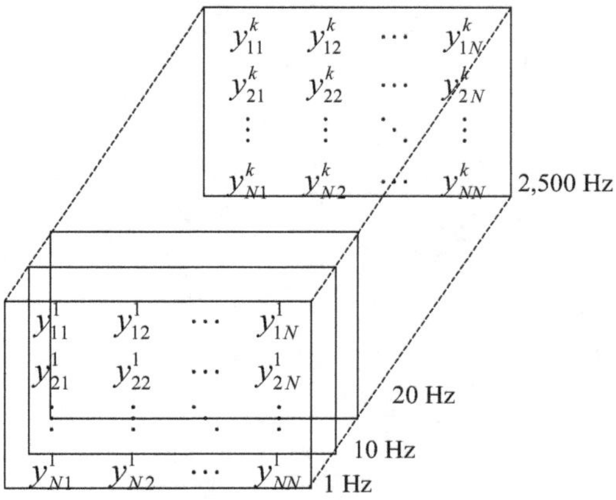

Figure 11.8 Concept of FDNE

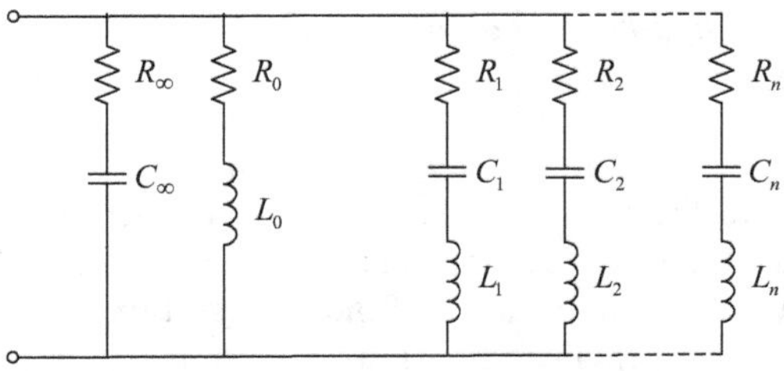

Figure 11.9 FDNE implemented in RLC circuit form

where $s = \text{j}2\pi f$, f is the frequency, and N is the port number. Every entry $y(s)$ is a rational function

$$y(s) = \sum_{i=1}^{n} \frac{c_i}{s - a_i} + d + sh \tag{11.2}$$

where poles $\{a_i\}$ and residues $\{c_i\}$ are either real or complex conjugate pairs; d and h are real; and n is the number of poles. In practice, all entries of $\mathbf{Y}(s)$ share the same set of poles $\{a_i\}$; therefore, n is called the order of the rational model.

11.4.2 Calculation of FDNE's rational model

Calculation of FDNE's rational model mainly concerns properly determining the aforementioned coefficients, so that $\mathbf{Y}(s)$ can represent the frequency characteristics of the original network within the considered frequency range.

The methodology of obtaining FDNE's rational model is fitting; hence, sampling values of the original AC network frequency response, observed form boundary, should be prepared first. This can be acquired by either measuring or

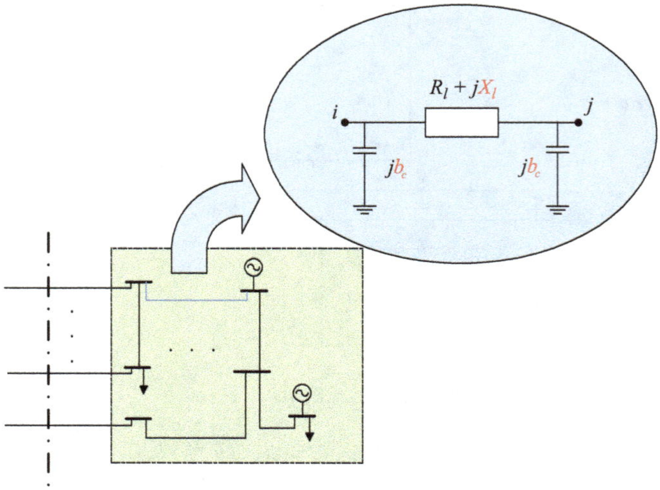

Figure 11.10 Simplified-model-based method to obtain network's frequency response

calculation. Here we briefly introduce a method based on simplified model, as illustrated in Figure 11.10. Take a transmission line as an example. It is represented by PI model, in which the impedance of inductor and the admittance of capacitor are frequency dependent. In a certain sampling frequency, this transmission line's frequency response can be established. Similarly, other devices are represented by suitable simplified models, as shown in Figure 11.11. Then using a standard network reduction method to compress the whole network to the boundary, the sampling value of network's frequency response in this particular frequency is obtained. By repeating the process we can get all the sampling values.

In fact, the method described here can be used to calculate AC network's frequency response (sampling values) directly from power flow data [16], such as PSS/e Raw file.

Next task is to obtain the coefficients of (11.2) based on a series of frequency sampling values, $y(s_i)$, $i = 1, 2, \ldots, N_0$, where N_0 is the number of samples. The most common used method for this part is vector fitting [20].

The basic concept is fitting, and the major problem is that the poles a_i in the denominator of (11.2) are unknowns, making the fitting problem nonlinear. The key point of the vector fitting is to somehow obtain the poles a_i. The remainder of the problem is then an ordinary linear fitting procedure, which can be easily solved.

Therefore, with vector fitting, we introduce an auxiliary rational function

$$\sigma(s) = \sum_{i=1}^{n} \frac{\bar{c}_i}{s - \bar{a}_i} + 1 \tag{11.3}$$

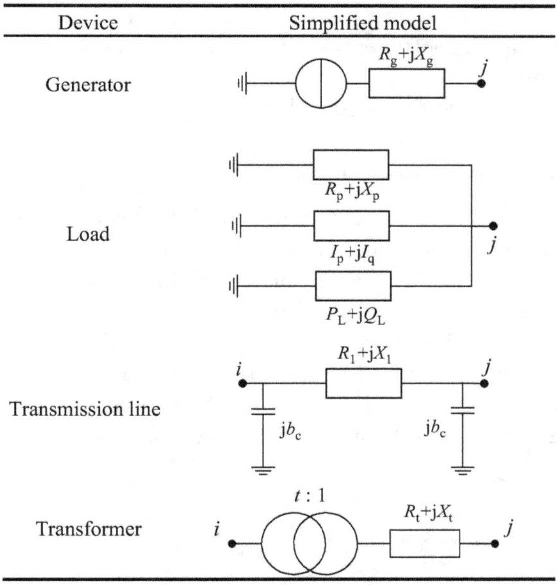

Figure 11.11 Simplified model adopted in the method

where the poles $\bar{a}_i$ are arbitrarily given, and the residues $\bar{c}_i$ are relaxed to meet the following relationship:

$$\sigma(s) \cdot y(s) = \sum_{i=1}^{n} \frac{\widetilde{c}_i}{s - \bar{a}_i} + \widetilde{d} + sh \tag{11.4}$$

where $\widetilde{c}_i$ and $\widetilde{d}$ are additional unknown coefficients.

An important feature of (11.4) is that $\sigma(s)y(s)$ has the same poles as $\sigma(s)$. This indicates that the zeros of $\sigma(s)$ are equal to the poles of $y(s)$. In other words, when $\sigma(s)$ is multiplied by $y(s)$, the poles of $y(s)$ are eliminated by the zeros of $\sigma(s)$, so that we retain the poles of $\sigma(s)$. Therefore, if the zeros of $\sigma(s)$ can be obtained, then the poles of $y(s)$ can also be obtained, and the problem is solved.

A brief description of the vector fitting procedure is given below, and a more detailed description can be found in Reference 20.

Step 1. Calculate $\bar{c}_i$

Since the poles $\bar{a}_i$ are given in (11.4), a linear fitting procedure can be used based on least-squares approximation with sampled values. The coefficients $\bar{c}_i$, $\widetilde{c}_i$, $\widetilde{d}$, and h are then obtained.

Step 2. Calculate a_i

Once $\bar{c}_i$ are found, $\sigma(s)$ can be determined. It is convenient to determine the zeros of $\sigma(s)$, which are also poles of $y(s)$. Then a_i can also be obtained.

Step 3. Calculate c_i, d, and h

With the poles of a_i known from (11.2), by solving a second ordinary linear regression problem, the coefficients c_i, d, and h can be estimated. By this way, $y(s)$ can be determined.

There are several additional considerations.

Remark 1. Starting poles and iteration

Initially, the poles of the auxiliary rational function $\sigma(s)$ are given arbitrarily. The choice of starting poles should be considered. In [20], the authors suggest choosing complex conjugate starting poles, with the imaginary parts linearly distributed over the frequency interval of interest.

In principle, the poles of $y(s)$ should be estimated directly when **Step 2** is completed. However, a more accurate result may be obtained by iterating this process, so that the a_i calculated during **Step 2** are used as the new poles of $\sigma(s)$, and **Step 1** is repeated. This iteration is a more elegant version of the Sanathanan–Koerner scheme [21].

Remark 2. Application to vector functions

Vector fitting can also be applied to vector functions, assuming that all elements have the same poles (see the Discussion section in Reference 20 for more details).

When using VF to calculate an FDNE rational model, all the elements of $\mathbf{Y}(s)$ are stacked into a single column vector and fitted using common poles. By considering the symmetry of $\mathbf{Y}(s)$, we found that the dimensions of the column vector are $N(N-1)/2$, rather than N^2.

It can be noticed that the number of poles n (the model order) must be given before using vector fitting. A fast method to appropriately choose model order can be found in Reference 22.

In what follows, we consider a typical FDNE rational model to demonstrate the calculation result.

Figure 11.12 shows a modified New England 39-bus system. Boundaries were chosen between internal buses #103 and #108, and external buses #3 and #8; hence, the left-hand side of the system was maintained as is, whereas the right-hand side was reduced to an FDNE $\mathbf{Y}(s)$. The elements of the 6×6 (two-port, three-phase) admittance matrix $\mathbf{Y}(s)$ were fitted in the range of 1–2.5 kHz using 48 poles by vector fitting, based on 500 evenly distributed samples.

It is worthy to note that the frequency range of interest was selected for practical reasons. The typical time step of EMT tools such as PSCAD/EMTDC® or RTDS® is 50 μs, which is the inverse of 20 kHz. It is a rule of thumb that the simulation is accurate to about 1/10 to 1/5 of this frequency, which gives 2–4 kHz. This time step has been widely used and continues to be used for line-commutated converters–high-voltage direct current (LCC-HVDC) systems and several protection studies. Hence the FDNE was selected to be accurate up to 2.5 kHz, which is the accuracy bandwidth of the EMT tool itself. For smaller time steps, the fitting bandwidth can be expanded as required.

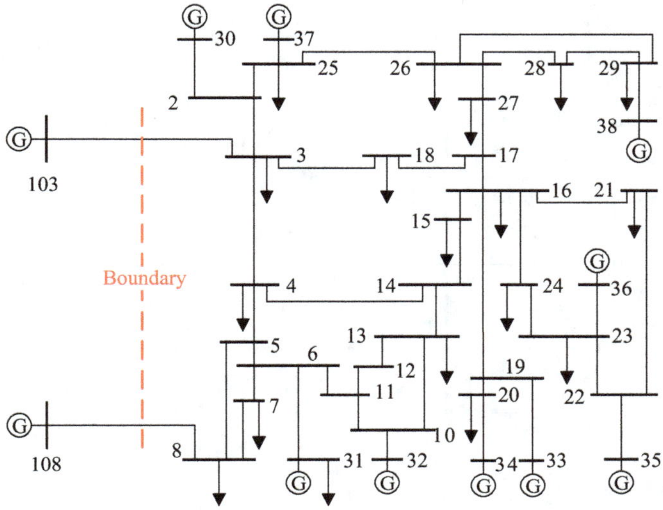

Figure 11.12 Modified New England 39-bus system

Figure 11.13 shows the fitting results of the FDNE rational model $\mathbf{Y}(s)$. For clarity without loss of generality, only one diagonal element $y_{11}(s)$ and one off-diagonal element $y_{12}(s)$ are illustrated: **Actual** denotes the sampling values; **Fitted** denotes the result of the rational model. A relatively good fit was obtained.

11.4.3 Passivity enforcement of FDNE's rational model

Rational models of FDNE must be passive to ensure numerical stability in time domain simulations. Therefore, passivity enforcement is an essential step in FDNE rational modeling procedures.

For clarity, we rewrite $\mathbf{Y}(s)$ into pole-residue form (since all entries share the same pole set):

$$\mathbf{Y}(s) = \sum_{i=1}^{n} \frac{\mathbf{R}_i}{s - a_i} + \mathbf{D} + s\mathbf{E} \tag{11.5}$$

where

$$\mathbf{R}_i = \begin{bmatrix} c_i^{11} & c_i^{12} & \cdots & c_i^{1N} \\ c_i^{21} & c_i^{22} & \cdots & c_i^{2N} \\ \vdots & \vdots & \ddots & \vdots \\ c_i^{N1} & c_i^{N2} & \cdots & c_i^{NN} \end{bmatrix} \tag{11.6}$$

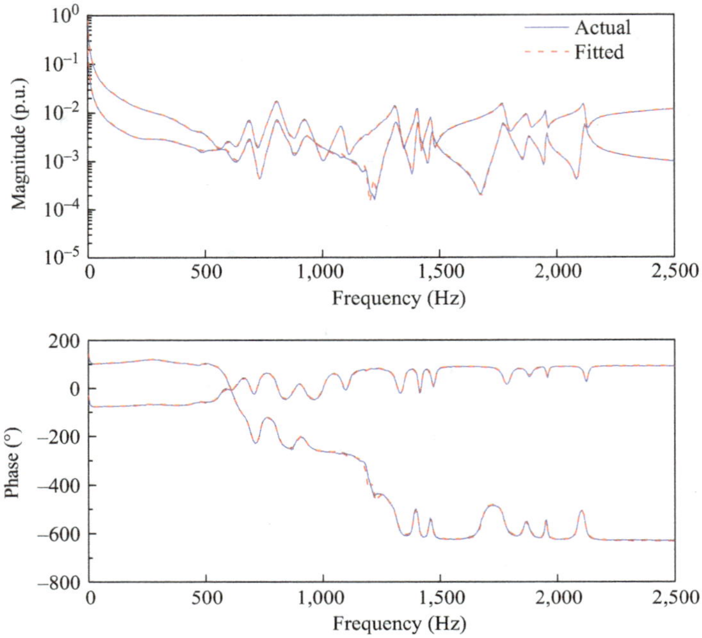

Figure 11.13 Fitting result

$$\mathbf{D} = \begin{bmatrix} d^{11} & d^{12} & \cdots & d^{1N} \\ d^{21} & d^{22} & \cdots & d^{2N} \\ \vdots & \vdots & \ddots & \vdots \\ d^{N1} & d^{N2} & \cdots & d^{NN} \end{bmatrix} \tag{11.7}$$

$$\mathbf{E} = \begin{bmatrix} h^{11} & h^{12} & \cdots & h^{1N} \\ h^{21} & h^{22} & \cdots & h^{2N} \\ \vdots & \vdots & \ddots & \vdots \\ h^{N1} & h^{N2} & \cdots & h^{NN} \end{bmatrix} \tag{11.8}$$

Passivity arises from the physical concept that the model cannot generate energy in all circumstances. Applying terminal voltage $\mathbf{u}$ of certain frequency on FDNE, it consumes energy

$$P = \mathbf{u}\mathrm{Re}\{\mathbf{Y}(s)\}\mathbf{u} \tag{11.9}$$

According to the definition of passivity, P should be non-negative, regardless of voltage's magnitude and frequency. Mathematically, it leads to the following constraints:

$$\mathrm{eig}(\mathrm{Re}\{\mathbf{Y}(s)\}) \geq 0 \tag{11.10}$$

In practice, **D** and **E** are also required to be positive semidefinite [23]:

$$\mathrm{eig}(\mathbf{D}) \geq 0 \tag{11.11}$$

$$\mathrm{eig}(\mathbf{E}) \geq 0 \tag{11.12}$$

Although vector fitting may fit the frequency response accurately, a good fit is not necessarily a passive fit, and in general (11.10)–(11.12) may not be satisfied.

However, even with a non-passive fit, the good fit and the fact that the initial frequency response is directly calculated from power flow data which is known to be passive lead to a consensus that the rational model calculated using vector fitting is only slightly non-passive, and only a small correction is required to make the model passive [23]. Therefore, the general idea of FDNE passivity enforcement is to adjust the elements of $\{\mathbf{R}_i\}$, **D**, and **E** slightly to make the model passive, while minimizing the corresponding change in **Y**. (Most approaches do not adjust the poles $\{a_i\}$, because changing poles complicates the problem significantly.) This leads to a constrained optimization problem:

$$\mathrm{Obj.}\ \Delta\mathbf{Y} = \sum_{i=1}^{n} \frac{\Delta\mathbf{R}_i}{s - a_i} + \Delta\mathbf{D} + s\Delta\mathbf{E} \cong 0 \tag{11.13}$$

$$\mathrm{s.t.}\ \mathrm{eig}(\mathrm{Re}\{\mathbf{Y} + \Delta\mathbf{Y}\}) \geq 0 \tag{11.14}$$

$$\mathrm{eig}(\mathbf{D} + \Delta\mathbf{D}) \geq 0 \tag{11.15}$$

$$\mathrm{eig}(\mathbf{E} + \Delta\mathbf{E}) \geq 0 \tag{11.16}$$

Many passivity enforcement methods have been proposed, and a summary and comparison can be found in Reference 24. The most commonly used method currently is described in [23,25,26], which converts the original optimization problem (11.13)–(11.16) into a quadratic programming model. There are still new methods in this area, such as those described in References 27 and 28.

We take the same example of FDNE defined by Figure 11.12 to show the effect of passivity enforcement. Figure 11.14 shows the change of eigenvalues of $\mathrm{Re}\{\mathbf{Y}(s)\}$, where **Raw** denotes the results before enforcement (referring to the same result as **Fitted**), and **Passive** denotes the results after enforcement. It is clear that **Raw** gave negative eigenvalues in some frequency range, whereas **Passive** did not.

Figure 11.15 shows the fit of $\mathbf{Y}(s)$ after passivity enforcement. Although it lost some accuracy (**Passive** was not as close to **Actual** as the **fitted** in Figure 11.13) to enforce the passivity, it was still a good fit.

Applying this FDNE rational model in PSCAD/EMTDC, the **Passive** model was stable numerically, whereas the **Raw** model could result in divergence in the time domain simulation, as shown in Figure 11.16.

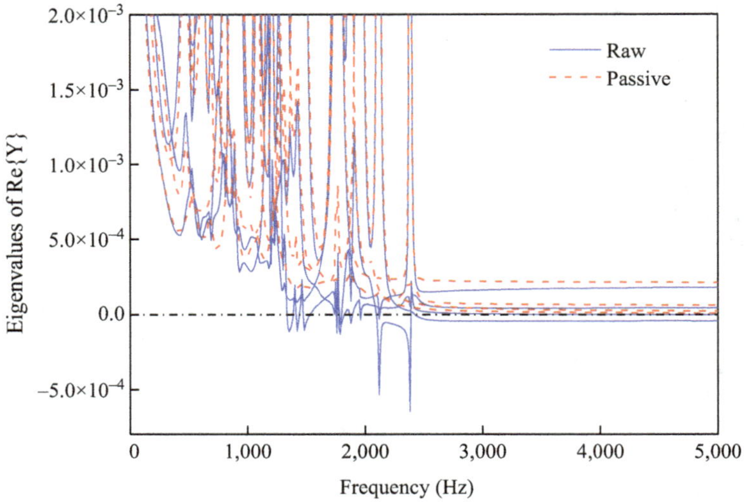

*Figure 11.14 Effect of passivity enforcement on **Y**(s)*

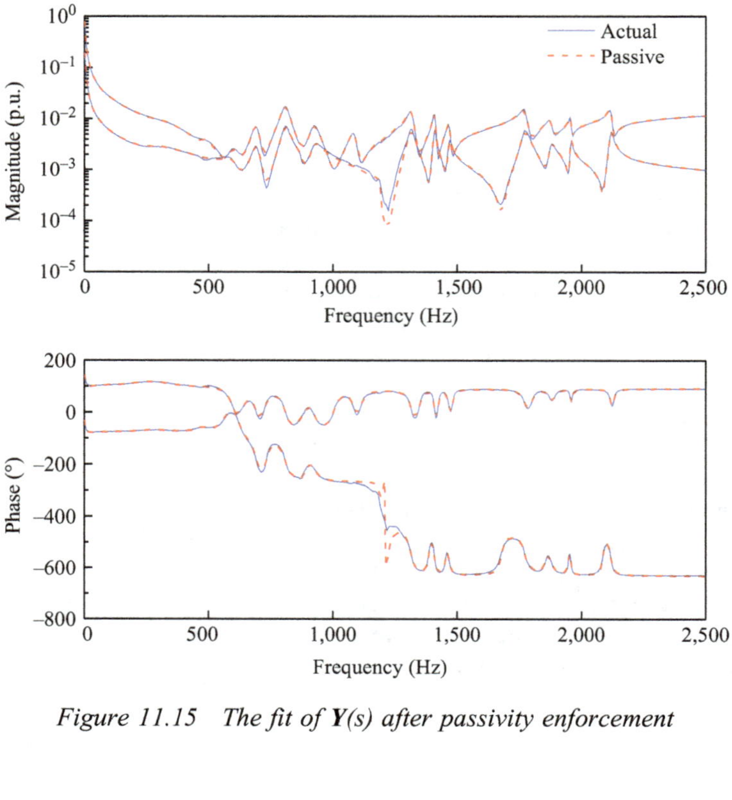

*Figure 11.15 The fit of **Y**(s) after passivity enforcement*

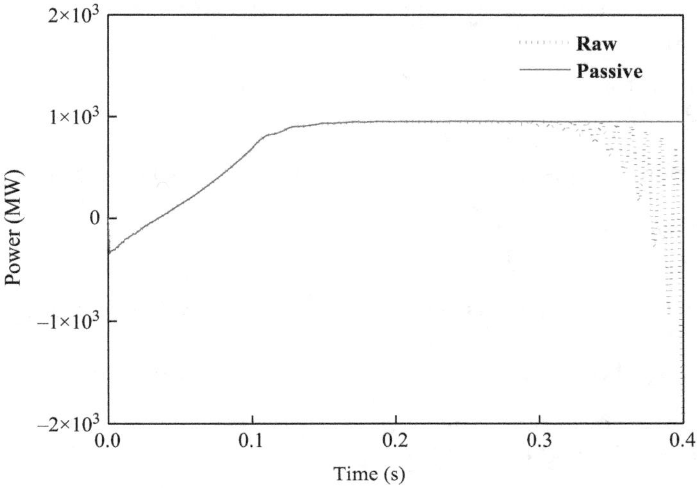

Figure 11.16 Time domain simulation

11.4.4 Implementation of FDNE's rational model in EMT

FDNE's rational models (11.1) and (11.2) can be rewritten in the following form:

$$\mathbf{Y}(s) = \mathbf{C}(s\mathbf{I} - \mathbf{A})^{-1}\mathbf{B} + \mathbf{D} + s\mathbf{E} \tag{11.17}$$

where $\mathbf{A} = diag(\mathbf{A}_1 \quad \cdots \quad \mathbf{A}_k \quad \cdots \quad \mathbf{A}_n), \mathbf{A}_k = diag\Big((a_k \quad a_k \quad \cdots \quad a_k)_{(1\times N)}\Big),$

$$\mathbf{B} = [\mathbf{B}_1 \cdots \mathbf{B}_k \cdots \mathbf{B}_n]^T, \ \mathbf{B}_k = diag\Big((1 \quad 1 \quad \cdots \quad 1)_{(1\times N)}\Big),$$

$$\mathbf{C} = [\mathbf{R}_1 \quad \cdots \quad \mathbf{R}_k \quad \cdots \quad \mathbf{R}_n], \ \mathbf{R}_k = \begin{bmatrix} c_k^{11} & c_k^{12} & \cdots & c_k^{1N} \\ c_k^{21} & c_k^{22} & \cdots & c_k^{2N} \\ \vdots & \vdots & \ddots & \vdots \\ c_k^{N1} & c_k^{N2} & \cdots & c_k^{NN} \end{bmatrix},$$

$$\mathbf{D} = \begin{bmatrix} d_{11} & d_{12} & \cdots & d_{1N} \\ d_{21} & d_{22} & \cdots & d_{2N} \\ \vdots & \vdots & \ddots & \vdots \\ d_{N1} & d_{N2} & \cdots & d_{NN} \end{bmatrix}, \ \mathbf{E} = \begin{bmatrix} h_{11} & h_{12} & \cdots & h_{1N} \\ h_{21} & h_{22} & \cdots & h_{2N} \\ \vdots & \vdots & \ddots & \vdots \\ h_{N1} & h_{N2} & \cdots & h_{NN} \end{bmatrix},$$

and $\mathbf{I}$ is the identity matrix.

In fact, the part $s\mathbf{E}$ in (11.17) is actually a pure capacitor part in EMT and can be implemented separately. This part can also be neglected since the absolute value of $\mathbf{E}$'s entries is quite small compared with other parameters in the model. The main question is how to implement the remaining part in EMT tool:

$$\mathbf{Y}(s) = \mathbf{C}(s\mathbf{I} - \mathbf{A})^{-1}\mathbf{B} + \mathbf{D} \tag{11.18}$$

$\mathbf{Y}(s)$ describes the relationship between terminal voltage $\mathbf{u}(s)$ and injection current $\mathbf{i}(s)$,

$$\mathbf{Y}(s) = \frac{\mathbf{i}(s)}{\mathbf{u}(s)} \tag{11.19}$$

Therefore, $\mathbf{Y}(s)$ is the transfer function of the following state-space:

$$\begin{cases} \dot{\mathbf{x}}(t) = \mathbf{A}\mathbf{x}(t) + \mathbf{B}\mathbf{u}(t) \\ \mathbf{i}(t) = \mathbf{C}\mathbf{x}(t) + \mathbf{D}\mathbf{u}(t) \end{cases} \tag{11.20}$$

where terminal voltage $\mathbf{u}(t)$ is input, current injection $\mathbf{i}(t)$ is output, and $\mathbf{x}(t)$ is state variable.

Applying the classic trapezoidal rule on (11.20), we get

$$\begin{cases} \mathbf{x}(t) = \boldsymbol{\alpha}\mathbf{x}(t-\Delta t) + \boldsymbol{\lambda}\mathbf{u}(t-\Delta t) + \boldsymbol{\lambda}\mathbf{u}(t) \\ \mathbf{i}(t) = \mathbf{C}\mathbf{x}(t) + \mathbf{D}\mathbf{u}(t) \end{cases} \tag{11.21}$$

where

$$\boldsymbol{\alpha} = \left(\mathbf{I} - \frac{\Delta t}{2}\mathbf{A}\right)^{-1}\left(\mathbf{I} + \frac{\Delta t}{2}\mathbf{A}\right) \tag{11.22}$$

$$\boldsymbol{\lambda} = \frac{\Delta t}{2}\left(\mathbf{I} - \frac{\Delta t}{2}\mathbf{A}\right)^{-1}\mathbf{B} \tag{11.23}$$

Substituting the first equation of (11.21) into the second one, it gives the EMT type model of FDNE as

$$\mathbf{i}(t) = \mathbf{I}_{\text{his}} + \mathbf{G}_{\text{eq}}\mathbf{u}(t) \tag{11.24}$$

where

$$\mathbf{G}_{\text{eq}} = \mathbf{C}\boldsymbol{\lambda} + \mathbf{D} \tag{11.25}$$

$$\mathbf{I}_{\text{his}} = \mathbf{C}\boldsymbol{\alpha}\mathbf{x}(t-\Delta t) + \mathbf{C}\boldsymbol{\lambda}\mathbf{u}(t-\Delta t) \tag{11.26}$$

Thus, the FDNE rational model can be implemented in EMT tools. More details, as well as techniques to improve model's computational efficiency, can be found in Reference 29.

11.5 Effect of FDNE in EMT–TSA hybrid simulation

11.5.1 FDNE-based RTDS–TSA hybrid simulation platform

We developed an FDNE-based RTDS–TSA hybrid simulation platform, and its hardware structure and network equivalent strategy are shown in Figures 11.17 and 11.18 separately. More details about the platform can be found in Reference 19 and here we focus on the effect of FDNE.

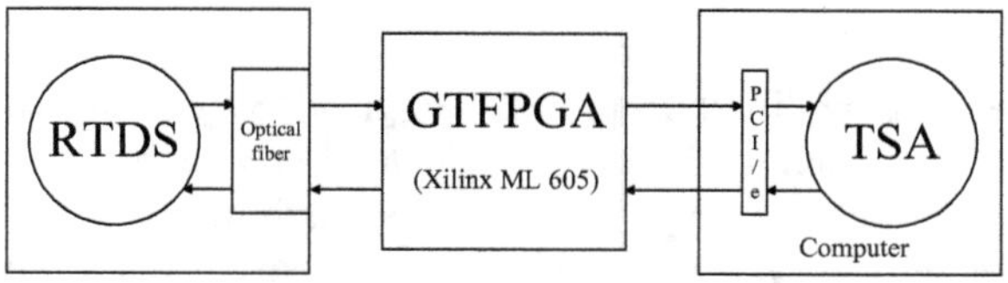

Figure 11.17 Hardware structure of the RTDS–TSA hybrid simulation platform

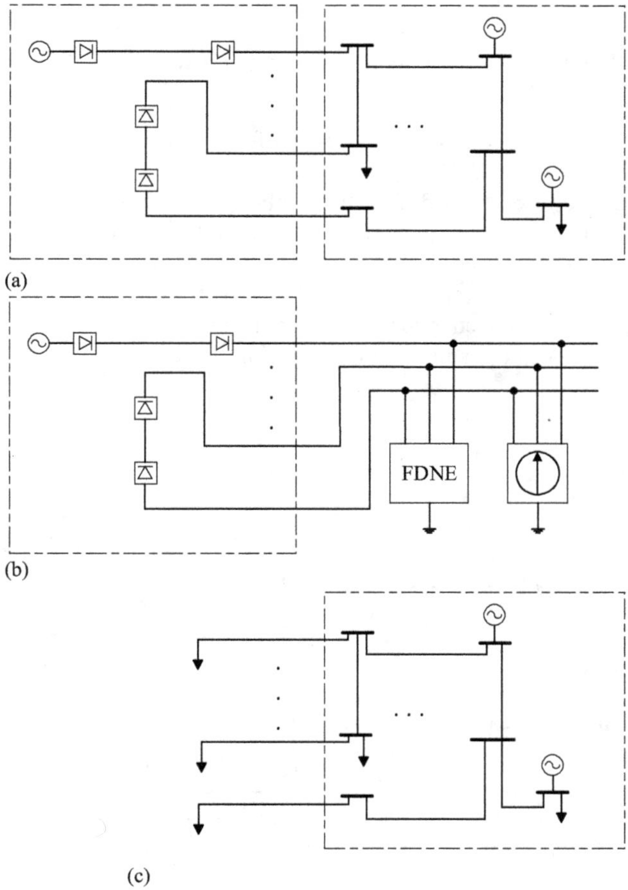

Figure 11.18 Network equivalent strategy adopted in the hybrid simulation platform. (a) System to be simulated; (b) part of system modeled in RTDS; (c) part of system modeled in TSA

11.5.2 Test system

The test system is modified from New England 39-bus system as shown in Figure 11.19. Two switches S1 and S2 are added to make the test system pure AC or AC–DC hybrid. When using the RTDS–TSA hybrid transient simulation platform to simulate the test system, the left part of system surrounded by dashed box is modeled in RTDS, while the rest is modeled in the TSA program. As boundary buses, bus #3 and bus #8 should be modeled both sides.

When using hybrid simulation, an FDNE rational model of the right part is needed. In fact, that FDNE has been presented in the previous sections in detail.

11.5.3 Simulating pure AC system

With switch S1 on and S2 off, the test system is a pure AC system. A 100-ms three-phase ground fault occurs at boundary bus #3 and it was simulated in four different ways:

1. using the proposed hybrid simulation platform with FDNE, the result is marked as **RTDS+TSA+FDNE**;
2. using the proposed hybrid simulation platform without FDNE, replaced with traditional admittance matrix, the result is marked as **RTDS+TSA+Y50**;
3. not using the proposed hybrid simulation platform, modeling the entire test system in RTDS, the result is marked as **Full RTDS** (benchmark);
4. not using the proposed hybrid simulation platform, modeling the entire test system in TSA, the result is marked as **Full TSA**.

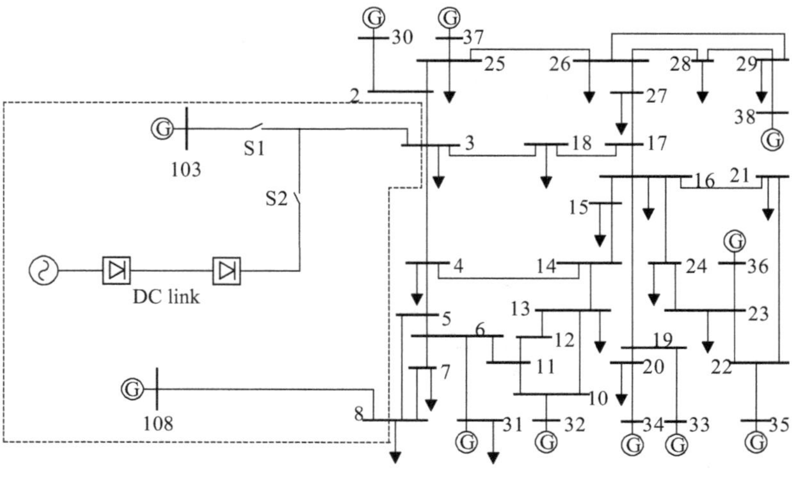

Figure 11.19 Test system

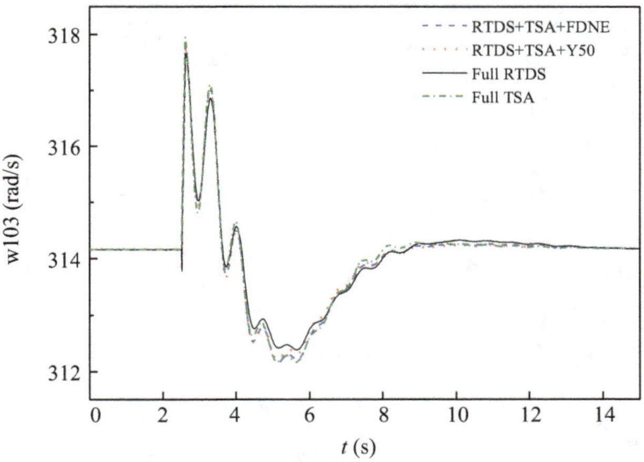

Figure 11.20 Generator #103 rotor speed

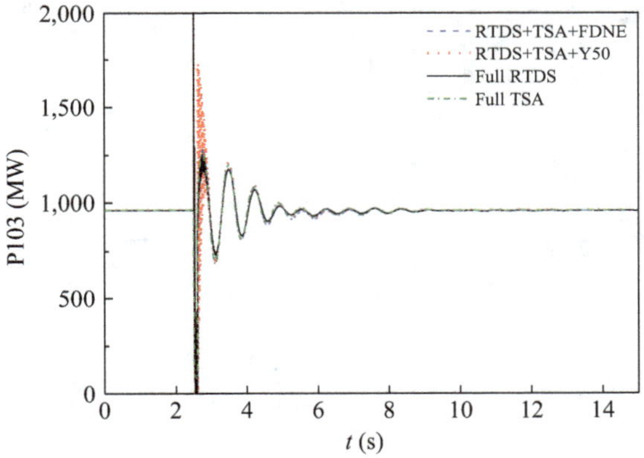

Figure 11.21 Generator #103 active power

The results are shown below, including the rotor speed of generator #103 (the generator connected to bus #103) (Figure 11.20), active power (Figure 11.21), and rotor speed of generator #37 (Figure 11.22). It can be noticed that, for the pure AC system, all four methods give almost the same results. In pure AC system, the high-frequency electromagnetic transients caused by fault are quickly damped within about one cycle, eliminating the advantage of FDNE.

11.5.4 Simulating AC–DC hybrid system

With switch S1 off and S2 on, a HVDC transmission line is inserted, making the system an AC–DC hybrid one.

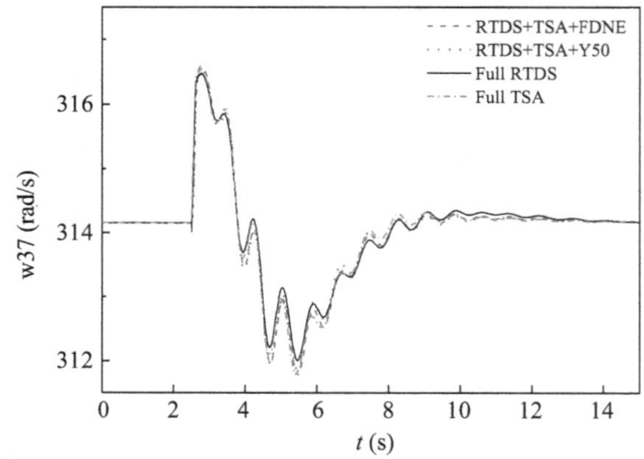

Figure 11.22 Generator #37 rotor speed

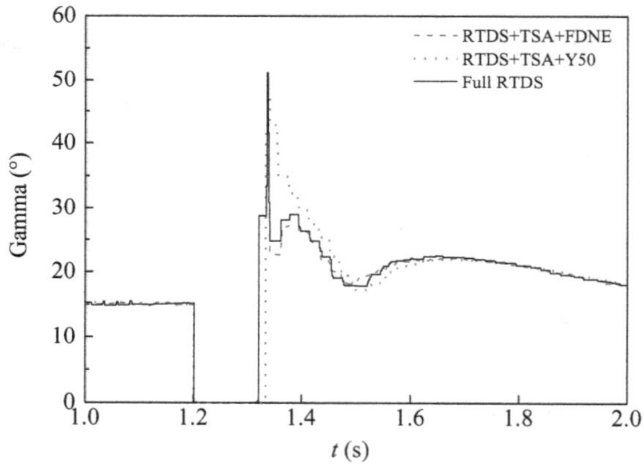

Figure 11.23 DC extinction angle

Case 1: A 100-ms three-phase ground fault occurs at boundary bus #3 (phase A voltage cross zero), and it was simulated in three different ways (**Full TSA** was not included in AC–DC hybrid system simulation since it cannot accurately model DC part).

Figures 11.23–11.26 are the results of DC extinction angle, DC voltage, DC active power, and Bus #3 phase A voltage (from RTDS side in hybrid simulation). It can be noticed that **RTDS+TSA+FDNE** was quite close to **Full RTDS**. However, there was a marked difference between these two and **RTDS+TSA+Y50**, especially in extinction angle and bus #3A voltage, which are closely related to HVDC commutation failure.

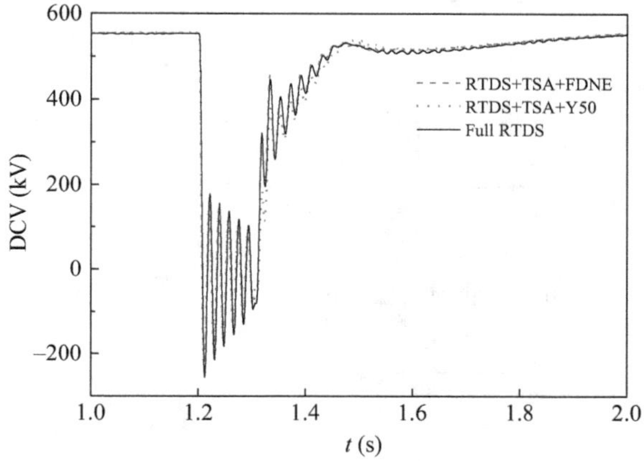

Figure 11.24 DC voltage

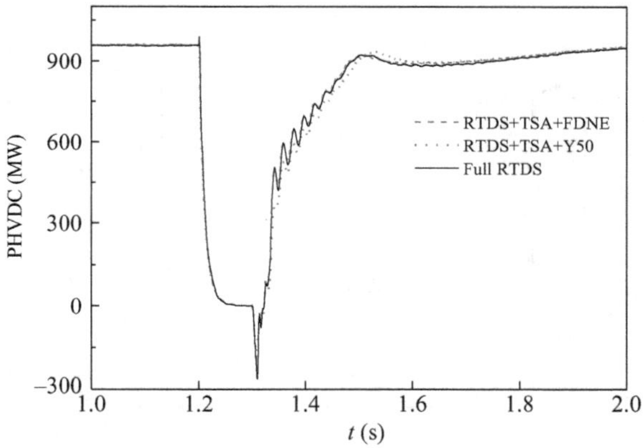

Figure 11.25 DC active power

Figure 11.27 shows the results of Generator #37 (the generator connected to bus #37) rotor speed (from TSA side in hybrid simulation). Three methods give almost the same result.

Case 2: A 100-ms phase A ground fault occurs at boundary bus #3 (phase A voltage cross zero) and it was simulated in three different ways.

The results are shown below. **Case 1** applied a three-phase ground fault, which means the DC system was completely disconnected with FDNE during the fault. On the other hand, this case applied a single-phase ground fault, so the advantage of FDNE was also reflected during fault (during fault, **RTDS+TSA+FDNE** and **Full RTDS** were close, but not the **RTDS+TSA+Y50**) (Figures 11.28–11.32).

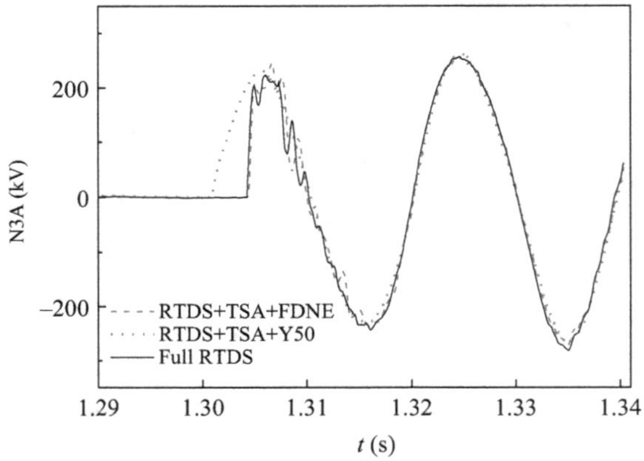

Figure 11.26 Bus #3 phase A voltage after fault

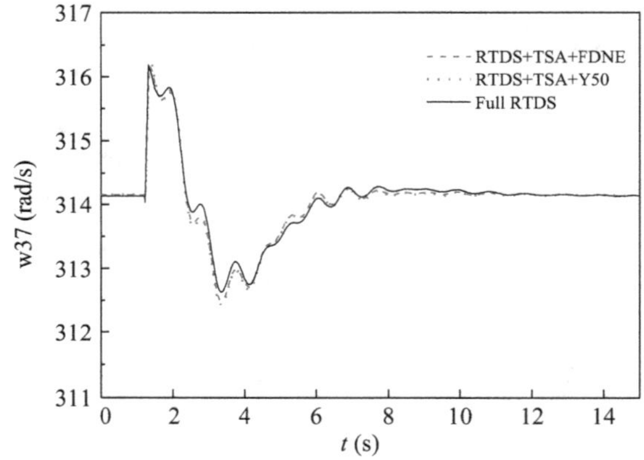

Figure 11.27 Generator #37 rotor speed

Case 3: Different commutation failure behavior given by different simulation methods.

Commutation failure is an important type of HVDC faults. Detection of a commutation failure may be done by comparing the DC current I_{DC} with the current flowing in converter transformer secondary windings. Under normal conditions, the maximum value of the absolute values of the converter transformer secondary windings

$$I_{wdn} = \max(|I_{wdnA}|, |I_{wdnB}|, |I_{wdnC}|) \tag{11.27}$$

will be equal to the DC current I_{DC}. During a commutation failure, I_{DC} will be much larger than I_{wdn} as defined above.

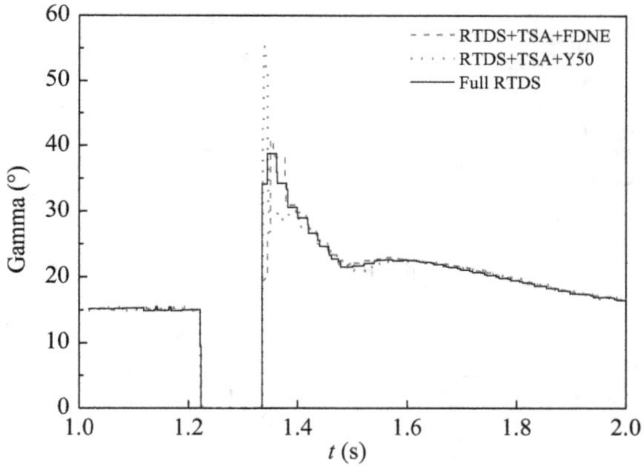

Figure 11.28 DC extinction angle

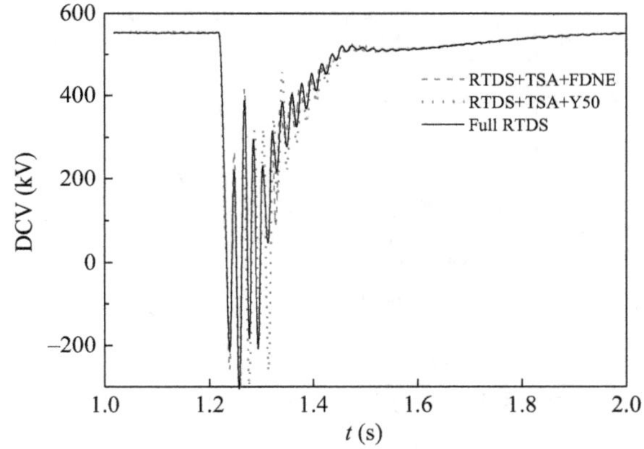

Figure 11.29 DC voltage

A 10-ms phase A ground fault occurs at boundary bus #3 (phase A voltage cross zero), and the results of I_{DC} and I_{wdn} given by three different methods are shown below.

RTDS+TSA+FDNE gave the same result as **Full RTDS** (one time commutation failure). However, **FDNE+TSA+Y50** gave a different result (three times commutation failure) (Figures 11.33–11.35). In this case, whether or not considering the frequency characteristics of the TSA network in EMT, leads different waveform distortion of the converter bus voltage, and gives different results of commutation failure.

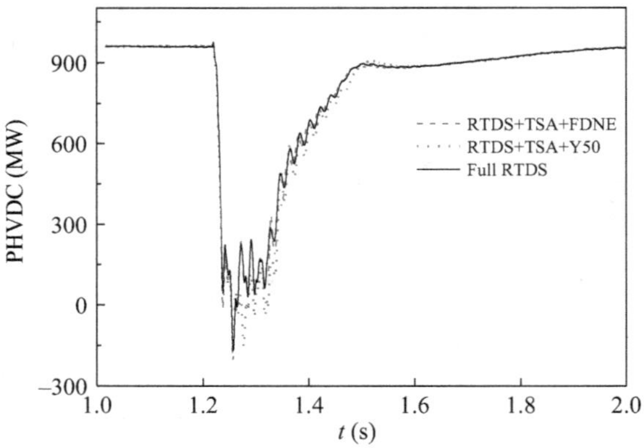

Figure 11.30 DC active power

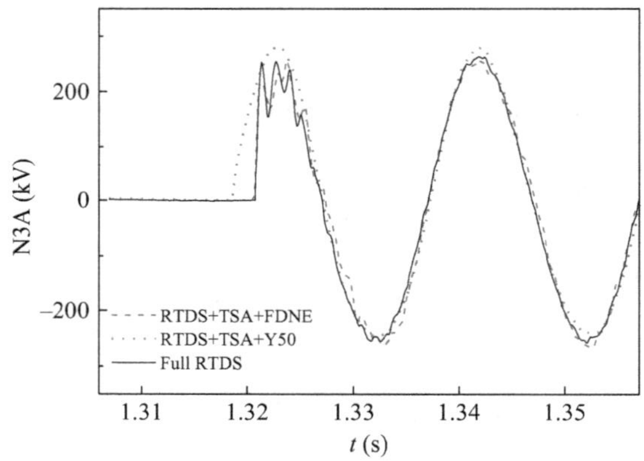

Figure 11.31 Bus #3 phase A voltage after fault

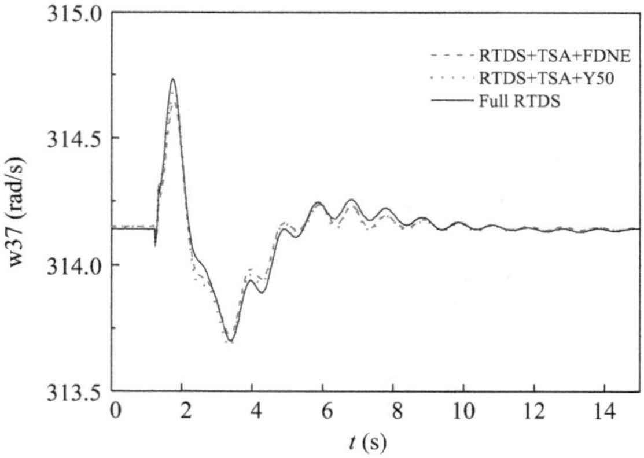

Figure 11.32 Generator #37 rotor speed

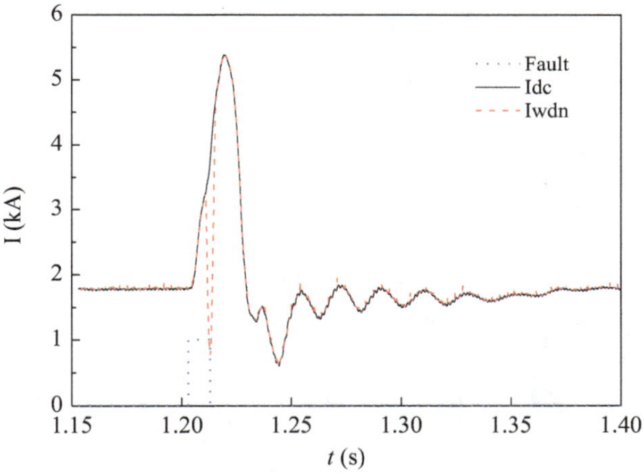

*Figure 11.33 Result given by **RTDS+TSA+FDNE***

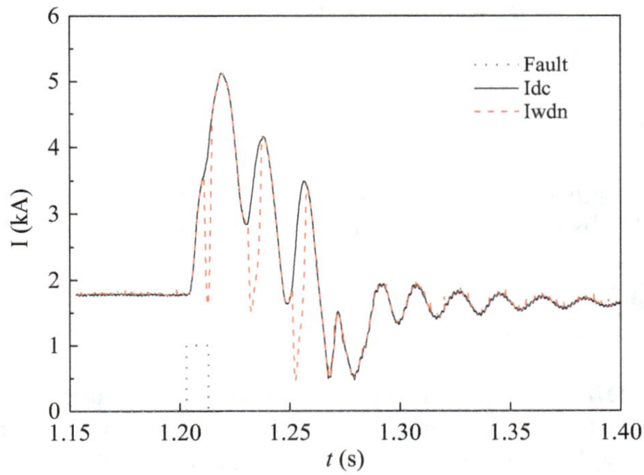

*Figure 11.34 Result given by **RTDS+TSA+Y50***

In AC–DC hybrid system, due to the non-neglected harmonics (or more accurately, non-fundamental frequency components) generated by DC part in the boundary, the traditional admittance matrix in Norton equivalent circuit is not accurate enough and is unable to give the correct result. This is exactly the advantage and meaning of FDNE.

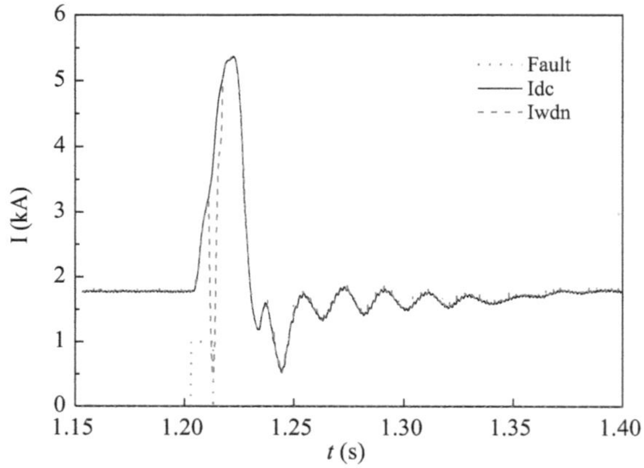

*Figure 11.35 Result given by **Full RTDS***

11.6 Summary

This chapter introduces the concept of EMT–TSA hybrid simulation for HVDC system and mainly focuses on the FDNE techniques. The advantage of FDNE is presented in detail.

References

[1] China Southern Power Grid Co., Ltd. (CSG) company profile [Online]. Available from http://eng.csg.cn/About_us/About_CSG [Accessed March 11, 2018].

[2] B. Kulicke. NETOMAC digital program for simulating electromechanical and electromagnetic transient phenomena in AC systems. Munchen: Siemens Co. Ltd., 1979.

[3] M. D. Heffernan, K. S. Turner, J. Arrillaga, *et al.* Computation of AC-DC system disturbances—Part I: interactive coordination of generator and convertor transient models. IEEE Transactions on Power Apparatus and Systems, 1981, 100(11):4341–4348.

[4] K. S. Turner, M. D. Heffernan, J. Arrillaga, *et al.* Computation of AC-DC system disturbances—Part II: derivation of power frequency variables from convertor transient response. IEEE Transactions on Power Apparatus and Systems, 1981, 100(11):4349–4355.

[5] K. S. Turner, M. D. Heffernan, J. Arrillaga, *et al.* Computation of AC-DC system disturbances—Part III: transient stability assessment. IEEE Transactions on Power Apparatus and Systems, 1981, 100(11):4356–4363.

[6] J. P. Bowles. AC system and transformer representation for HVDC transmission studies. IEEE Transactions on Apparatus and Systems, 1970, PAS-89(7):1603–1609.

[7] N. G. Hingorani and M. F. Burbery. Simulation of AC system impedance in HVDC system studies. IEEE Transactions on Apparatus and Systems, 1970, PAS-89(5-6):820–828.
[8] J. Reeve and K. Adapa. A new approach to dynamic analysis of AC networks incorporating detailed modeling of DC systems. Part I: Principles and implementation. IEEE Transactions on Power Delivery, 1988, 3(4):2005–2011.
[9] J. Reeve and K. Adapa. A new approach to dynamic analysis of AC networks incorporating detailed modeling of DC systems. Part II: application to interaction of DC and weak AC systems. IEEE Transactions on Power Delivery, 1988, 3(4):2012–2019.
[10] G. W. J. Anderson and N. R. Watson. A new hybrid algorithm for analysis of HVDC and FACTS systems. The 1995 International Conference on Energy Management and Power Delivery, Invercargill, 1995:462–467.
[11] G. W. J. Anderson. Hybrid simulation of ac-dc power systems. Christchurch: University of Canterbury, 1995.
[12] N. R. Watson and J. Arrillaga. Frequency-dependent AC system equivalents for harmonic studies and transient convertor simulation. IEEE Transactions on Power Delivery, 1988, 3(3):1196–1203.
[13] X. Wang, P. Wilson, and D. Woodford. Interfacing transient stability program to EMTDC program. The 2002 International Conference on Power System Technology, Kunming, 2002:1264–1269.
[14] C. Yue, F. Tian, X. Zhou, *et al.* Principle of interfaces for hybrid simulation of power system electromagnetic-electromechanical transient process. Power System Technology, 2006, 30(1):7–11.
[15] Y. Liu, X. Liang, Y. Min, *et al.* An interface algorithm in power system electromechanical transient and electromagnetic transient hybrid simulation. Automation of Electric Power System, 2006, 30(11):44–48.
[16] X. Lin, A. M. Gole, and M. Yu. A wide-band multi-port system equivalent for real-time digital power system simulators. IEEE Transactions on Power Systems, 2009, 24(1):237–249.
[17] Y. Liang, X. Lin, A. M. Gole, *et al.* Improved coherency-based wide-band equivalents for real-time digital simulators. IEEE Transactions on Power Systems, 2011, 26(3):1410–1417.
[18] Y. Zhang, A. M. Gole, W. Wu, *et al.* Development and analysis of applicability of a hybrid transient simulation platform combining TSA and EMT elements. IEEE Transactions on Power Systems, 2013, 28(1):357–366.
[19] Y. Hu, W. Wu, B. Zhang, and Q. Guo, "Development of an RTDS-TSA hybrid transient simulation platform with frequency dependent network equivalents," 2013 4th IEEE/PES Innovative Smart Grid Technologies Europe (ISGT EUROPE), Copenhagen, Denmark, 2013.
[20] B. Gustavsen and A. Semlyen, "Rational approximation of frequency domain responses by vector fitting," IEEE Transactions on Power Delivery, 1999, 14(3):1052–1061.

[21] W. Hendrickx and T. Dhaene. A discussion of rational approximation of frequency domain responses by vector fitting. IEEE Transactions on Power System, 2006, 21:441–443.

[22] Y. Hu, W. Wu, and B. Zhang. A fast method to identify the order of frequency-dependent network equivalents. IEEE Transactions on Power Systems, 2016, 31(1):54–62.

[23] B. Gustavsen and A. Semlyen. Enforcing passivity for admittance matrices approximated by rational functions. IEEE Transactions on Power System, 2001, 16(1):97–104.

[24] S. Grivet-Talocia and A. Ubolli. A comparative study of passivity enforcement schemes for linear lumped macromodels. IEEE Transactions on Advanced Packaging, 2008, 31(4):673–683.

[25] B. Gustavsen. Fast passivity enforcement for pole-residue models by perturbation of residue matrix eigenvalues. IEEE Transactions on Power Delivery, 2008, 23(4):2278–2285.

[26] A. Semlyen and B. Gustavsen. A half-size singularity test matrix for fast and reliable passivity assessment of rational models. IEEE Transactions on Power Delivery, 2009, 24(1):345–351.

[27] Y. Hu, W. Wu, and B. Zhang. A semidefinite programming model for passivity enforcement of frequency-dependent network equivalents. IEEE Transactions on Power Delivery, 2016, 31(1):397–399.

[28] Y. Hu, W. Wu, A. M. Gole, *et al.* A guaranteed and efficient method to enforce passivity of frequency dependent network equivalents. IEEE Transactions on Power Systems, 2017, 32(3):2455–2463.

[29] Y. Hu, W. Wu, and B. Zhang. Compacting and partitioning-based simulation solution for frequency-dependent network equivalents in real-time digital simulator. IET Generation Transmission & Distribution, 2015, 9 (16):2526–2533.

Chapter 12

Real-time modeling and simulation for HVDC systems

Yi Zhang[1]

12.1 Converter real-time digital simulator (RTS) models [line commuted converter (LCC) and voltage source converter (VSC)]

This chapter describes the modeling and simulation methodologies for high-voltage direct current (HVDC) systems on RTSs. In the past 25 years, RTSs have been successfully used in factory and dynamic testing of HVDC controllers by HVDC manufactures and electrical power utilities. It is widely accepted by today's power industry that the RTSs based on ElectroMagnetic Transient (EMT) theory are the mainstream simulation platforms for closed-loop testing of HVDC controllers. These RTSs have sufficient accuracy, while they are more affordable and more flexible compared with previous analogue simulators. As the EMT theory and algorithms are well-documented in many classical publications [1–6], and also in the previous chapters of this book, there is no need to duplicate the EMT methodology in this chapter. This chapter focuses on the special techniques used in real-time modeling of HVDC equipment and systems [7,8].

12.1.1 LCC valve group model

The most important and commonly used model in HVDC simulation is the valve group model for LCC-HVDC converters [5,8]. It is well-known that RTSs have been widely used as the main test platforms for LCC-HVDC converters. The six-pulse valve group model as shown in Figure 12.1 is the basic element of HVDC converter stations (Figure 12.1) and can be developed with mathematically rigorous algorithm to ensure its accuracy and stability. There are two different approaches to represent and solve the HVDC valve group. One approach is to write an integrated valve group model in the simulation software, in the way as writing a model for a synchronous machine or a transformer. The other approach is to represent the valve group with a set of individual branches. These branches are then included in the

[1]RTDS Technologies Inc., Winnipeg, MB, Canada

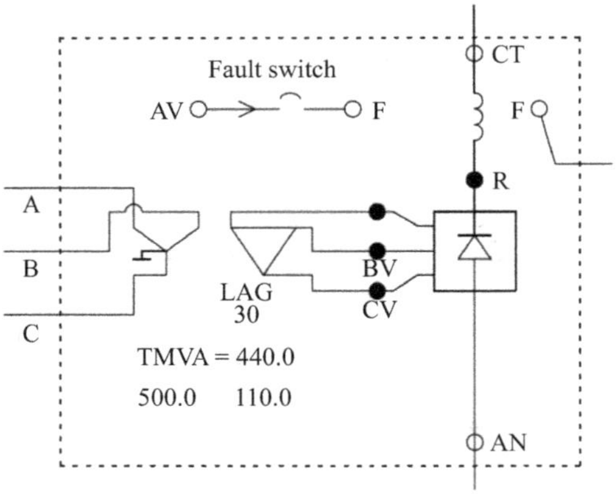

Figure 12.1 Six-pulse valve group

equation of the entire circuit. Both approaches have been used in research and commercial real-time simulators.

The approach to write an integrated valve group model has the advantages of minimizing the computation resources. It treats the entire valve group as a sub-network and absorbs several internal nodes. The number of nodes has always been a dominant concern for real-time simulation, especially early on when processors were not very powerful [9,10]. Writing an integrated valve group model also makes it easier to include some advanced features, such as the compensation algorithm, measurement of firing angle and distinction angle, etc. The disadvantage of this approach is that the development workload is higher because each model requires a customized design, including the derivation of the circuit equations and specific programing.

One key technique in real-time simulation is how to connect the models to the entire power system network. In earlier years of real-time digital simulation, the interface technique with one time-step delay was widely used in power system models, especially in models with heavy computation load, such as HVDC valve group. However, the numerical instability due to one time-step delay causes many difficulties. One of the problems is that the delay brings inaccuracy to the precise modeling of harmonics. With the increasing processing power of modern computer technology, the interface models with one time-step delay are no longer used. A superior approach is to solve the models as a part of the entire admittance matrix. This approach is called the embedded modeling technique. The terminology "embedded" means that the equations of the individual model are embedded as part of the nodal analysis equations of the entire network. Using the embedded algorithm, the valve group equation is solved simultaneously with the main network equations; therefore, completely eliminating the interfacing delay. The embedded algorithm significantly enhanced the numerical stability and the accuracy of the models. As a result, with the embedded valve group models, the accuracy of harmonics modeling of the HVDC converter station has been significantly improved.

Alternatively, the valve group can be solved as a normal part of the entire circuit without pre-grouping. For instance, it can be treated as an independent state space group, where the circuit equations are pre-computed for all possible combinations of switch positions before entering the real-time loop. Matrix manipulation techniques such as partitioning and re-ordering can be used in organizing the circuit with six switches per group. In addition, the different switch permutations can be pre-calculated and stored with various optimal memory management techniques. An advantage of this approach is easy programing because the valve group is handled as a set of generic branches. The disadvantage is that efficiency and flexibility will be lower. For example, the number of nodes and branches of the main network will increase, if nodes and branches of the valve group are counted. The large admittance matrix requires more computation load, resulting in slower speed. Although the faster processing power from the semiconductor industry in recent years has offset many of the speed issues, efficient is still a major concern in real-time simulation. Therefore, this approach is not recommended to be used in modern RTSs.

The six-pulse valve group model has worked very well throughout the past 15 years. However, recently, when ultra HVDC (UHVDC) systems must be modeled, each pole needs four valve groups with some additional nodes and breakers. This requires excessive computation resources and brings new challenges to the real-time simulation. In order to reduce the resource requirement, more six-pulse valve groups can be packed into one model. An example is the new 12-pulse valve group model which is designed to improve simulation quality and efficiency for UHVDC systems [11].

Figure 12.2 illustrates the 12-pulse HVDC valve group model prepared for use in one of the often used RTSs [8]. Ten internal nodes are labeled as R, P, AV1, BV1, CV1, M, AV2, BV2, CV2 and N. Five perimeter nodes are labeled as A, B, C, CT and AN. The model can also be expanded to optionally support four windings on the three single-phase transformers to enable filters and reactive power support to be connected to the converter transformer. In that case, the model has ten internal nodes and eight external nodes, resulting in a conductance matrix of dimension 18. This new model significantly reduces the processing resources required for modeling UHVDC systems. In addition, the model has a more efficient solution sequence with less time delay, therefore, increasing the accuracy of the simulation, especially for control hardware in loop testing.

The valve group models use snubbers to damp and absorb the numerical spikes during valve transitions. This is necessary for real-time simulation because the real-time constraint prevents interpolation from being used. The impedance of the resistor–capacitor (RC) snubber should be significantly larger than Ron and significantly smaller than Roff. In order to get reasonable simulation results, the time constant of the snubber (Rsn $\times$ Csn) is recommended to be at a duration equals to or above 2.0 time steps.

In real-time simulators, thyristor valves are represented by ideal switches. Often, internal logic is provided to measure the extinction angle of an inverter valve

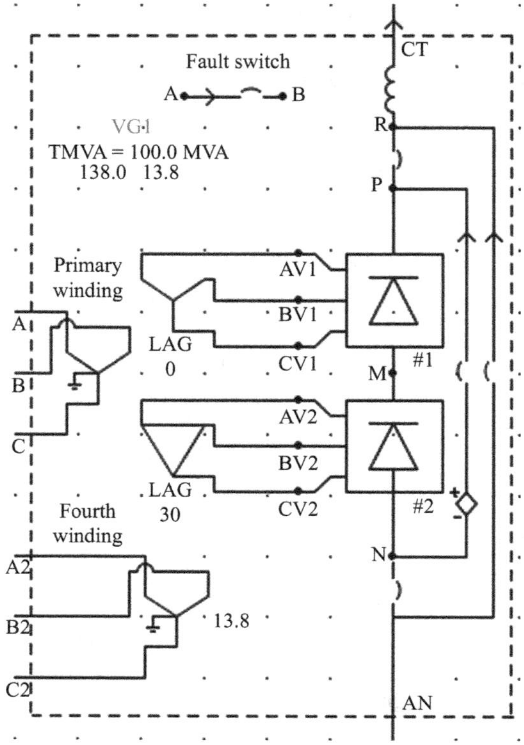

Figure 12.2 Twelve-pulse valve group

following current extinction. The thyristors will be re-fired if the measured extinction angle is less than the user-defined minimum setting. This type of model has been widely accepted and provides an adequate but slightly optimistic simulation of a commutation failure. A minimum extinction angle can be specified to offset the optimistic trends. In an actual thyristor valve, following current extinction, a negative voltage of a certain magnitude and duration is required to ensure that the valve will withstand positive voltage without conducting. It is the voltage-time area (not just the time) of the negative voltage appearing across the valve following current extinction that determines whether the thyristor can withstand positive voltage.

Alternatively, the valve group model can output the valve voltage zero crossing pulses, based on which the users can calculate the extinction angle in an external controller or using internal controls on the simulator. Some valve groups also provide the alpha angle, so that manufactures can calculate the overlapping angle and derive the extinction angle rather than relying on direct measurement. One of the advantages of digital simulation is that any new idea or approach can be implemented and tested with minimum effort compared to analogue simulators or field testing.

12.1.2 VSC-MMC converter model

Compared to other two- or three-level VSC topologies, the modular multilevel converter (MMC) has been considered as a promising solution for HVDC applications. However, this topology presents some different challenges for real-time simulation. The arm of an MMC valve is made up of multiple sub-modules (SMs) connected in series. Each SM is formed by capacitors and switches, for instance, IGBTs (insulated-gate bipolar transistors) and their corresponding free-wheel diodes, connected with different configuration to obtain several topologies: half-bridge (two switches with one capacitor), full-bridge (four switches with one capacitor) and double clamped (five switches, two diodes and two capacitors). Figure 12.3 illustrates an MMC valve containing several SMs, but in reality, a valve may contain hundreds of SMs. The number of SMs is determined according to the voltage rating of the DC grid and employed switches.

The large number of SMs creates an excessive computational burden for the EMT simulation of MMC-HVDC systems, especially in a real-time environment.

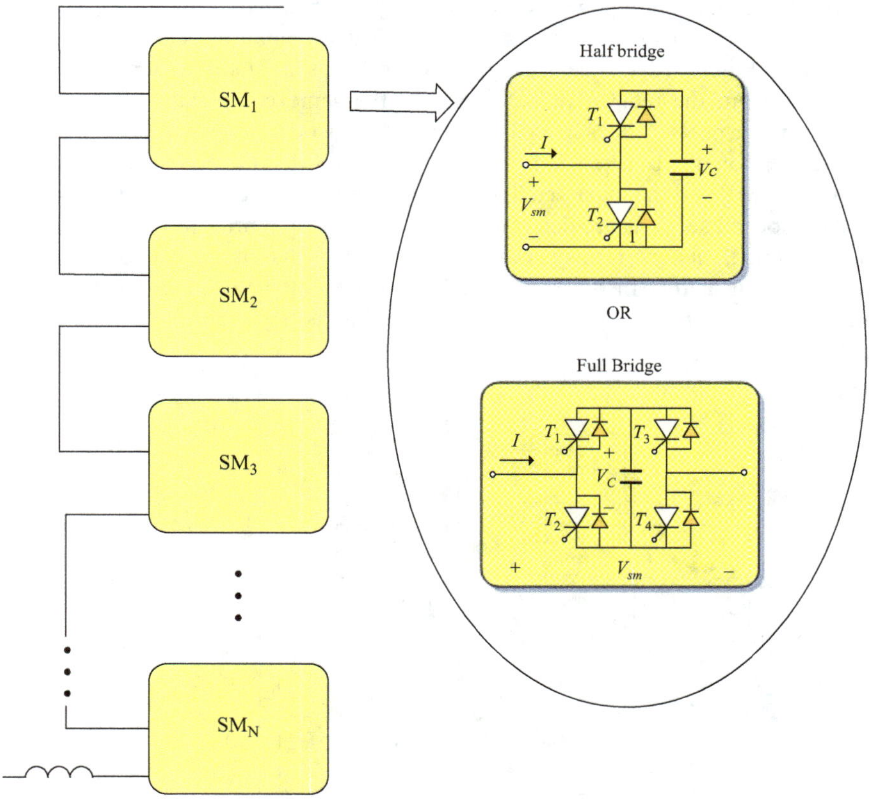

Figure 12.3 Half-bridge MMC valve

For hardware-in-the-loop testing, the exchange of large capacitor voltages and individual firing pulses presents a communication challenge for the simulation model and the external physical controller. In order to meet the requirements of MMC-HVDC simulation and testing, various types of models, which can model MMCs with up to thousands of SMs in real time [12], have been developed on both processors and field programmable gate arrays (FPGAs). For high-level control testing, the voltage balance of SMs can be assumed as ideal without modeling the firing of individual SMs. Under this condition, an efficient MMC model with ideal voltage balance assumption can be developed on the processor of the real-time simulator. On the other hand, low-level control testing needs to model the individual firing and internal faults of the MMC valves. Sometimes, manufactures require the modeling of different parameters of individual SMs, i.e., the capacitor and the discharging resistor of each SM. The purpose of this requirement is to examine whether a slight difference among SM parameters causes a difference in the performance of voltage balance and control stability. A feature allowing the input of individual parameters has been developed on the RTDS real time simulator to allow users to specify parameters in a table or a normal distribution function. These requirements require a significant amount of computation power, which can only be achieved by FPGAs.

A surrogate network [12] for an MMC valve containing half-bridge SMs is shown in Figure 12.4. The surrogate network has a modified topology that is different from the real valve but produces the same computational result in all significant aspects. The surrogate network consists of four series sections: the reactor section, the blocked SM section ('Blocked SM'), the deblocked SM section ('In +ve SM') and the bypassed SM section. The SMs to be included in the blocked and deblocked sections are determined in each time step prior to the calculation of new Dommel history terms, and their selection is based on the external firing control input at that instance. The SMs that are not included in the blocked or deblocked sections during a time step, referred to as 'Bypassed SMs', are allowed

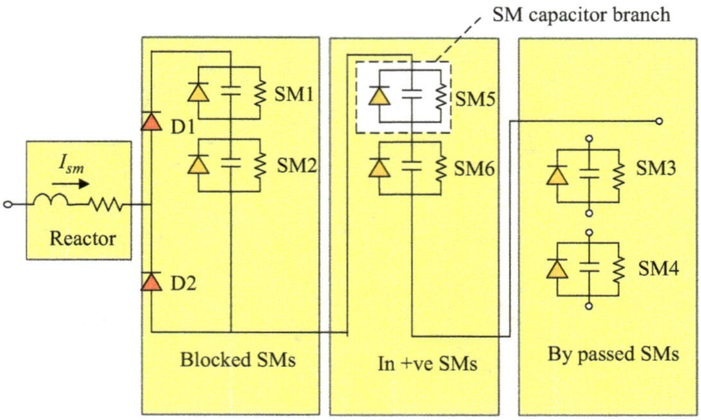

Figure 12.4 Half-bridge surrogate network topology

to discharge in isolation according to the RC time constant of the particular SM. The RC time constant is determined by the physical resistor and capacitor of the SM. In some circumstances, the representation of random parameter deviation is required, such as all the discharge resistors of the SMs can be very close to an average number but still have a small deviation.

The surrogate network and other equivalent methods facilitate the efficient real-time simulation of MMC valves with large numbers of SMs on both processors and FPGAs. Various MMC models on processors or FPGAs have been developed on the RTDS simulator based on the surrogate network's algorithm for various purposes of simulation and study. The most recent development is the generic MMC model shown in Figure 12.5, which has three terminals (GMT3) on the FPGA. This model is capable of modeling the different parameters of individual SMs, and the internal faults inside the legs and between the legs and ground. The feature of individual parameter input allows users to specify a parameter in a table or a normal distribution function, by which the impact of the parameter differences on the operation of MMC HVDC systems can be discovered. The capability to model the internal faults provides a complete mechanism to examine the performance of the protection and control systems in responding to all scenarios possible occurring inside the MMC HVDC stations.

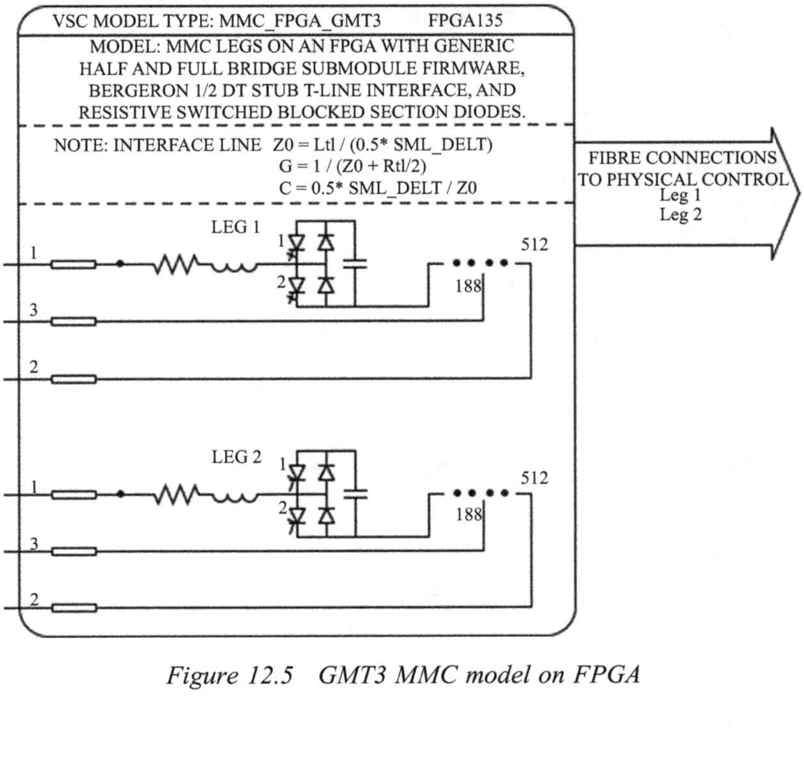

Figure 12.5 GMT3 MMC model on FPGA

12.1.3 Other VSC models

Although MMC is becoming the mainstream configuration of VSC-HVDC projects, other types of VSC-HVDC still exist. The other typical VSC configurations include two-level, three-level T type, three-level neutral-point-clamped (NPC), etc. These VSC configurations are often used in FACTS (flexible alternate current transmission system) devices and small-size HVDC links, as depicted in Figure 12.6.

We can still use the integrated model approach and the compensation algorithms as we do for the LCC valve model to simulate the two- and three-level VSC converters. This method requires significant heavy computation cost, therefore, and can only be done with large time steps of around 30–50 μs. Also, this method can be done but are limited to certain topologies, such as two-level VSC converters. In addition, the compensation only responds to one switching per time-step, so only works for slow switching events. However, VSC-HVDC sometimes uses high-frequency pulse width modulation (PWM) control which may result in multiple switching operations in one large time step. The local compensation becomes invalid and less accurate for this situation. This leads to the development of small time-step solutions.

A more general approach to simulate power electronics is by representing a closed switch as a small inductance and an opened switch as a small capacitor. By choosing the proper small L and small C, we can make the Dommel equivalent impedances R_L equal to R_{RC} as in Figure 12.7. As a result, the conductance matrix of the global circuit remains unchanged with the changing of switching states. This approach avoids the decomposition of the nodal admittance matrix and achieves a time step of less than 2 μs for a doubly fed induction generator (DFIG) wind generation circuit. It can also be

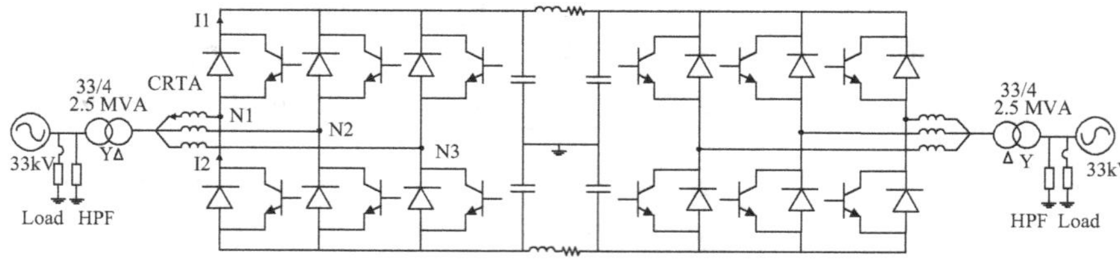

Figure 12.6 Back-to-back two-level VSC-HVDC

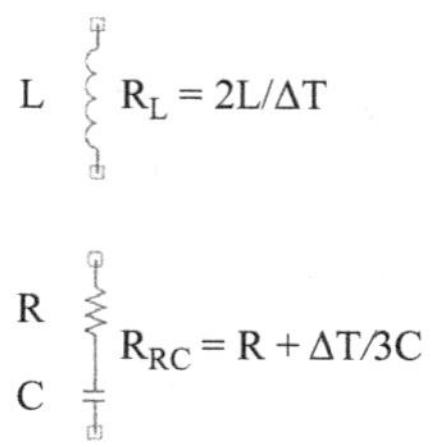

Figure 12.7 Valve representation with small L and small C

implemented on an FPGA to simulate the VSC and surrounding circuits in HVDC stations using a time step below 1 μs. It is hoped that the small time step makes the in-step dynamic negligible for simulating most HVDC systems.

The L/C approach has been used in modeling power converters in system-level real-time simulations since 2003. Despite its frequent use, several issues were discovered in the results of L/C based simulations: (1) unrealistically high virtual loss (especially at high PWM frequencies), (2) fictitious current oscillations between the L and C represented devices and with the external network, and (3) limitations in the impedance ratio between OFF and ON switch representations. It was then concluded that, for general circuits, this approach is acceptable only when the switching frequency is less than 3,000 Hz.

Therefore, there are potential benefits to use switched-resistance representations of electronic switch devices. This type of simulation has recently become feasible due to the higher performance of computer processor cores. However, the switched-resistance representation of switches requires reliable prediction of the ON/OFF status of switch devices before each simulation time step. Recently, a new method has been developed for highly reliable prediction of ON/OFF switch statuses for VSCs with switched-resistance representations in fixed time-step simulation. Real-time simulation results of systems containing two- and three-level VSCs showed this new method is very promising [13]. This method is based on two factors: first is the high-performance computing hardware which can decouple the admittance matrix in real time, and second is the successful prediction of the switching sequence in the power electronics device. It is reported in [13] that the new method can model the circuit in Figure 12.6 at a switching frequency of 10,000 Hz with a time step of 2.0 μs. The simulation results and loss are shown in Figure 12.8 and Table 12.1.

12.2 Converter transformer model (including saturation)

The converter transformer model is combined with valve group in Figures 12.1 and 12.2 for two reasons, namely, to save computation resources and to provide gamma measurement to the controllers which require it.

In most of the traditional EMT theory, transformer saturation is implemented by either a variable inductor or an injection connected on one terminal of a transformer. It is a common understanding that the method with variable admittance is numerically more stable than the method with injection. When the integrated valve group is used, the nodes at the secondary side of the converter transformer are not accessible. As in Figure 12.9, the saturation block can be added at the transformer primary side. In most cases, this has been an acceptable solution.

In some situations, the accuracy of this approximation causes concerns. For example, when the converter transformer is charged from the primary side, the inrush current can be different depending on whether the saturation block is connected at the primary or secondary side. It is more accurate to use a T-type

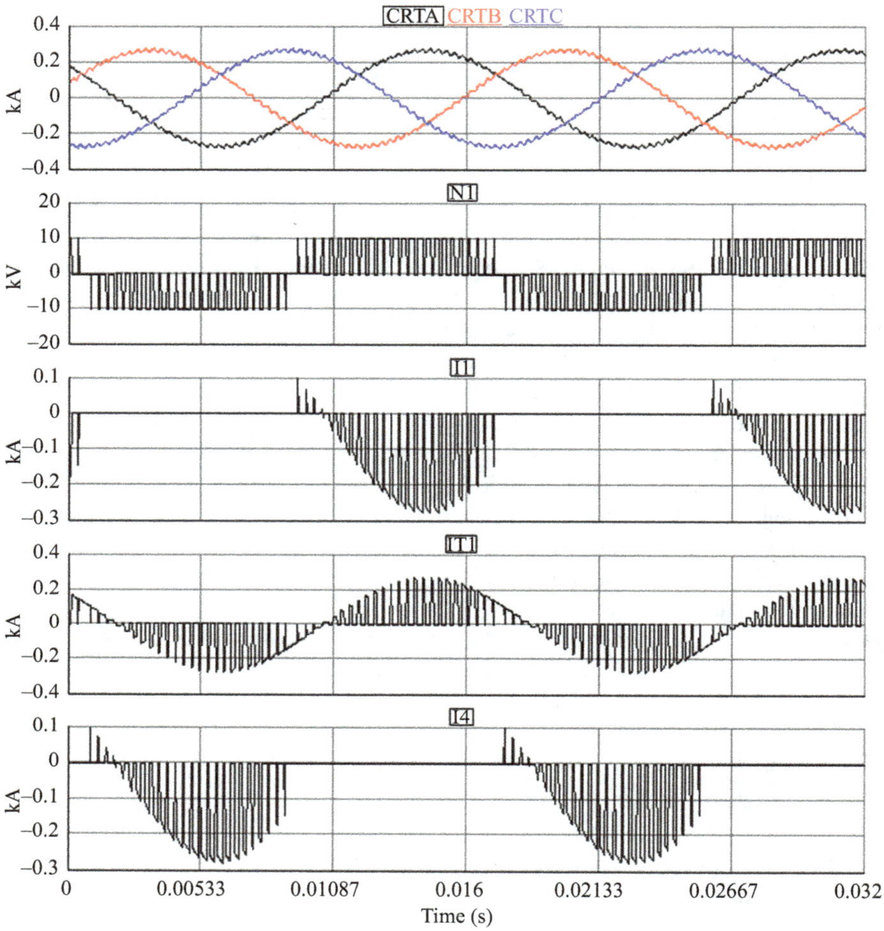

Figure 12.8 Simulation result using predicted resistive switching

Table 12.1 The loss of simulation using predicted resistive switching

Switching frequency (Hz)	Losses (%)
3,060	0.259
9,900	0.333

connection in which the saturation block is placed at the middle of the transformer winding. An equivalent method has been developed to move the saturation branch to the middle point, which was recently implemented on RTDS real-time simulator.

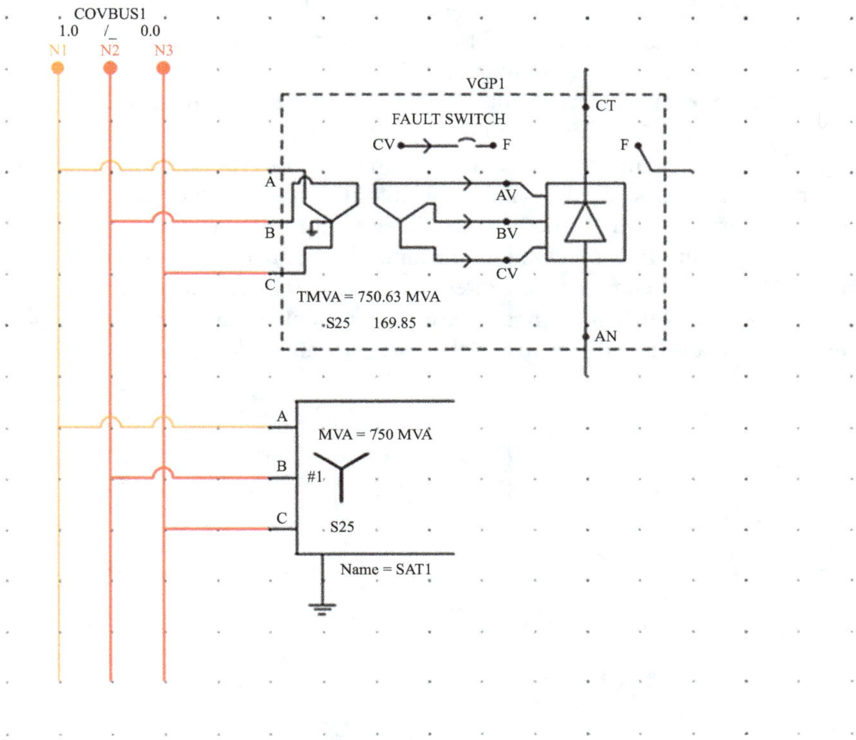

Figure 12.9 Saturation modeling of converter transformer

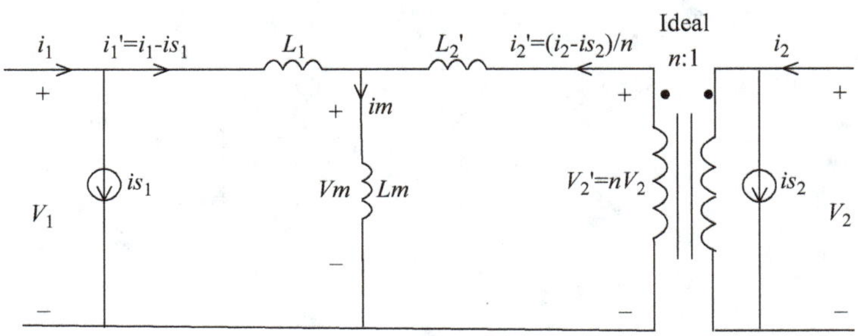

Figure 12.10 Mathematical equivalent model of transformer electric circuit

The advantages of this method are that it does not significantly increase computation cost, and that it does not require access to the middle-point nodes. Two injections at the terminal are used to achieve the same effect as the saturation branch at the middle point as in Figure 12.10.

The default transformer in the valve group model is not designed to model the internal faults of a transformer. In order to simulate the internal faults on the transformer winding, special transformers with winding to ground faults and turn-to-turn faults are needed as in Figure 12.11.

Recently, demands to model the DC bias in transformer saturation have led to a more sophisticated transformer modeling algorithm. The algorithms based on magnetic circuit theory have advantages to model this type of phenomenon with higher accuracy. One of the successful algorithms called the unified magnetic equivalent circuit (UMEC) model is available on RTDS simulator.

The following introduces the multi-limb transformer model based on UMEC theory [14]. The conventional three-phase transformer model in EMT-type programs assumes that three phases are independent, meaning that three phases are exactly balanced and the flux in the core legs and yokes are the same as they

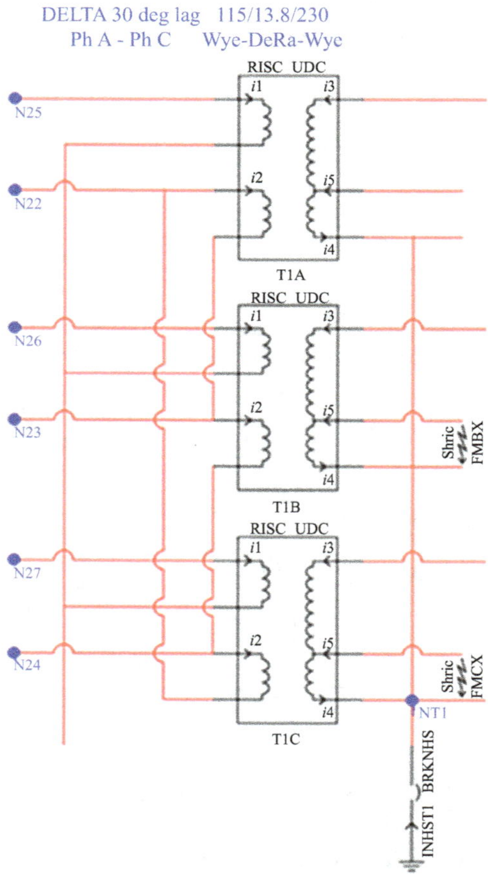

Figure 12.11 Transformer with internal faults

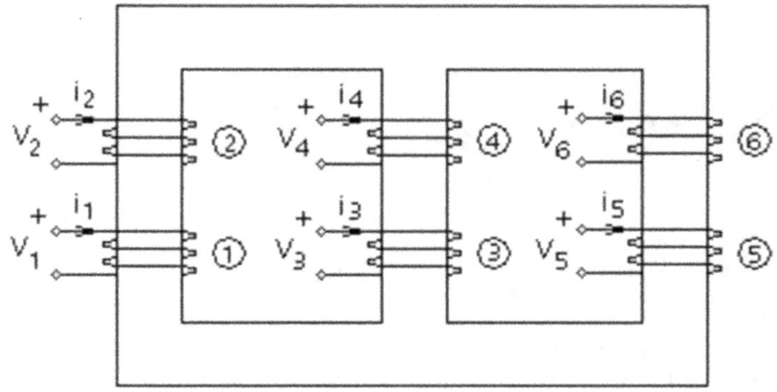

Figure 12.12 Three–phase, three-limb transformer model

would be in three single-phase transformers. Moreover, the primary and secondary winding leakages are combined, and the magnetizing current is placed at one side. As a result, the conventional model largely ignores physical structure. The UMEC transformer model has been developed based on magnetic circuit theory, which represents the core physical structure as well as the mutual coupling of the electrical windings. It is, therefore, a more accurate model to represent the three-phase transformer, especially when it has a three-limb construction as in Figure 12.12.

In reality, the saturation in a three-phase transformer is not uniform. Particular limbs may saturate more or less depending on the currents flowing in the windings. This makes it difficult to provide a more detailed representation of three-phase transformer saturation. Fortunately, considerable effort has been placed on providing these more detailed models. The UMEC transformer models have been developed on the RTDS simulator based on the literatures and have been compared against the same models in PSCAD [14,15]. At present, the RTDS simulator supports the UMEC transformer models in various configurations of single-phase two-winding, single-phase three-winding, three-phase two-winding and three-phase three-winding, as shown in Figure 12.13.

The advantages of the UMEC transformer are that it can model the coupling between three phases and that it can model the saturation of individual part of different limbs in the transformer. Therefore, it is more accurate than the conventional transformer in some circumstances such as modeling the inter-phase coupling and DC bias. However, the UMEC transformer has much heavier computation load than conventional transformers, so it is only recommended for use when analyzing some particularly sophisticated scenarios.

12.3 Transmission line/cable model using in HVDC

This section introduces the phase-domain frequency-dependent transmission (FDPD) line model for use in real-time simulation [18]. The accuracy of the modeling of transmission lines is a very important part for the close-loop testing of HVDC systems.

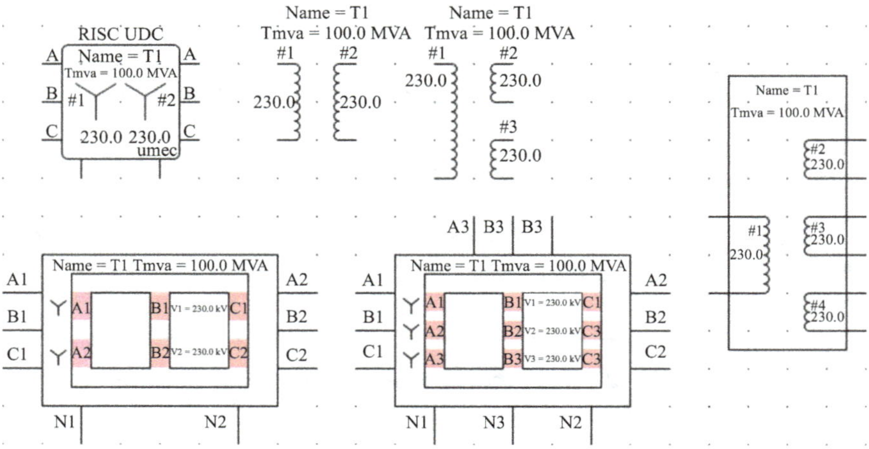

Figure 12.13 UMEC transformer models on RTDS simulator

The most commonly used frequency-dependent transmission line model is the J. Marti model [4] based on the modal domain concept. The J. Marti model works well in most cases; however, it lacks accuracy when modeling certain transients, such as the parallel operation of AC and DC transmission lines. Using the modal domain model, it is assumed that the transformation matrix is constant, which is not always the case. To eliminate the assumption of a constant modal transformation matrix, the transmission line equations need to be solved directly in the phase domain.

12.3.1 Transmission line model

Transmission lines have been modeled in EMT-type programs using the modal analysis approach since the beginning of electromagnetic transient simulation of power systems. The first line model in EMT simulation was the classical Bergeron model, in which the capacitance and inductance were calculated at only the fundamental frequency. However, the depth of penetration of current into the ground and the conductor changes with frequency and resistivity. Consequently, the capacitances and inductances of a transmission line are dependent on frequency. This led to the development of frequency-dependent transmission line models for use in EMT-type programs. The introduction of the popular J. Marti modal-domain frequency-dependent transmission line model marked the arrival of the first generation of these frequency-dependent models. More recently, FDPD line models have been introduced [16–18].

The modal-domain frequency-dependent transmission line models use constant phase-to-mode and mode-to-phase transformation matrices for all frequencies. The frequency dependence of the modal-domain models is provided only in the representation of the characteristic impedance and the attenuation of the separate modes. This was a very successful development; but in reality, it can be shown that the transformation matrices should also be frequency dependent. Unfortunately, in some cases, when frequency-dependent transformations were included in EMT simulations, the simulations were numerically unstable. This led to the development of FDPD line models. These models completely avoid using phase-to-mode and mode-to-phase transformations to avoid the problems with the frequency dependence of the transformations.

Bergeron line models are widely used in the modeling of AC systems. Despite their frequent usage, these models cannot properly model the coupling between AC and DC lines on the same right of way [8]. Figure 12.14 shows a very simple circuit including a 100-km, 230-kV AC line and a parallel mono-polar DC line separated by 50 m on a right of way.

For illustration purposes, the parallel circuits were modeled on one RTS first with a Bergeron line model and then with an FDPD line model. The DC current was ramped from 0.0 to 1.0 kA over a period of 0.1 s, as shown in Figure 12.15. The zero-sequence current in the AC line according to the Bergeron model is shown in Figure 12.16. The zero-sequence current in the AC line according to the FDPD model is shown in Figure 12.17. Clearly, the low-frequency (in this case DC) coupling provided by the Bergeron line model is not accurate. In reality, when the ramping of DC current is completed and the *di*/*dt* of the DC line is zero, then the zero-sequence current in the parallel AC line should also go back to zero. The zero-sequence current in the parallel AC line should be temporary. Figure 12.17 shows a more accurate result produced by the FDPD line model for the zero-sequence current in the AC line.

The FDPD line model can model up to 12 parallel conductors at the typical time step of 50 μs on RTDS simulators. Ground wires are in addition to the 12 insulated conductors. Recently, a new small time-step frequency-dependent transmission line model that can run a transmission line with up to 12 coupled conductors at a time step in the range of 2.5–3.0 μs has been developed on FPGA on RTDS. It should be noted as shown in Figure 12.18 that there will be an interface between the FDPD line on FPGA and the rest of the network. The interface is

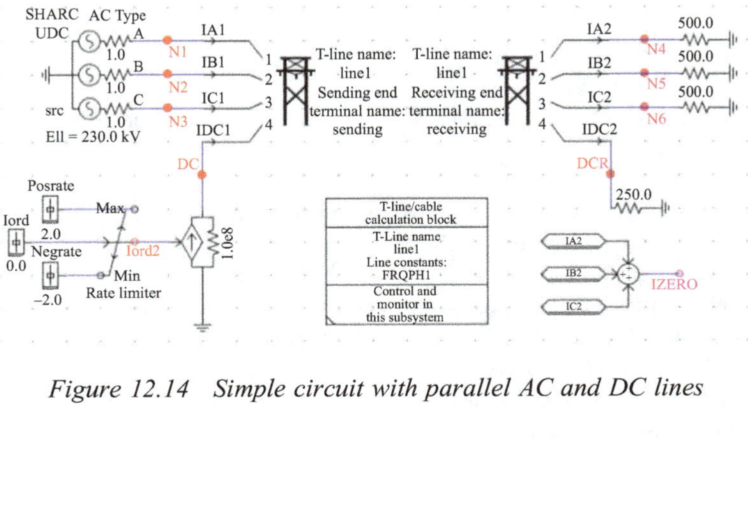

Figure 12.14 Simple circuit with parallel AC and DC lines

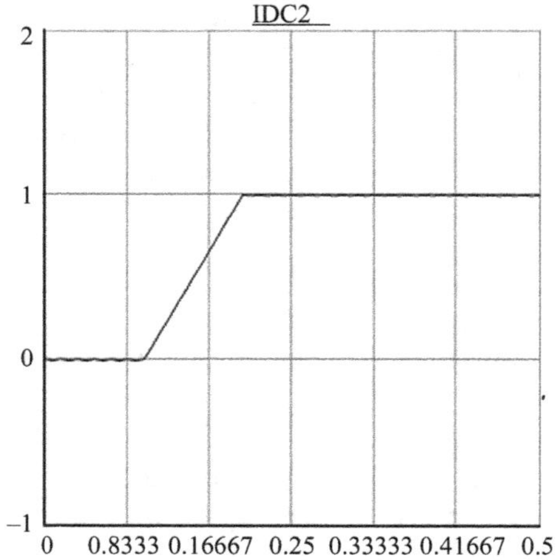

Figure 12.15 Ramped DC current

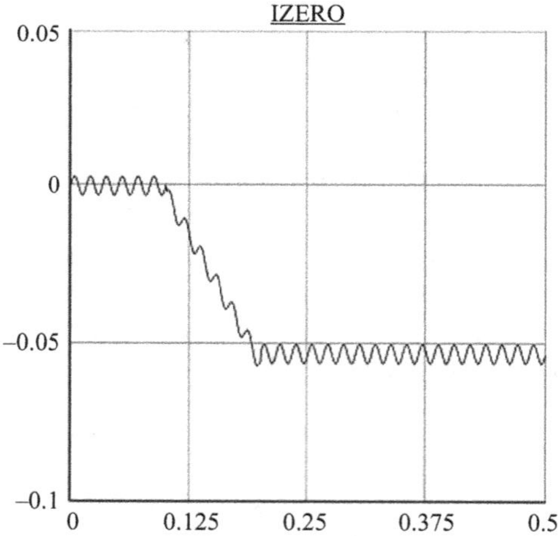

Figure 12.16 Zero-sequence current in AC line (Bergeron mode)

normally realized by a Bergeron transmission line, which is physically comprised of metallic wires of about 600–750 m.

It should be pointed out that special techniques are often used to enhance the numerical stability of transmission line models at very low frequency (0 Hz). The FDPD line model (also called the universal line model) is often used to simulate the coupling between AC and DC parallel lines. Its curve fitting process near 0 Hz needs special care to avoid non-passivity issues. In PSCAD and RTDS, a robust correction technique is used to realize the passivity enforcement [16].

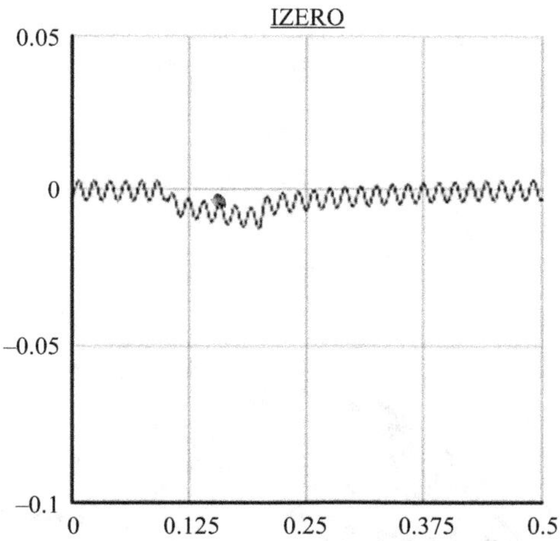

Figure 12.17 Zero-sequence current in AC line (frequency-dependent phase domain)

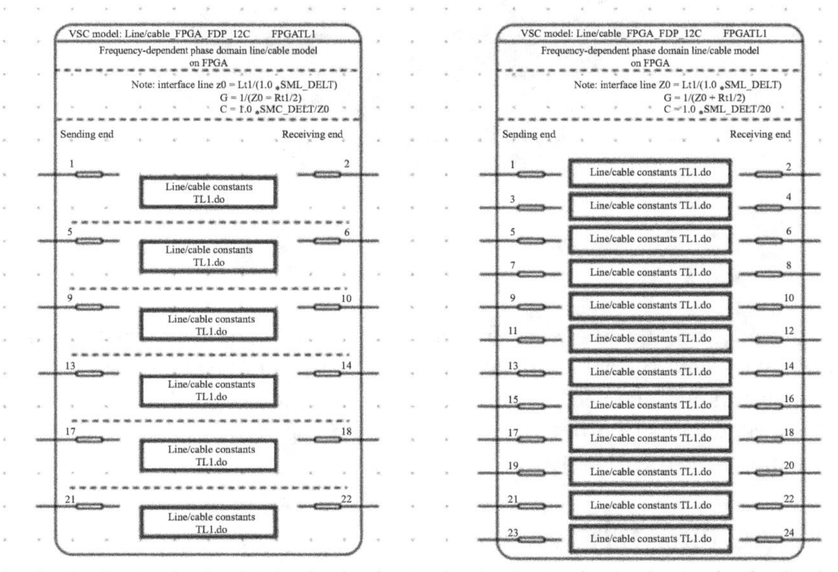

Figure 12.18 Frequency-dependent phase-domain line model on FPGA

12.3.2 Cable model

In EMT simulation, the theoretical base of the cable model is the same as that of the transmission line. The difference is that the configuration of conductors in the cable is different, so that the calculation of the parameters is different accordingly.

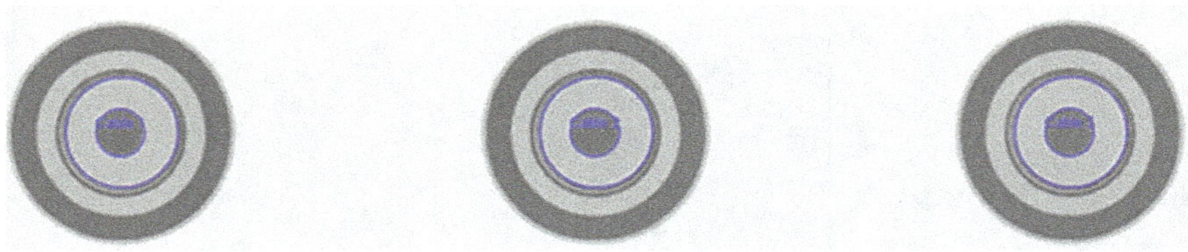

Figure 12.19 Cable with cross coupling between phases

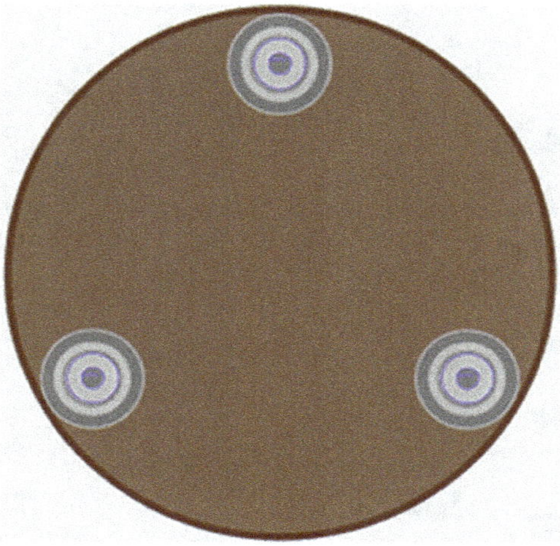

Figure 12.20 Pipeline cable

Recently, significant efforts have been dedicated to expanding the capability to model cables with special configurations. In PSCAD and RTDS, cables with up to 12 conductors are allowed with the FDPD line model [20]. The cross couplings between conductors of different cables as well as pipe type cables are included as shown in Figures 12.19 and 12.20.

12.4 AC filter RTS model

There are many passive AC filters in every LCC-HVDC converter station. These filters are designed to limit harmonics to an acceptable level to ensure the normal operation of HVDC systems and the connected AC systems. The filters also participate in the compensation and control of reactive power for the converter stations. They normally provide approximately 40–60% per unit reactive power with regard to the rated active power of the HVDC system. The filters are basically some combinations of RLC branches. However, filters at HVDC converter stations consist of many RLC branches; therefore, they will need excessive computation resources if they are modeled by individual RLC components.

The switchable filter model is designed to resolve this specific problem. The switchable filter model is a filter bank consisting of 12 various types of filter and 1 additional shunt arrestor. Each filter and arrestor are independently switchable in real time during the simulation as shown in Figure 12.21.

Each branch of the switchable filter model can be defined as any of the following types, as shown in Figure 12.22: RLC branch, high pass, C type, double damped, D2 type, D type, triple tuned, quadruple tuned and T2 type. These configurations of filter branches were developed for the RTDS simulator based on the recommendations from HVDC manufacturers over the past 20 years; therefore, reflecting real filter designs existing in HVDC systems. Given different parameters, almost all possible configurations of filters in HVDC systems can be achieved by the circuits in Figure 12.22.

The advantage of the switchable filter model is that it combines multiple filters into one model. This eliminates all internal nodes and breakers and, therefore, significantly reduces the computation burden. The solution of the switchable filter model is based on the embedded algorithm mentioned in Section 12.1; as such, the equation of the filter is solved simultaneously with the entire network. There is no additional delay between the switchable filter model and the entire network solution.

The benefit of the switchable filter model is mainly on resource saving. A single-phase filter may contain 3–10 internal nodes. A converter station has ten groups of three-phase filters, and there could be 90–300 additional nodes if simple

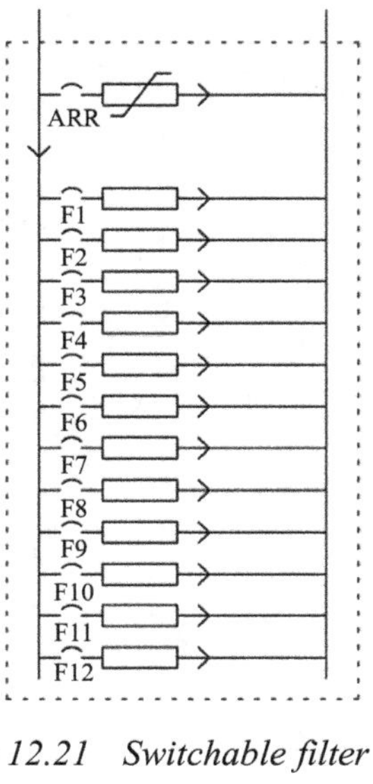

Figure 12.21 Switchable filter model

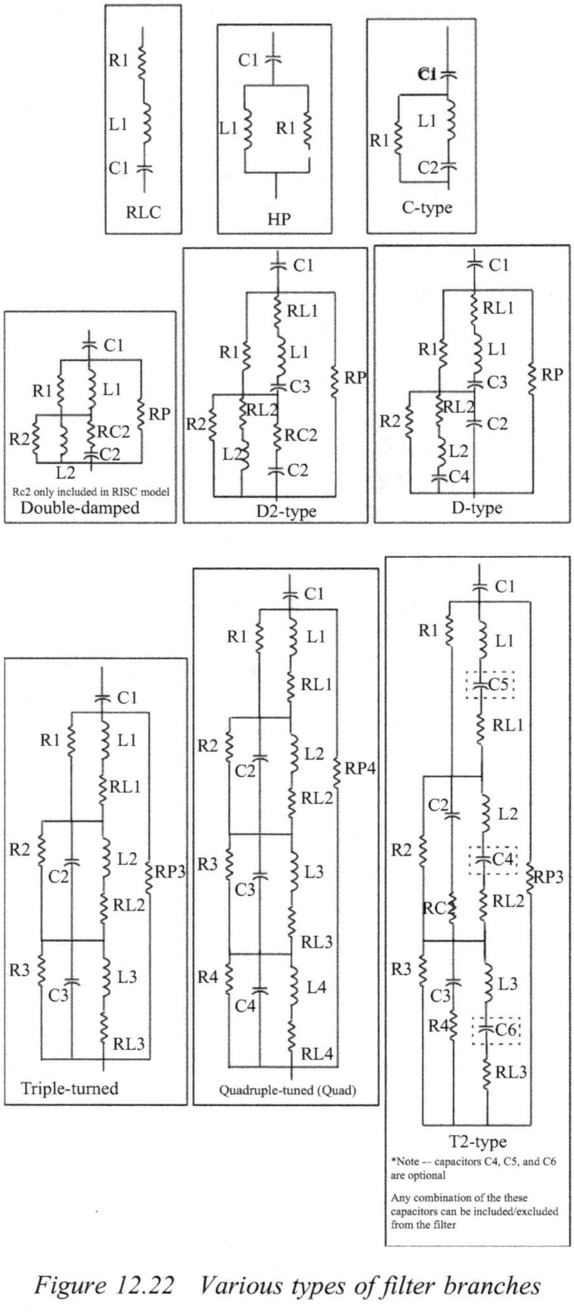

Figure 12.22 Various types of filter branches

RLC connections are used. It is neither possible nor necessary to use the direct RLC connection. With the switchable model, no additional nodes are needed. Therefore, the computation load of the network solution is significantly reduced by 50–70%.

The switchable filter approach also has a disadvantage, that is, it is not easy to create internal faults, so it is necessary to use individual RLC circuits when the modeling of filter internal faults is required. However, most filters can be modeled with the switchable filter model, leading to significant savings in computation resources.

12.5 DC filter RTS model

The modeling of DC filters is relatively easy because the configurations of DC filters are usually simpler than AC filters and are not required to be constantly switched. Sometimes, DC filters can be modeled as individual components, so that arbitrary internal faults can be modeled. Certainly, the methods used to model AC filters in the previous sections are also fully applicable to DC filters.

12.6 Smoothing reactor model

Based on the practical experiences, we have concluded that there is no special requirement needed to model the smoothing reactor. Normal linear indictors are recommended to model the smoothing reactor.

12.7 DCCT and DCVT

Theoretically, DCCT and DCVT can be modeled with a magnetic circuit with saturation modeling capability. However, the response of both DCCT and DCVT is fairly fast and linear. Based on the experiences obtained during the testing practices of many previous HVDC projects, we understand that the dynamic behaviors of DCCT and DCVT are negligible in the effective frequency range of EMT simulation. Therefore, we recommend both DCCT and DCVT to be modeled by linear gain.

12.8 DC breaker model/bypass switch (BPS) model

DC breakers have been become a popular topic in the recent years. Several DC breaker configurations consisting of many electronic and mechanical components have been published. The fast switching characteristics of DC breakers require the use of a small simulation time step of around 2–4 μs.

A DC breaker is a key component in the operation of a DC grid. In a full-bridge MMC converter, power electronics are the primary source of protection. However, in a DC grid based on a half-bridge MMC converter, a fast and reliable DC breaker is necessary. A hybrid DC breaker model has been developed for the small time-step VSC sub-networks of the RTDS simulator as in Figure 12.**23**. This

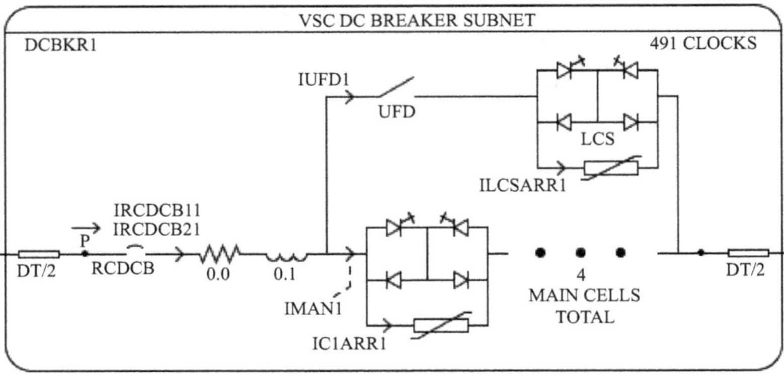

Figure 12.23 A DC breaker model

DC breaker operates based on the technique of creating a back-voltage using arrestors to force the current to zero. The topology of this model is based on the hybrid HVDC breaker model presented in [19].

The bypass switches (BPSs) in HVDC converters provide a current path during non-conventional operation or transition. It is known that DC current does not have a zero-crossing point, so it is very difficult to be cleared with simple switching components. The BPS never operates with DC current; they only operate after the DC currents are driven to zero by the controller. Therefore, we can model the BPSs with ideal switches, i.e., by a small resistor for ON and a big resistor for OFF. It should be mentioned that some valve group models (such as on RTDS) contain BPSs internally, avoiding the use of additional computational resources. It is recommended to use internal BPSs, so the best efficiency can be achieved.

12.9 Improved firing algorithm

It is commonly understood that conducting interpolation in real-time digital simulation is not possible. This is due to a fundamental property of real-time digital simulation: firing pulses from an external controller can arrive at any time in a time step. The design of such an interpolation method fundamentally conflicts with the requirement of rigorous hard real time. It is impossible to design the same interpolation used in the non-real-time EMT tool in real-time digital simulation. As an alternative, several local compensation algorithms, such as the improved firing algorithm of the RTDS valve-group models, have been developed to enhance the simulation accuracy.

One breakthrough advancement on HVDC valve group modeling is the improved firing algorithm developed in 1998 by RTDS Technologies Inc. The improved firing algorithm was developed based on the idea of linear interpolation, which is widely used in most non-real-time simulation tools. By using interpolation, the moment a circuit switches can be accurately predicted, therefore achieving much higher simulation resolution compared to simulations with the same time step but without interpolation. The interpolation algorithm provides an alternative to achieve adaptive variable time-step simulation with minimum computation cost.

Unfortunately, the interpolation algorithm cannot be directly used in real-time digital simulation because it requires repeated computation, therefore making the time step hard to predict. In real-time simulation, the computation time of each time step is rigorously restricted. It is impossible to ensure that there is sufficient time to conduct interpolations in all circumstances. Since the late 1990s, a variety of developmental works have been done to overcome this problem. One successful example is a compensation method called the improved firing algorithm, which approximates the result of interpolation by compensating for some local variables. Figure 12.24 shows the principle of the improved firing algorithm for a thyristor when its current crosses zero. An artificial voltage source is inserted into the valve branch to offset the overshooting of valve current. By using the improved firing technique, the accuracy of the model was increased to approximately 2 μs, even when the simulation time step is typically set at 50 μs [8,20].

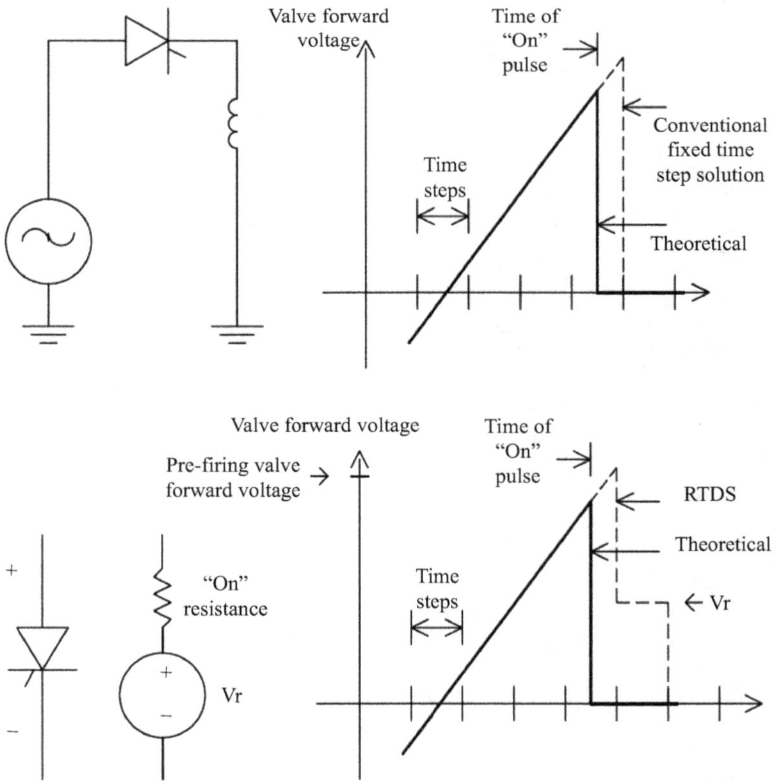

Figure 12.24 Improved firing algorithm

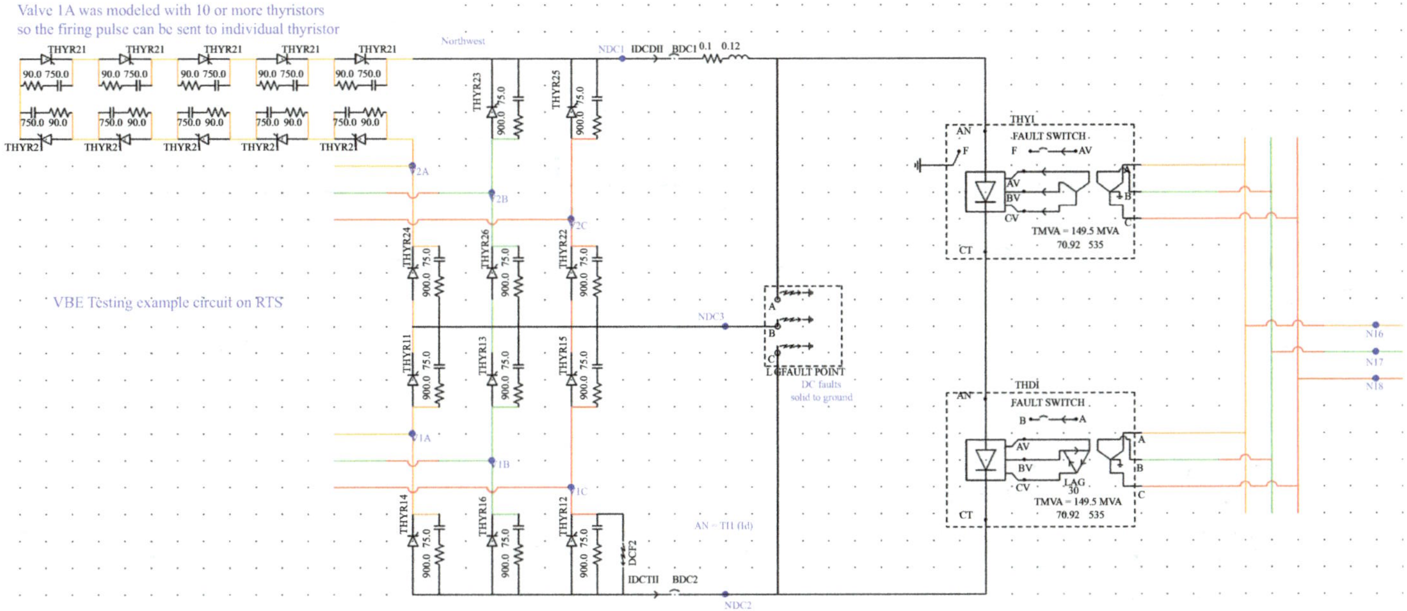

Figure 12.25 HVDC VBE testing circuit

12.10 Test of valve-base electronics (VBE) (LCC only)

In normal function or dynamic testing, the HVDC converters are modeled in the unit of six-pulse valve group. In reality, each arm of the thyristor valve consists of many thyristors in series, and each thyristor component receives an individual firing pulse. For a long time, there was no need to test the valve-base control of the firing details. Each arm was modeled with a single thyristor which represents many thyristors in series, and it is assumed that they were ON and OFF simultaneously. Recently, some manufactures and HVDC operators proposed to test the valve-based controller. The most straightforward approach is to model the valve group with individual components, in which one arm is modeled with multiple thyristors in series as shown in Figure 12.25. The external VBE devices can be tested arm by arm with this setting. If the RTS has enough computation power, all the arms can be modeled with multiple thyristors. In practical cases, testing VBE arm by arm is acceptable to both manufactures and the HVDC project owners.

Although the VBE testing scheme in Figure 12.25 simplifies the testing circuit to one arm, it still requires a large amount of computation resources, especially when there are a large number of thyristors in the arm. An alternative way is proposed in Figure 12.26 to test the multiple firing pulse with control logic. This approach can handle many individual firing pulses with very low computation resources.

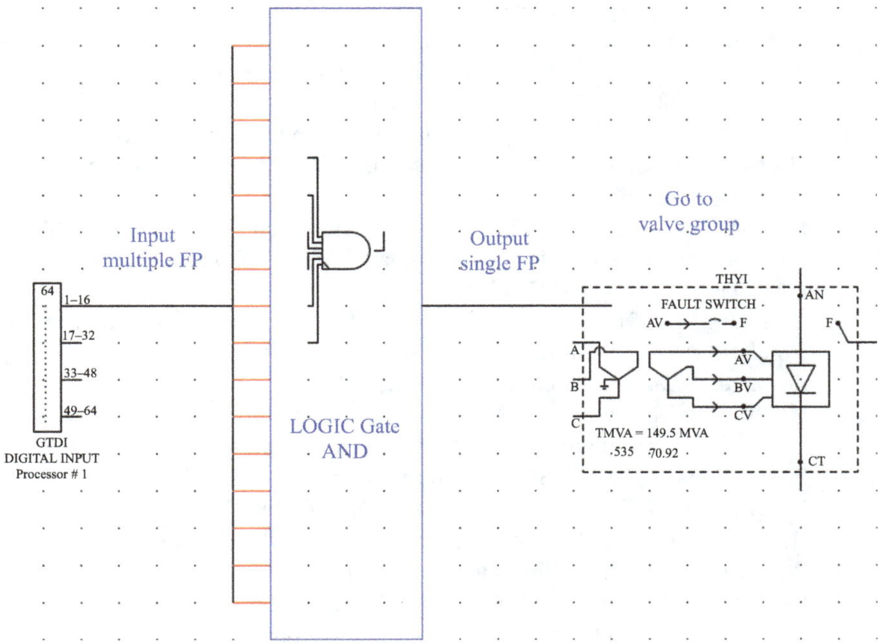

Figure 12.26 HVDC VBE testing bench with control logic

12.11 Limitations in real-time simulation for HVDC systems

The limitations in real-time simulation are listed as following:

1. Interpolation cannot be implemented with current technology.
2. The physical characteristics of the valve cannot be modeled in real time. All the valves are modeled with ideal switches.
3. The network size that can be modeled is limited by available simulation hardware resources.

References

[1] H.W. Dommel, *'EMTP Theory Book'*, 2nd ed. Microtran Power System Analysis Corporation.

[2] H. Dommel, 'Digital computer solution of electromagnetic transients in single-and multiphase networks', *IEEE Transactions on Power Apparatus and Systems*, 1969, PAS-88 (4).

[3] H. Dommel, 'Computation of electromagnetic transients', *Proceedings of the IEEE*, 1974, 62 (7), 983–93.

[4] J. Marti, 'Accurate modelling of frequency-dependent transmission lines in electromagnetic transient simulations', *IEEE Transactions on Power Apparatus and Systems*, 1982, PAS-101 (1), 147–57.

[5] D.A. Woodford, A.M. Gole, and R.W. Menzies, 'Digital simulation of DC links and AC machines', *IEEE Transactions on Power Apparatus and Systems*, 1983, PAS-102 (6), 1616–23.

[6] T.L. Maguire, and A.M. Gole, 'Digital simulation of flexible topology power electronic apparatus in power systems', *IEEE Transactions on Power Delivery*, 1991, 6 (4), 1831–40.

[7] P.G. McLaren, R. Kuffel, R. Wierckx, J. Giesbrecht, and L. Arendt, 'A real time digital simulator for testing relays', *IEEE Transactions on Power Delivery*, 1992, 7 (1), 207–13.

[8] Y. Zhang, H. Ding, and R. Kuffel, 'Key techniques in real time digital simulation for closed-loop testing of HVDC systems', *CSEE Journal of Power and Energy Systems*, 2017, 3 (2), 125–30.

[9] H. Duchen, M. Lagerkvist, N. Lovgren, and R. Kuffel, 'Validating the real time digital simulator for HVDC dynamic performance studies', In Proceedings of the ICDS '97, Montreal Canada, May 1997, pp. 245—50.

[10] Y.-B. Yoon, T.-K. Kim, S.-T. Cha, R. Kuffel, R. Wierckx, and M. Yu, 'HVDC control and protection testing using the RTDS simulator', In Proceedings of the 4th International HVDC Transmission Operation Conference, Yichang PRC, September 2001, Paper No. 17, pp. 101—6.

[11] T. Maguire, 'Multi-processor Cholesky decomposition of conductance matrices,' IPST, Delft, The Netherlands, July 2011, Paper No. 11IPST106.

[12] T. Maguire, B. Warkentin, Y. Chen, and J.P. Hasler, 'Efficient techniques for real time simulation of MMC systems', IPST, Vancouver, Canada, July 2013, Paper No. 13IPST143.

[13] T. Maguire, S. Elimban, E. Tara, and Y. Zhang, 'Predicting switch ON/OFF statuses in real time electromagnetic transients simulations with voltage source converters', In 2018 2nd IEEE Conference on Energy Internet and Energy System Integration (EI2), October 2018.

[14] Y. Zhang, T. Maguire, and P. Forsyth, 'UMEC transformer model for real time digital simulator', IPST 2005, Montreal, QC, Canada.

[15] Y. Liang, Y. Zhang, and A. Dehkordi, 'Key aspects in modelling transformers for real time simulation: Open phase, DC bias and internal faults', In 2019 IEEE 8th International Conference on Advanced Power System Automation and Protection (APAP), October 2019, Xi'an, China.

[16] H.M. Jeewantha De Silva, A.M. Gole, J.E. Nordstrom, and L.M. Wedepohl, 'Robust passivity enforcement scheme for time-domain simulation of multi-conductor transmission lines and cables', *IEEE Transactions on Power Delivery*, 2010, 25 (2), 930–8.

[17] B. Gustavsen, A. Semlyen, 'Simulation of transmission line transients using vector fitting and modal decomposition', *IEEE Transactions on Power Delivery*, 1998, 13 (2), 605–14

[18] B. Gustavsen and A. Semlyen, 'Rational approximation of frequency domain responses by vector fitting', *IEEE Transactions on Power Delivery*, 1999, 14 (3), 1052–61.

[19] J. Häfner, and B. Jacobson, 'Proactive hybrid HVDC breakers – a key innovation for reliable HVDC grid,' Presented in Cigré Symposium, September 2011, Bologna, Italy, pp. 13–15.

[20] RTDS User Manuel, Available from https://www.rtds.com/contact/RTDS Technologies Inc.

Index

www.ingramcontent.com/pod-product-compliance
Lightning Source LLC
LaVergne TN
LVHW020055220926
848484LV00002BB/383